Origins of Life

Robert M. Hazen, Ph.D.

THE
GREAT
COURSES®

PUBLISHED BY:

THE GREAT COURSES
Corporate Headquarters
4840 Westfields Boulevard, Suite 500
Chantilly, Virginia 20151-2299
Phone: 1-800-832-2412
Fax: 703-378-3819
www.thegreatcourses.com

Printed in the United States of America

Robert M. Hazen, Ph.D.

Clarence Robinson Professor of Earth Science, George Mason University

Robert M. Hazen, research scientist at the Carnegie Institution of Washington's Geophysical Laboratory and Clarence Robinson Professor of Earth Science at George Mason University, received the B.S. and S.M. in geology at the Massachusetts Institute of Technology (1971) and the Ph.D. at Harvard University in earth science (1975). After studies as NATO Fellow at Cambridge University in England, he joined Carnegie's research effort in 1976.

Professor Hazen has authored 250 articles and 19 books on science, history, and music. He received the Mineralogical Society of America Award (1982), the American Chemical Society Ipatieff Prize (1986), the ASCAP-Deems Taylor Award (1989), the Educational Press Association Award (1992), Fellowship in the American Association for the Advancement of Science (1994), and the American Crystallographic Association's Science Writing Award (1998). Professor Hazen is the President of the Mineralogical Society of America and serves on advisory boards of the Committee for Public Understanding of Science and Technology (AAAS), NOVA (WGBH TV Boston), Earth & Sky radio programs, Encyclopedia Americana, and the Carnegie Council. He appears frequently on radio and television programs on science.

Professor Hazen's recent research focuses on high-pressure organic synthesis and the origin of life, as well as the many roles of minerals in the origin of life. He is also active in the study of Precambrian microscopic fossils. He is a co-investigator and team leader of the NASA Astrobiology Institute, which supports his research.

Professor Hazen is active in presenting science to a general audience. His articles on science for a general readership have appeared in *Newsweek, The New York Times Magazine, Smithsonian Magazine, Technology Review*, and *Scientific American.* At George Mason University, he has developed courses and companion texts on scientific literacy. His books include the bestselling *Science Matters: Achieving Scientific Literacy* and *The Sciences: An Integrated Approach.*

Robert Hazen is also a professional trumpeter. He has performed with numerous ensembles, including the National Symphony Orchestra, the Boston Symphony Orchestra, and orchestras of the Metropolitan Opera, the

New York City Opera, the Kirov Ballet, and the Royal Ballet. He is currently a member of the National Philharmonic and the National Gallery Orchestra.

Table of Contents

Origins of Life

Professor Biography		i
Course Scope		1
Lecture One	The Grand Question of Life's Origins	4
Lecture Two	The Historical Setting of Origins Research	21
Lecture Three	What Is Life?	38
Lecture Four	Is There Life on Mars?	54
Lecture Five	Earth's Oldest Fossils	72
Lecture Six	Fossil Isotopes	90
Lecture Seven	Molecular Biosignatures	110
Lecture Eight	Emergence	130
Lecture Nine	The Miller-Urey Experiment	150
Lecture Ten	Life from the Bottom of the Sea	169
Lecture Eleven	The Deep, Hot Biosphere	189
Lecture Twelve	Experiments at High Pressure	208
Lecture Thirteen	More Experiments Under Pressure	227
Lecture Fourteen	Deep Space Dust, Molten Rock, and Zeolite	247
Lecture Fifteen	Macromolecules and the Tree of Life	267
Lecture Sixteen	Lipids and Membrane Self-Organization	287
Lecture Seventeen	Life on Clay, Clay as Life	309
Lecture Eighteen	Life's Curious Handedness	330
Lecture Nineteen	Self-Replicating Molecular Systems	351
Lecture Twenty	Günter Wächtershäuser's Grand Hypothesis	371
Lecture Twenty-One	The RNA World	390
Lecture Twenty-Two	The Pre-RNA World	409
Lecture Twenty-Three	Natural Selection and Competition	426
Lecture Twenty-Four	Three Scenarios for the Origin of Life	445

Table of Contents
Origins of Life

Chronology—Key Events in Research on Life's Origins 463
Chronology—Key Events in Life's Origins 469
Glossary 470
Biographical Notes 475
Bibliography 479

Origins of Life

Scope:

The story of life's origins on Earth is truly epic in scale, both in time and in space. Scientists surmise that 4 billion years ago the Earth's surface was desolate and lifeless. The hostile surface was constantly disrupted by volcanic eruptions and meteorite bombardment, yet evidence locked in Earth's most ancient rocks suggests that by about 3.8 billion years ago, tiny bacteria had emerged to establish a firm foothold. Life would persist in an unbroken lineage to the present day.

What happened during that ancient time interval between about 4 billion and 3.8 billion years ago that led to life's remarkable origin? An answer to this fundamental question will certainly alter the way we think about ourselves and about our place in the cosmos. For one thing, if life could begin on Earth by a natural process, then it could also begin by similar processes on Earth-like planets circling other stars throughout our galaxy and beyond. Perhaps we are not alone in the universe.

Of all the great unanswered questions, none is more fascinating and none has deeper resonance than the question of where we came from—the question of life's origins and evolution. The origin of life is a subject of immense complexity, and we do not yet know how life began. We are in the midst of a remarkably dynamic stage of research, during which advances are made every month. Consequently, the ultimate objective of these lectures is not to provide a complete picture of how life emerged. Rather, it is an opportunity to paint a picture of our present state of knowledge and ignorance and to provide a perspective of where we are going and how we are getting there.

The first two lectures set the stage with a review of the historical and philosophical context of this age-old question of life's origins. Because men and women have wondered about the origins of life for thousands of years, a rich context informs the scientific question. Lecture Three examines recent efforts to understand one of the most basic of biological questions: "What is life?" An understanding of what distinguishes living systems from the nonliving world is essential before exploring the origins of life.

Much of the lecture series focuses on two complementary strategies that scientists have adopted to clarify the gap between the lifeless geochemistry of 4 billion years ago and the time of Earth's earliest life-forms. On the one hand, Lectures Four through Seven look at the so-called "top-down" approach to the origins of life. This strategy scrutinizes modern living organisms and ancient fossils to work backward in time toward the "last common ancestor" of today's life. A small army of biologists and paleontologists tackles the question of origins from this perspective. This is a field filled with exciting discoveries and lots of controversy—it can be exceedingly difficult to tease out evidence of life from ancient rocks.

On the other hand, the so-called "bottom-up" approach starts with the primitive lifeless Earth of air, water, and rocks and maps the gradual increase in chemical complexity that must have occurred. In fact, this increase in chemical complexity must be the centerpiece of any model for life's origins. Lecture Eight, therefore, provides a conceptual framework for understanding this stepwise approach by presenting an overview of the new and exciting field of *emergence*. Emergence is now widely recognized as the process by which simple systems of many interacting particles spontaneously become more complex. Emergence thus provides a unifying theme for origins research.

Most of the rest of the course examines the growing body of "bottom-up" experimental evidence for the emergence of chemical complexity on the path to life. Life's origins can be simplified by examining a sequence of three broad steps, each of which adds complexity to the system. Lectures Nine through Fourteen review the many ways that simple organic molecules were synthesized in prebiotic environments. Competing ideas about organic molecules formed on Earth's surface, versus deep underground, versus in deep space, add richness to this topic. In fact, making potential biomolecules turns out to be the most well understood part of life's origins.

Lectures Fifteen through Eighteen deal with the knottier problem of life's essential macromolecules, which are large structures made of just the right arrangements of hundreds to millions of smaller molecular building blocks. The processes by which just the right mix of molecules was selected and concentrated remain a central problem in origins research.

The last six lectures in this series address the final stages of life's emergence. At some point, a collection of molecules must have begun to make copies of itself. Then, those self-replicating cycles of molecules must have experienced competition, which quickly drove the evolution to even more complex assemblages. Lecture Twenty-Three, in particular, introduces some remarkable new experiments on molecular evolution that give us hints as to how this process might have happened.

In sum, this series of 24 lectures provides an overview of the current understanding of life's origins and establishes a foundation for sharing in the many new discoveries that are sure to follow in the coming months and years.

Lecture One

The Grand Question of Life's Origins

Scope: This first lecture is an introduction to our exploration of the age-old mystery of life's origins. As we'll see, this is a deeply rooted question that is still a long way from being solved. As we move through these lectures, you'll learn about some amazing discoveries in chemistry, biology, astronomy, geology, and other scientific fields. We'll also meet some remarkable people—mainstream researchers and controversial iconoclasts, famous Nobel Prize winners and unknown investigators struggling to gain acceptance. This is an intense and fascinating field, filled with debate, controversy, rivalries, and animosity. There is as much drama here as in any scientific subject.

Outline

I. What exactly does it mean to study the origins of life? This is a story that is truly epic in scale.

A. Imagine yourself back on the ancient Earth, some 4 billion years ago; the surface is unbelievably hostile. The temperature is boiling hot; giant volcanoes are everywhere, belching noxious gases into the atmosphere. Meteors and comets are constantly bombarding Earth's surface. Just about the only landmasses present are large volcanic islands not much bigger than Hawaii or Iceland. There are no continents.

B. That world of 4 billion years ago was utterly lifeless. Imagine yourself standing on one of the active volcanic islands, perhaps on a black lava beach with shallow tidal pools. Even the most powerful microscope would detect no sign of life. Chemical analysis would show nothing more than a dilute solution of organic compounds and salts.

C. Now, skip forward to about 3.8 billion years ago. If you could then come to the same spot, not much will appear to have changed.

1. The islands are still violently volcanic, though perhaps a little less intense than before. Meteorite impacts still pepper the surface, though at a somewhat slower rate. The surface of the Earth is a little bit cooler, as well.

2. When you scrutinize the tidal pools, however, you now see a kind of cloudiness that wasn't there before. If you examine the water under a microscope, tiny bacteria swarm in the pool. In the space of a couple of hundred million years, life has begun.

D. What happened during that ancient time interval between about 4 billion and 3.8 billion years ago that led to life's remarkable origin? An answer to this fundamental question will alter the way we think about ourselves and our place in the cosmos. Consider the fact that if life could begin on Earth by a natural process, then it could also begin by similar processes on Earth-like planets throughout our galaxy and beyond. Perhaps we are not alone in the universe.

II. We're about to begin an incredible adventure, and this first lecture sets the stage, giving you a sense of where we're going and what we will discover. This will not be a standard science course for at least three reasons.

A. First, this course is unusual because at this point in time, there is much that we don't know about how life emerged on Earth.

1. Of all the great unanswered questions, none is more fascinating, and none has deeper resonance than the question of the origin and evolution of life. The origin of life is a subject of immense complexity and one that we have just begun to explore.
2. Over the next 24 lectures, I will tell you what we think we know about life's origins, but more importantly, I will tell you much about what we *don't* know.

B. The second reason that this course is unusual is that, unlike most courses in physics, chemistry, and biology, this one will emphasize the *process* of science. We will see how unscripted and creative the scientific process is, and you will gain a sense of how scientists around the world are trying to fill in the gaps of our ignorance.

C. The third unusual aspect of this course is that the scientific study of life's origins is controversial in a way that other scientific studies are not. There are controversies in our society today about the scientific approach to life's origin and evolution versus the revealed truth of sacred texts, especially the Old Testament of the Bible.

1. This course focuses exclusively on the scientific approach to the question of life's origins. In this lecture series, I make an assumption that life emerged from basic raw materials through a sequence of events that was completely consistent with the natural laws of chemistry and physics.
2. Even with this scientific approach, there is a possibility that we'll never know how life originated. It is possible that life emerged by an almost infinitely improbable sequence of difficult chemical reactions. If that's true, then living planets must be rare in the cosmos. It is even possible that Earth is the only living planet.
3. If life is the result of an infinitely improbable succession of chemical steps, then any scientific attempt to understand life's origin is doomed to failure; such a succession could not be duplicated in a program of lab experiments.
4. The other possibility is that the universe is pregnant with life. Perhaps nature is organized in such a way that life emerges inevitably as a consequence of chemistry—what chemistry Nobelist Christian de Duve calls a "cosmic imperative." If that's true, then scientists can fruitfully study life's origins in the lab.
5. Not surprisingly, virtually all origins-of-life researchers adopt the philosophical view that life is indeed a cosmic imperative. However, we cannot prove that idea until we find a second, independent origin of life, either on another world or in the lab.

III. I have organized this course in the following general outline.

A. The first two lectures establish the historical and philosophical context of the age-old question of life's origins. The third lecture examines recent efforts to understand one of the most basic of biological questions: "What is life?"

B. Lectures Four through Seven look at the so-called "top-down" approach to the origin of life. This strategy scrutinizes modern living organisms and ancient fossils to work backward in time toward the "last common ancestor" of today's life.

C. Lecture Eight provides a conceptual framework for understanding the "bottom-up" approach by presenting an overview of the new and exciting field of *emergence*—the process by which simple systems of many interacting particles spontaneously become more complex.

D. Lectures Nine through Fourteen review the many ways that simple organic molecules were synthesized in prebiotic environments, whether on Earth's surface, deep underground, or in deep space.

E. Lectures Fifteen through Eighteen deal with the knottier problem of life's essential macromolecules—large structures made of just the right arrangements of hundreds to millions of smaller molecular building blocks. We'll look at a significant problem in origins research, that is, how just the right mix of molecules was selected and concentrated.

F. The last six lectures in this series address the final stages of life's emergence. At some point, a collection of molecules must have begun to make copies of itself. Then, those self-replicating cycles of molecules must have experienced competition, which quickly drove the evolution to even more complex assemblages. In Lecture Twenty-Three, we'll look at remarkable new experiments on molecular evolution that give us hints about how this process might have happened.

IV. To conclude this introduction, let's look briefly at the sociology of the origins research community. Who does this research, and where is it conducted? How is the research funded? And how do origins researchers communicate with one another?

A. The origins-of-life community has one principal scientific society, the International Society for the Study of the Origin of Life (ISSOL). Total membership is only about 500 scientists from three dozen countries.

B. The ISSOL promotes research, bestows awards, encourages the exchange of scientific data, holds international meetings, and publishes a journal entitled *Origins of Life and Evolution of the Biosphere*.

C. Important breakthroughs are also published in *Science* and *Nature*, as well as *The Journal of Molecular Evolution* and *The Journal of Molecular Biology*.

D. Week-long Gordon Research Conferences are restricted to about 120 scientists, who sign promises not to reveal what is discussed—

thereby encouraging consideration of wild new ideas and "active debate."

E. Finally, we should note one other important aspect of the sociology of origins research: Funding drives science. No matter how compelling a scientific question might be, it won't receive much attention unless someone pays the bills.

1. In the early days of origins research, funding was spotty, but the past decade has seen a major change in the origins-of-life community, primarily because of new interest and funding from NASA. In the mid-1990s, NASA began its Astrobiology Institute and started pouring about $25 million per year into the field.
2. NASA's funding increase also changed the character of origins research. NASA defines *astrobiology* as "the search for the origin, distribution, and future of life in the universe." As we'll see throughout this course, NASA is now sponsoring research on Earth life in extreme environments, telescope observations of distant planets, and a variety of other studies. In this context, the study of life's origins becomes one facet of a much larger inquiry.

V. The question of life's origins is squarely in the mainstream of science, and only with the methods of science can we hope to explain the chemical details of how life arose. In the next lecture, we'll look backward and place the modern scientific pursuit of origins in a historical context.

Essential Reading:

Dick and Strick, *The Living Universe: NASA and the Development of Astrobiology.*

Hazen, *Genesis: The Scientific Quest for Life's Origin*, Preface.

Supplemental Reading:

Fry, *The Emergence of Life on Earth*, chaps. 13–14.

Lahav, *Biogenesis: Theories of Life's Origin*, part I.

Schopf, *Cradle of Life: The Discovery of Earth's Earliest Fossils*, chs. 2 and 4.

Questions to Consider:

1. "How did life arise?" is one of the great unanswered questions in science. What other scientific questions are of comparable importance?
2. Is it reasonable for NASA to spend tax dollars funding astrobiology? What other sources of funding might be used to support this research?

Lecture One—Transcript

The Grand Question of Life's Origins

Hi. I'm Robert Hazen, Clarence Robinson Professor of Earth Science at George Mason University in Fairfax, Virginia. At George Mason, I teach a variety of undergraduate courses that integrate the sciences with other areas of study. I recently taught classes on scientific literacy and on scientific ethics. I also offer a really fun course on symmetry and art in science, and several classes in the dynamic new field of astrobiology.

In addition, since 1976, I've been a research scientist at the Carnegie Institution of Washington's Geophysical Laboratory in northwest Washington, DC. At the Carnegie Institution, my colleagues and I spent most of the last decade focusing on one of science's greatest unanswered questions: How did life originate? We started small, with just four of us working together in 1996. But that effort has now expanded to about 30 scientists in DC, and dozens of other collaborators in institutions around the world.

I am absolutely thrilled that The Teaching Company has given me this opportunity to tell you about this age-old mystery of life's origins and to explore with you how scientists around the world are attempting to find answers. In the process, I'll tell you about some amazing discoveries in many scientific fields in chemistry, biology, astronomy, geology and more. We'll also meet some remarkable people: mainstream researchers and controversial iconoclasts, famous Nobel Prize winners, and unknown investigators struggling to gain acceptance. This is an intense and fascinating field, filled with debate, controversy, rivalries, and animosity. There is as much drama here as in any scientific subject.

So what exactly does it mean to study the origins of life? This is a story that is truly epic in scale. I want you to imagine yourself back to the ancient Earth, some four billion years ago. The Earth's surface back then is unbelievably hostile. The temperature is boiling hot, much hotter than anyplace on Earth today. Giant volcanoes are everywhere, belching noxious gases into the atmosphere. Meteors the size of city blocks and even larger are constantly bombarding our surface. Comets also crash through the atmosphere and pulverize the surface. Just about the only land masses present are large volcanic islands not much bigger than Hawaii or Iceland There were no big continents.

That world of four billion years ago was utterly lifeless. Now imagine yourself standing on one of the active volcanic islands, perhaps in a black, lava beach with shallow tidal pools. You can examine those pools with the most powerful microscope, but there's nothing to see. Chemical analysis would show nothing more than a dilute solution of organic compounds and salts.

Now, skip forward to about 3.8 billion years ago. If you could then come to the same kind of spot, not much would appear to have changed. The islands are still violently volcanic, though perhaps a little less intense than before. Meteorite impacts still pepper the surface, though at a somewhat slower rate. The surface of the Earth is a little bit cooler as well, but when you scrutinize the tidal pools, you now see a kind of cloudiness that wasn't there before. If you examine the water under a microscope—lo and behold! Tiny bacteria now swarm in the water. In the space of a couple of hundred million years, life has begun.

What happened during that ancient time interval between about 4 billion and 3.8 billion years ago that led to life's remarkable origin? An answer to this fundamental question will certainly alter the way we think about ourselves and about our place in the cosmos. For one thing, if life could begin on Earth by a natural process, then it could also begin by similar processes on Earth-like planets circling the stars throughout our galaxy and beyond. Perhaps we are not alone in the universe.

We are about to begin an incredible adventure. In this first lecture, I want to set the stage to give you a sense of where we are going and what we are about to discover. I have to be honest with you right up front that this will not be your usual science course for at least three reasons.

First, and perhaps foremost, this course is unusual because at this point in time, there is so much that we do not know about how life emerged on Earth. There is a popular misconception that scientists know it all, that we have answers to every possible question. I think that this erroneous view has been exacerbated by the silly claim, notably by science writer John Horgan, that we have come to the "end of science," that researchers have discovered almost everything of significance. "All that's left to do," Horgan says, "is fill in a few blanks, like completing a postage-stamp collection." Well, that is just absurd. There are huge gaps in our understanding. Incredibly important and exciting unanswered questions remain. That is what drives us as scientists. Humans are curious, and we are compelled to discover what we do not know. Science would be deadly dull if we had all the answers.

Of all the great, unanswered questions, none is more fascinating, and none has deeper resonance than the question of where we came from, the question of the origin and evolution of life. The origin of life has been a subject of immense complexity, and I have to tell you right up front we do not know how life began. It is as if we are trying to assemble a huge jigsaw puzzle. We have a few pieces clumped together here and there, but most of the puzzle pieces are missing. What is more, we have lost the box top, so we are not even sure what the complete picture is supposed to look like.

Stewart Kaufman, who is a distinguished theorist at the Santa Fe Institute in New Mexico, put it succinctly. Let me read you his quote. He says, "Anyone who tells you that he or she knows how life started on the Earth is a fool or a knave."

Well, that ignorance might seem to pose a problem. After all, how can I tell you about the origin of life when we are so woefully ignorant of that history? Actually, I see this as a tremendous opportunity rather than a problem. You see, science is all too often presented as a dry recitation of facts we know to be true, or concepts we think we know to be true. In fact, the most exciting aspect of science is the process of discovery; the challenge of wresting from nature, slowly but surely, an understanding of what we do not know. The origin of life is a perfect subject to reveal that ongoing adventure that is the very essence of science.

So over the next 24 lectures I will tell you what we think we know about life's origins. But more important, I'll tell you a whole lot about what we don't know.

That leads me to a second reason why this course is unusual. Unlike most courses in physics or chemistry or biology, I'm going to emphasize the process of science. I want to take you into the field to see what geologists see and to puzzle with them as they sort out the meanings of ancient rocks. I want to take you into the laboratory and show you how we do origin experiments. These aren't simple and straightforward procedures. We make mistakes. We follow false leads. We fool ourselves. I want you to see how unscripted and creative the scientific process really is. I want you to get a sense of how scientists around the world are trying to fill in the blanks of our ignorance.

We are in the midst of a remarkably dynamic stage of research into the origins of life. One consequence of this rapid rate of discovery is that there are numerous competing ideas and lots of controversy. In fact, I have never seen a scientific field with so many wild, new ideas. We will see how amazingly creative people keep coming up with new models and hypotheses. They just present them to the community and see if they take root. Then there are endless debates and conflicts among competing ideas. There is a tremendous amount of give and take, as the best ideas went out in a kind of scientific natural selection.

The ultimate objective of these lectures is not to provide a complete picture of how life emerged. Rather, it is an opportunity to give you a vivid picture of our state of knowledge and ignorance, to give you an insider's perspective of where we are going and how we are getting there. By the time we are through, you will be in a great position to understand new discoveries and advances in the field; that is one thing I can absolutely guarantee. Every month we learn new things. Every month we are closer to solving the mystery of life's origins.

There is a third unusual aspect to this course, one that has probably already crossed your mind; most scientific topics are not controversial. Studies of the laws of motion or experiments on chemical reactions or fieldwork on rocks or minerals or observations in the structures of cells for that matter; they are all mainstream research topics. No one I know feels uneasy about scientists tackling those questions. However, the scientific study of life's origins has an edge. There are real controversies in our society today about the scientific approach to life's origin and evolution versus the revealed truth of the sacred text, especially the Old Testament of the Bible.

Let me say now, for the record, I am a scientist; I am not a theologian, nor am I a philosopher. This course focuses exclusively on the scientific approach to the question of life's origins. In this lecture series I make a basic assumption that life emerged by some kind of natural process. I propose that life arose by a sequence of events that are completely consistent with the natural laws of chemistry and physics. In this assumption I am like most other scientists. I believe in a universe that is ordered by these natural laws. Like other scientists, I rely on the power of observations and experiments and theoretical reasoning to understand how the cosmos came to be the way it is.

Using this kind of reasoning, I assume that life arose in the ancient scorched Earth from the simplest of basic raw materials: air, water, and rock. Thus, life emerged nearly four billion years ago by a series of chemical and physical processes that were completely in accord with natural laws. Though I have to confess, the nitty-gritty details of that transformation remain a deep mystery. But just because we do not know all the details does not mean it did not happen by lawful natural process.

I have to be honest, even with a scientific approach there is a possibility that we will never know—in fact, that we cannot ever know—how life emerged. That is because it is always possible that life emerged by an almost infinitely improbable sequence of difficult chemical reactions. If that is true, then living planets will be very rare in the cosmos. It is even possible that Earth is the only living planet in the entire universe. If that is true, then any scientific attempt to understand life's origins is doomed to failure. You see, an infinitely improbable succession of chemical steps can not be duplicated in a program of lab experiments. The origin of life was an infinitely improbable accident. There is absolutely nothing you or I or anyone else could do to figure out how it happened. I must tell you, that is a depressing thought to someone like me who's devoted a decade to studying the origin of life.

But there is another possibility; perhaps the universe is pregnant with life. Perhaps nature is organized in such a way that life emerges inevitably as a consequence of chemistry—what chemistry Nobel winner Christian de Duve calls "a cosmic imperative." Perhaps, given water and molecules and energy in an appropriate environment plus sufficient time, life is extremely likely to emerge from non-life. If so, then scientists can approach this big question with optimism. If life is likely to begin on just about any planet or moon where favorable conditions occur, then we scientists can fruitfully study life's origins in the lab.

I guess it is not too surprising then that virtually all origin-of-life researchers adopt a philosophical view that life is indeed "a cosmic imperative." We trust that it is only a matter of time before we know how it happened. It is wonderful to think that, given the scenario, genesis occurs throughout the universe all the time. However, we cannot prove that idea until we find a second independent origin of life either on another world or in the laboratory.

I need to pause here and provide you with a little truth in advertising. My scientific background is in geology. In fact, my specialty is mineralogy and crystallography. That might seem a far cry from the emergence of life. Why should an Earth scientist enter what sounds like a field of biology?

Well, the fact is I was seduced. In the mid-1990s I had been doing research at the Carnegie Institution of Washington's Geophysical Laboratory for about 20 years. My studies were in the specialized field of high-pressure crystal chemistry. My colleagues and I had a steady stream of federal funding. We published lots of articles. Scientific life was good. However, I have to confess, something was lacking in that stage of my career. The excitement of science lies in the search for answers to questions about the natural world. We had gotten into a kind of rut where we had learned the basics of how crystals compress, and there were fewer and fewer scientific surprises. I was really eager to try something different. So when Harold Morowitz, a longtime friend and colleague, asked me to try an origin of life experiment, I jumped at the chance. I'll tell you more about that experience in Lecture Twelve.

There is another important reason why I entered the origins research field. Since 1989, I worked closely with Jim Trefil, at my college, George Mason University. We had been trying to foster changes in undergraduate science education. As you probably know, most science requirements for college students are discipline based. They are intensive courses in biology or in physics or in chemistry. Jim and I contend this approach is fundamentally flawed. Discipline-bound science courses just don't provide the kind of useful knowledge with which non-science majors can relate. Narrow courses also fail to give a foundation on which students can build and grow as lifelong science enthusiasts. We believe that most non-science majors receive tremendous benefit from integrated science courses. Over-arching scientific principles and their application to daily life should provide the core of undergraduate curriculum. By the way, my other Teaching Company course, *The Joy of Science*, was a direct outgrowth of this effort in science education reform.

It turns out that origin-of-life research is a perfect example of why integrated science in education is important. The study of life's origins fundamentally integrates many branches of science: astronomy, biology, chemistry, geology, physics; they are all important. It has been tremendously exciting for me to incorporate the full spectrum of scientific ideas into my research. That is why, when the chance to change research directions occurred, I jumped.

The outline for this lecture series, as well as for the companion book, *Genesis: The Scientific Quest For Life's Origin*, began to come together in September of 2000 at a fascinating conference in the beautiful little town of Modena, in Northern Italy. The conference addressed one seemingly simple question, "What is life?" I was amazed at the range of opinions of the 100 scientist, philosophers, and theologians who were present. Surprisingly, the most heated debates seemed to be within the ranks of scientists themselves. There was one researcher who studied molecules that make cell membranes who claimed that life began with a first-enclosing cell membrane. Another authority on metabolism countered that life began with the first metabolic cycle, which is the process all cells use to change energy and atoms into useful molecules. A group of experts on cell-genetics were certain that the first living entity must have been a RNA-like genetic system. Believe it or not, there is even a famous mineralogist who argues that life began as a self-replicating mineral.

As you can imagine, for a neophyte in the origin of life field, this discussion sounded a lot like the classic tale, "The Blind Men and The Elephant." Recall, each of the blind men was asked to describe an elephant based only on whatever part was close at hand. So one blind man says, "The elephant is like a tree"; that's the massive leg. Another says, "The elephant is like a rope," based on the trunk. A third says, "The elephant is like a snake," and that's the tail, and so forth. Each of the blind men is partly correct, but none has the whole picture. It struck me that the story of life's origin might likewise require a multiplicity of viewpoints. So that's the strategy I used in this lecture series.

Let me give you a basic outline of the course. My first two lectures set the stage: the historical and philosophical context of the age-old question of life's origins. Men and women have wondered about the origin of life for thousands of years. So there's a rich context to the scientific question. Then, in Lecture Three, I'll examine recent efforts to understand one of the most basic of biological questions, "What is life?" That's an important topic if we are going to explore the origins of life.

Much of the lecture series will focus on two complimentary strategies that scientists use to zero in on the gap between lifeless geochemistry, on one hand, and the earliest life form on the other. In Lectures Four through

Seven, I will look at the so-called "top-down" approach to the origin of life. This strategy scrutinizes modern living organisms and ancient fossils, working backward in time to the last common ancestor. As we will see, a small army of biologists and paleontologists tackle the question of origins from this perspective. This is a field filled with exciting discoveries and lots of controversy. It's exceedingly difficult to tease out evidence of life from ancient rocks.

By contrast, the so-called "bottom-up" approach starts with a primitive, lifeless Earth of air, water, and rocks, and maps the gradual increase in chemical complexity that must have occurred. In fact, this increase in chemical complexity must be the centerpiece of any model for life's origins. So in Lecture Eight, I provide a conceptual framework for understanding this stepwise approach, by presenting an overview of the new and exciting field of emergence. Emergence is now widely recognized as the process by which simple systems of many interacting particles spontaneously become more complex. Emergence thus provides a unifying theme for origins research.

For most of the rest of the course, I will examine the growing body of experimental evidence for the emergence of chemical complexity on the path to life. My thesis is that life's origin can be simplified by examining a sequence of three broad steps, each of which adds complexity to the system. In Lectures Nine through Fourteen, I'll review the many ways that simple organic molecules were synthesized in pre-biotic environments. We'll see competing ideas about how organic molecules form on Earth's surface versus deep underground versus in deep space. In fact, making potential biomolecules turns out to be the best understood part of life's origins.

Lectures Fifteen through Eighteen deal with the knottier problem of life's essential macromolecules. Macromolecules are large structures made of just the right arrangements of hundreds to millions of smaller molecular building blocks. We'll see that how just the right mix of molecules was selected and concentrated remains a huge problem in origins research.

The last six lectures of this series address the final stages of life's emergence. At some point, a collection of molecules must have begun to make copies of itself. Then those self-replicating cycles of molecules must have experienced competition, which quickly drove the evolution to even more complex assemblages. In Lecture Twenty-Three, I'll tell you about remarkable new experiments on molecular evolution that gives us hints as to how this process might have happened.

I hope you now have a sense of where we are going. I want to conclude this introduction by telling you about the sociology of the origin's research community. Who does this research? Where are they based? How is this research funded? How do we communicate with each other?

It is actually pretty easy to keep track of the origins of life community because there is one principal scientific society: The International Society for the Study of the Origin of Life, or ISOL. The total membership of ISOL is about 500 scientists from roughly three-dozen countries. The United States is by far the biggest contingent with more than 250 members, but there are also significant groups from France, Germany, Japan, and Russia. Even so, these are really tiny numbers in the context of global science, only 500 members out of perhaps a couple of million scientists worldwide.

For the most part, ISOL does just what other scientific societies do. Indeed, its mission statement could be that of any such organization. It promotes research, it hands out awards, it encourages the exchange of scientific data, and it holds international meetings every few years. In fact, recent meetings have been held in Trieste, Mexico City, and Beijing.

But ISOL also has its quirks. Most scientific societies have open memberships, but not ISOL. Early on, when the society was small, there was a real concern that the membership could be taken over by crackpots and fringe scientists with their own agendas. The origins of life field has, from time to time, attracted people with ideas who are, to say the least, a bit odd. There is one contingent, for example, that is convinced that life on Earth was "planted" by aliens and that we are all just one big experiment. So to prevent an "alien takeover," so to speak, ISOL says you can't become a member without proof of a scientific track record in the field and a nomination by two members in good standing.

Another historical oddity of ISOL is the name of its journal: *The Societies for the Studies of the Origin of Life*, with "Origin" singular. But the journal, which first appeared in 1974, is titled *Origins of Life*, with "Origins" plural. This difference reflects a fascinating shift in attitudes that occurred in the 1970s and 1980s. That is when many researchers began to argue that life has arisen frequently in the universe. Indeed, as I have indicated, without such an assumption, the scientific study of life's origin is probably a waste of time. By the way, that is why this Teaching Company course is called *Origins of Life*.

Then, there was another significant change in 1985. The society began primarily as an organization of chemists who were engaged in the "bottom-up" approach to life's origins. But in the 1980s ISOL increasingly became the home for geologists and paleontologists who use the "top-down" approach to study the oldest rocks on Earth. So the journal was renamed "Origins of Life and Evolution of the Biosphere."

Of course, that is not the only journal for origin studies. The most important breakthroughs in origin-of-life research, as in any other scientific field, usually appear in the big two prestigious weekly periodicals. Those are *Science* and *Nature*. In fact, research that advances our understanding of life's origins and evolution probably has a better chance of getting into these journals than almost any other topic.

Then there are a number of specialized journals such as the *Journal of Molecular Evolution* and the *Journal of Molecular Biology*, and lots of others. There are also other scientific meetings on the origin of life, most notably, the Gordon Research Conferences on the Origin of Life. There have been about fifteen of these, one held every year and a-half in the United States. These weeklong meetings are restricted to about 120 scientists, and they are quite unusual in that everything is strictly said off the record. If you attend, you actually have to sign a statement that you will not reveal what was discussed. That gives people a lot of freedom to talk about wild new ideas and engage in what might be politely called "active debate." Gordon Conferences are also unusual because each meeting has a single organizer who puts together the scientific program and invites all the speakers. Then at each meeting, the attendees elect the next organizer. In fact, the January 2008 Conference in Ventura, California, I became the organizer on the theme of "Steps in Life's Emergence."

I think one of the most amusing aspects of the recent Gordon Conferences is their struggle to decide on whether to call it "Origin of Life" or "Origins of Life." The official registration forms still use the singular name from the 1970s, but in the 2005 meeting, I noticed that the fine print on about half the name badges read "Origin" while the other half read "Origins."

There is one more aspect of the sociology of origins research that I want to tell you about: funding drives science. No matter how compelling a scientific question might be, you are not going to get much progress unless someone pays the bills. In the early days of origins research, funding was spotty at best. In the U.S., the National Science Foundation and NASA both put perhaps a million dollars a year into investigations by small groups, but there was nothing large-scale.

The past decade or so has seen a major change in the origin-of-life community, primarily because of new interest in funding by NASA. Now, NASA has long been a principal source of research funding for origins research; after all, one of their central missions is to find life on other worlds. But the stakes were raised in the mid 1990s when NASA began their Astrobiology Institute and started pouring about 25 million dollars per year into the field. That money supports a lot of researchers as well as the new NASA-sponsored astrobiology meetings and two new journals; one is *Astrobiology,* another is the *International Journal of Astrobiology*.

Now, NASA's funding also increased the character of origins research. For one thing, NASA's mission is a lot bigger than just the origin of life. NASA defines astrobiology as "the search for the origin, distribution, and the future of life in the universe." So, as we will see throughout this lecture series, NASA is now sponsoring research on Earth life in extreme environments, telescope observation of distant planets, field studies of the oldest rock on Earth, investigation of meteorites, and a host of chemical synthesis studies including my own. Of course, NASA still sponsors missions to other planets and moons at a price tag on the order of a billion dollars a year. That's in addition to the astrobiology budget, and that is big science. In this context, the study of life's origins becomes one facet of a much larger inquiry. It is really an effort to understand our place in the universe.

I will tell you a lot more about NASA's astrobiology institute in Lecture Thirteen, but suffice it to say, with a network of over a thousand researchers, it is now a major player in the origins game.

To summarize, in this introductory lecture we have seen that the mystery of life's origins is a deeply rooted question that is still a long way from being solved. It is, however, a question squarely in the mainstream of science. Only with the methods of science can we hope to explain the chemical details of how life arose.

In the next lecture, we will look backward, and place the modern scientific pursuit of origins in a historical context.

Lecture Two

The Historical Setting of Origins Research

Scope: The mystery of life's earliest beginnings represents one of the oldest questions asked by humans; indeed, the scientific investigation of life spans almost three millennia. In this lecture, we'll review some of the history of origins research, and we'll see that early efforts to answer this great question were hampered by the *absence* of three things: relevant evidence, appropriate experimental equipment, and a theoretical notion of emergence.

Outline

I. Before the discovery of living cells in the 17th century, there was little experimental evidence to inform the question of life's origins.

A. Thales of Miletus (c. 625–547 B.C.) anticipated centuries of thought by citing water as the ultimate substance of the world and of life, and his student Anaximander (c. 611–547 B.C.) echoed this view. Both Thales and Anaximander relied on a spontaneous, naturalistic explanation of life. They invoked a natural process, not a creation by the gods.

B. Later Greek scholars, notably Empedocles (c. 492–432 B.C.), expanded the list of fundamental elements to four: earth, water, air, and fire.

C. By the 4th century B.C., a significant debate regarding the nature of matter arose between the schools of Democritus (c. 460–370 B.C.) and Aristotle (384–322 B.C.).

1. Democritus argued for a world of atoms—tiny particles that combine to form the matter in our world. In his view, life arose spontaneously through the combination of atoms of soil and fire.
2. Aristotle opposed this atomic hypothesis, but he embraced the spontaneous naturalistic generation of life on Earth. At the core of his belief was the doctrine of *vitalism*—the idea that every organism is imbued with a "life force," intrinsically different from the forces that act on inanimate matter.

D. In the 17th century, the theory of life's spontaneous generation was accepted knowledge, disputed by only a few theologians.

E. The first experiment to challenge the conventional wisdom of spontaneous generation was performed by the Tuscan physician Francesco Redi (1627–1697). Using raw meat, Redi discovered that flies, not spontaneous generation, cause maggots.

F. One important factor in the spontaneous generation controversy was the 17th-century invention of the microscope and the surprising discovery that microscopic life is everywhere, but even the existence of microorganisms failed to resolve the controversy.

G. Noted Italian physiologist Lazzaro Spallanzani (1729–1799) was opposed to the idea of spontaneous generation. He explored the subject by comparing the results of boiling nutrient-rich water in sealed and unsealed flasks. Spallanzani believed that his results discounted the idea of spontaneous generation, but he was challenged in this belief by the Englishman John Needham (1713–1781).

II. By the 19th century, experiments had clearly discounted the spontaneous generation of larger organisms, such as mice and flies, but the origin of microbes was still a matter of debate. The great French chemist Louis Pasteur (1822–1885) finally resolved the issue.

A. In the 1850s and 1860s, Pasteur helped to abolish belief in vitalism and the theory of spontaneous generation with his own series of experiments on sterilized solutions.

1. He prepared a nutrient-rich sugar solution and poured it into several beakers; one set of beakers was tightly sealed to prevent any contact with the ambient air.
2. Pasteur prepared other beakers so that they were left open to the air but had narrow twisting necks or cotton stoppers. Thus, the sugar solution was in contact with the ambient air, but microbes were unable to traverse the long glass passage or cotton barrier. Pasteur also ran a series of control experiments in which he left beakers open or deliberately contaminated them with dust.
3. Over the course of several years, he showed that boiled water, if isolated from airborne microbes, remained sterile indefinitely. His conclusion was that only microbial

contamination causes new growth. Cells do not arise by spontaneous generation.

B. In the course of this work, Pasteur also contributed in a significant way to the study of life's origins. By verifying the dictum that no cellular life can occur without prior cellular life, his findings seemed to push back life's origins to an inconceivably remote time and place.

C. With Pasteur's refutation of spontaneous generation, the door was opened for new ideas. Ironically, the only valid scientific response to Pasteur's work was to propose that life arose by—spontaneous generation!

1. In 1871, for example, Charles Darwin speculated that natural chemical processes, in a plausible environment, might lead to an increase in complexity and, ultimately, to life.
2. Charles Darwin wrote to Joseph Hooker, "If (and Oh! What a big if!), we could conceive in some warm little pond, with all sorts of ammonia and phosphoric salts, light, heat, electricity, etc., present, that a protein compound was chemically formed ready to undergo still more complex changes."

D. Darwin's insight indicated that life's origin depended on three key resources.

1. First, all terrestrial life depends on liquid water, which is the essential growth medium for all living cells.
2. Life also depends on a ready and reliable source of energy. The radiant energy of the Sun provides the most obvious energy source for life today, but the energy of lightning bolts, asteroid impacts, and other phenomena have also been invoked as the energy sources that triggered life.
3. The third essential input for life is a variety of chemical elements. Every known living organism incorporates atoms of carbon, oxygen, hydrogen, nitrogen, and quite a few other elements as well. Countless atoms of these essential elements combine in graceful geometries to form myriad essential biomolecules.

III. In spite of Pasteur's work and the intrinsic importance of the topic, Darwin's musings about a chemical origin of life flew in the face of conventional theology. Indeed, it was more than half a century after

Darwin, in the 1920s, that scientific speculation on life's origin took a more formal guise.

A. The most notable pioneer of the modern school of origin theorists was the Russian biochemist Aleksandr Oparin (1894–1980). In 1922, Oparin elaborated on the idea that life arose from a body of water, dubbed the *Oparin Ocean* or the *primordial soup*, which gradually became enriched with organic molecules.

B. Many of Oparin's postulates were echoed in 1929 in the independent ideas of British biochemist and geneticist J. B. S. Haldane (1892–1964). Haldane speculated on the production of large carbon-based molecules under the energetic influence of the Sun's ultraviolet radiation. He also proposed that the first living objects were self-replicating, specialized molecules that accumulated in the ocean.

IV. All origins-of-life researchers have to deal with the vast gap between the lifeless geochemical world on the one hand and the vibrant living Earth on the other. How did all of the mind-boggling biochemical complexity arise?

A. It is virtually impossible to conceive of this history as a single step that might be replicated in a single experiment. Rather, today's conventional wisdom is that living cells emerged from the barren, primordial geochemical world through a sequence of chemical steps.

1. Thus, the only reasonable experimental approach is to reduce biological complexity to a series of simple chemistry experiments.

2. How do we begin? Even the simplest known living cell depends on the intricate interplay of millions of molecules engaged in hundreds of different, yet interdependent, chemical reactions.

B. One approach to documenting life's origin lies in reducing the living cell to the much simpler level of life's basic chemical components—that is, the small carbon-based organic molecules, as well as the collective structures that they form. A fascinating theoretical field, called *emergence*, points to a promising research strategy in this regard.

C. To set the stage, let's review the natural laws that deal with how energy behaves—the laws of thermodynamics.

1. The first law of thermodynamics states a principle called the *conservation of energy*. According to this law, energy, which is simply a measure of a system's ability to do work, can appear in diverse forms; thus, heat is a form of energy, as is light. The first law of thermodynamics also tells us that energy can change from any one form to another over and over again, but the total amount of energy does not change.
2. The second law of thermodynamics places severe limitations on how energy can change from one form to another. Heat energy, for example, always flows from warmer to cooler regions, never the other direction. Note that dissipated heat energy still exists, but it cannot be put to any practical use.
3. Another way of stating the second law is that all natural systems tend spontaneously to become disordered. Scientists measure this disorder with a characteristic called *entropy*. Any collection of atoms, including a car or the body, gradually deteriorates.
4. If we look around, however, we see any number of remarkably ordered systems, such as houses, trees, and birds. In spite of the second law's pronouncement that entropy inexorably increases, it is obvious that disorder is not the only end point in the universe.
5. Observations of familiar phenomena reveal that the two laws of thermodynamics don't seem to provide a complete picture. Indeed, some scientists suspect that a fundamental law of nature, the law that describes the emergence of complex ordered systems, including every living cell, has yet to be formulated.
6. The second law of thermodynamics states that systems *as a whole* become more disordered with time, but at the *local scale* of a single living cell, or an ant colony, or the brain, we don't see disorder—we see the complexity that emerges.

D. A true pioneer in this work was the chemist Ilya Prigogine (1917–2003). Prigogine recognized that all sorts of complex emergent systems arise when energy flows through a collection of interacting particles.

1. Familiar examples include the arms of spiral galaxies that are formed from hundreds of millions of stars and sand dunes formed from myriad sand grains. Living cells, the conscious

brain, and bustling cities, are all ordered structures that emerge when many interacting particles are subjected to an energy flow.

2. The formation of patterns in emergent systems speeds up the dissipation of energy. For this reason, a sand surface covered with ripples or dunes is more efficient at breaking up the energetic flow of water or wind than a flat surface.
3. If we can elucidate the principles that govern emergent systems, then we should be able to apply those insights to designing an experimental program to investigate the origins of life.

E. Only recently has the conceptual framework of emergence helped to frame the origin of life as a sequence of chemical steps. In this scenario, each step in life's origins increases the complexity of the system. Yet before we explore the bottom-up chemical steps leading to life, we need to understand more about life after it emerged from the prebiotic soup. And that's the subject of the next lecture: What, exactly, is life?

Essential Reading:

Hazen, *Genesis: The Scientific Quest for Life's Origin*, ch. 6.

Lahav, *Biogenesis: Theories of Life's Origin*, part I.

Supplemental Reading:

Oparin, *The Origin of Life.*

Strick, *Sparks of Life: Darwinism and the Victorian Debate over Spontaneous Generation.*

Wills and Bada, *The Spark of Life: Darwin and the Primeval Soup*, chaps. 1–2.

Questions to Consider:

1. Which side would you have taken in the debate between Spallanzani and Needham? Why?
2. The writings of Darwin, Oparin, and Haldane presented a scenario for life's origin that was testable in the laboratory, yet it was decades before such experiments were attempted. Why do you think it took so long?

Lecture Two—Transcript

The Historical Setting of Origins Research

The mystery of life's earliest beginnings represents one of the oldest questions asked by humans. Creation myths lie at the heart of some of humanity's oldest writings, while the scientific investigation of life spans almost three millennia. In this, the second lecture of our series, I want to review some of the long and varied history of origins research. Both in this lecture and in those to come, we will see that early efforts to answer this great question were hampered by the absence of three things: relevant evidence, appropriate experimental equipment, and a theoretical notion of emergence.

Prior to the discovery of living cells in the 17th century, there was very little experimental evidence to inform the question of life's origins. Nevertheless, philosophers of ancient Greece were not shy about proposing theories. Thales of Miletus, who lived from about 625 to 547 B.C., and who is widely credited as the first of the Greek philosophers, anticipated centuries of thought by citing water as the ultimate substance of the world and of life. Thales's student Anaximander, who lived from about 611 to 547 B.C., echoed this view. He claimed that animals were created from water by the evaporative power of the sun. Of special note is that Thales and Anaximander relied on a spontaneous, naturalistic explanation of life. They invoked a natural process, not a creation by the gods.

Subsequent Greek scholars, notably Empedocles, who lived from about 492 to 432 B.C., expanded the list of fundamental elements to four—earth, water, air, and fire. He proposed that life arose from an association of water and earth under the influence of heat.

By the 4th century B.C., a significant debate regarding the nature of matter arose between the schools of Democritus, who lived from about 460 to 370 B.C., and Aristotle, who lived a bit later, from 384 to 322 B.C. Democritus argued for a world of atoms, tiny particles that combine to form the variety of matter in our world. In his view, life arose spontaneously through the combination of atoms of soil and fire.

Aristotle opposed this atomic hypothesis, with its supposed chance mixing of elements. Nevertheless, he too embraced the spontaneous naturalistic generation of life on Earth. At the core of this belief was the doctrine of "vitalism;" that is the idea that every organism is imbued with a "life force," intrinsically different from the forces that act on inanimate matter. That is the way things remained for almost two thousand years.

We fast forward to the 17th century, which is a time when the theory of life's spontaneous generation was accepted knowledge. In fact, the only real challenge to the Aristotelian view came from a few theologians who argued that all living things were formed by God during the first days of creation.

A sense of the times is provided by a famous quotation from the influential Belgian naturalist Jan Baptista van Helmont, who lived from 1577 to 1644. He wrote, "If one stuffs a soiled shirt into the mouth of a jar containing grains of wheat, the ferment released from the soiled shirt, combining with the odor of the grain, transmutes the mixture into mice in about 21 days." Therefore, van Helmont was describing the spontaneous generation recipe for mice.

The first experiment to challenge this conventional wisdom was performed by the Tuscan physician Francesco Redi, who lived from 1627 to 1697. Redi tackled the problem of preserving raw meat, which deteriorates rapidly in air. In the first part of the experiment, he used open containers. If he let flies land on the meat, then maggots would invariably appear a few days later.

In the second part of the experiment, he covered the containers with muslin to protect it from the flies. In that case, no maggots grew. He described all these studies in his influential book *Experiments in the Generation of Insects*, published in 1668. Redi's discovery that maggots appear only in meat contacted by flies led him to conclude that the flies, not spontaneous generation, cause maggots. Redi's experiment was brilliantly conceived and executed, and it provides us with one of the earliest examples of the scientific method. Even so, Redi appears to have accepted the idea that microscopic organisms may well arise spontaneously all around us, all the time.

For the better part of the next two centuries, until well into the 19th century, the doctrine of vitalism held sway. What is more, many naturalists felt that the "life force" was not only different from the well studied forces and motions of the non-living world, but that the life force was inherently unknowable. The influential German philosopher Immanuel Kant, who lived from 1724 to 1804, wrote in his *Critique of Judgment* the following telling quote, "It is absurd for men…to hope that another Newton will arise in the future, who shall make comprehensible by us the production of a blade of grass according to natural laws which no design has ordered. We must abundantly deny this insight to men." In other words, the origin of life is an ongoing, supernatural phenomenon, a process beyond the study and understanding of humans.

Why did so many people accept this idea of spontaneous generation? Think for a moment about your own experience, and I bet you will get a feeling for why this idea of spontaneous generation is not such a strange point of view. After all, mold appears to grow spontaneously on cheese and on bread. Worm-like maggots form in old meat, and every spring new leaves sprout and flowers blossom in the season of renewal. Therefore, it is not so surprising that many people accepted unquestioningly that life arises through spontaneous generation.

As we will see, this view continued well into the 19th century, though there were an increasing number of respected dissenters, and the topic fostered lots of debate. One important factor in the spontaneous generation controversy was the 17th-century invention of the microscope, and the surprising discovery that microscopic life is everywhere. Nevertheless, microorganisms failed to resolve the controversy. After all, anyone who favored spontaneous generation could cite microbes as just one more manifestation of a life force, while those opposed pointed to microscopic life as an obvious source of contamination in any experiment.

As anyone who does experiments will tell, the interpretation of data is seldom unambiguous. This fact of scientific life was highlighted by a marvelous 18th-century exchange. On the one hand, the noted Italian physiologist Lazzaro Spallanzani, who lived from 1729 to 1799, was opposed to spontaneous generation. He explored the subject by comparing the behavior of flasks that had been filled with nutrient-rich water. In some experiments, he boiled flasks that were sealed, and he observed that those flasks remained sterile indefinitely, as long as they remained sealed. However, when he boiled unsealed flasks, they rapidly took on a cloudy appearance due to the growth of microbes. Spallanzani concluded that ubiquitous microscopic life forms must have contaminated all of his unsealed flasks.

Spallanzani's conclusions were challenged by Englishman John Needham, who lived from 1713 to 1781. Needham was an astute amateur experimenter who performed his own fragrant experiments on hot mutton gravy. Needham agreed that boiling killed microbes, but he found that microbes soon reappeared in abundance when his gravy cooled. That result led him to a very different conclusion. He claimed that this new population of cells arose by spontaneous generation.

Naturally, Spallanzani countered that Needham's new microbes came from air-borne contamination. To prove his point, he undertook a new set of experiments in which he pumped air out of his flasks and then boiled the water. No microbes appeared in these trials. However, Needham countered with a new argument: It was a property of the air, not the water, that must carry the life force, he said.

Today, we are likely to react to this historical incident with a biased world view. It seems obvious to us that Spallanzani's conclusions about microbial contamination were correct. Moreover, it seems just as obvious that Needham's support of spontaneous generation was misguided. However, put yourself in the place of a naïve and impartial observer of the time. You would not be that familiar with the nature of bacteria, nor would their ability to replicate rapidly be common knowledge. Therefore, faced with these conflicting claims, you would have had a hard time choosing between invisible microbes on the one hand and an invisible life force on the other. Indeed, both arguments were internally consistent, so doubts must have remained.

Let us fast forward again, this time to the 19th century. By then, experiments had clearly discounted the spontaneous generation of larger organisms such as mice and flies, but the origin of microbes was still a matter of debate. It was the great French chemist Louis Pasteur, who lived from 1822 to 1885, who resolved the issue once and for all. Pasteur was a giant of experimental research, and he contributed a host of fundamental ideas to biology.

In the 1850s and 1860s, Pasteur helped to abolish belief in vitalism and the theory of spontaneous generation once and for all with his own brilliant series of experiments on sterilized solutions. He prepared a nutrient-rich sugar solution and poured it into several beakers. As with previous researchers, one set of beakers was tightly sealed to prevent any contact with the ambient air.

However, Pasteur's innovation was to prepare other beakers so that they were left open to the air, but with a narrow twisting neck, or a cotton stopper. Thus, his sugar solution was in contact with the ambient air, but microbes were unable to traverse the long glass passage or cotton barrier. Pasteur also ran a series of control experiments, by leaving beakers wide open or deliberately contaminating them with ordinary dust. Over the course of several years, he showed that boiled water, if isolated from airborne microbes, remained sterile indefinitely. His conclusion: Only microbial contamination causes new growth. Cells do not arise by spontaneous generation.

This unambiguous result had more than purely intellectual appeal. Pasteur's discoveries, along with his subsequent perfection of techniques to sterilize sealed containers of food and beverage, proved to be of tremendous practical as well as theoretical consequence. His work helped to reduce the incidence of infectious diseases, while his development of the pasteurization process is still a mainstay of the food industry.

In the course of this perceptive and elegant work, Pasteur also contributed in a curious and significant way to the study of life's origins. By experimentally verifying the dictum that no cellular life can occur without prior cellular life, his findings seemed to push back life's origins to an inconceivably remote time and place. Think about that. If life does not arise spontaneously, then where and when did it come from? How could anyone make useful observations or perform relevant experiments to study an event so ancient and inaccessible?

The experimental investigation of life's origins is a surprisingly recent game. As we have seen, as recently as two centuries ago, many scientists accepted the intuitively reasonable idea of spontaneous generation. Two hundred years ago, most naturalists accepted the concept that a life force permeates the cosmos, and a central precept of vitalism is that life arises spontaneously all around us, all the time. Therefore, the question of life's origins was not even asked, at least it was not asked in the modern experimental sense.

However, with Pasteur's experimental refutation of spontaneous generation, the door was flung wide open for new ideas. Ironically, it turns out that the only valid scientific response to Pasteur's work was to propose that life arose by—what else? Spontaneous generation!

In 1871, for example, Charles Darwin penned a speculative letter to his friend, the botanist Joseph Hooker. He wrote: "If—and Oh! What a big if!—we could conceive in some warm little pond, with all sorts of ammonia and phosphoric salts, light, heat, electricity, etc., present, that a protein compound was chemically formed ready to undergo still more complex changes..."

What Darwin was actually proposing is the spontaneous generation of life, but this origin of life did not occur by some mystical life force. He suggested instead that natural chemical processes, in a plausible natural environment, led to an increase in complexity and ultimately to life. Darwin's insightful letter also points to critical aspects of that ancient process. For example, he indicates that life's origin, wherever and however it occurred, depended on three key resources.

First, all terrestrial life depends on liquid water, which is the essential growth medium for all living cells. Every living cell, even those that survive in Antarctic ice or the driest desert ecosystem, is formed largely of water. That is true for multi-cellular life as well. Our bodies are typically 70 percent water; and many foods are normally even more water-rich, as you will know if you have ever dealt with the dehydration of freeze-dried products. Therefore, Darwin must be correct; surely, the first cells arose in a watery environment like his "warm little pond."

Life also depends on a ready and reliable source of energy. Darwin speaks of "light, heat, lightning, etc." and today's origin researchers would agree. The radiant energy of the sun provides perhaps the most obvious energy source for life today, but the energy of lightning bolts, asteroid impacts, the steady flow of heat from the Earth's interior, or the chemical energy of a variety of unstable minerals have also been invoked as the energy sources that triggered life.

The third essential input for life is a variety of chemical elements. Every known living organism incorporates atoms of carbon, oxygen, hydrogen, and nitrogen, with a bit of sulfur, phosphorus, and quite a few other elements as well. Countless atoms of these essential elements combine in graceful geometries to form myriad essential bio-molecules.

I may have given a false impression of agreement here that Pasteur laid to rest the question of spontaneous generation, so everyone agreed that life arose by a natural process long ago. In spite of Pasteur's work and the intrinsic importance of the topic, Darwin's private musings about a chemical origin of life flew in the face of conventional theology. "In the beginning God created…," was sufficient for most people, including a majority of the most distinguished scholars of the time. Consequently, Darwin, whose theory of evolution by natural selection met with a great deal of controversy following the 1859 publication of *The Origin of Species*, never published his thoughts on life's origins. Indeed, it was more than half a century later, in the 1920s, that scientific speculation on life's origin took a more formal guise.

The most notable pioneer of the modern school of origin theorists was the Russian biochemist Aleksandr Oparin, who lived from 1894 to 1980. In 1922, while he was still in his twenties and under the scrutiny of the rigid Soviet hierarchy, Oparin elaborated on the idea that life arose from a body of water that gradually became enriched in organic molecules. This body of water has subsequently been dubbed as the "Oparin Ocean" or the "primordial soup." Somehow, Oparin suggested, life emerged as these primordial molecules clumped together and they organized spontaneously into a chemical system that could make copies of themselves. In many other countries, where religious doctrine more pervasively colored thinking on origins, such revolutionary ideas would have been seen as heretical. However, Oparin's ideas arose in Communist Russia, and they resonated with the materialist philosophical worldview of the Soviet leadership.

Oparin's speculative ideas were not alone. Though Oparin's treatise was unknown to western researchers—it didn't appear in English translation until 1938—many of his postulates were closely echoed in 1929 in the independent ideas of British biochemist and geneticist J. B. S. Haldane, who was born two years before Oparin, in 1892, and died in 1964. Haldane, who was a Marxist activist in addition to his scientific work, wrote a brief, perceptive, and, many contemporaries might have said, radical essay entitled "The Origin of Life." His article appeared in the eclectic secularist British publication called *The Rationalist Annual*. That was a periodical largely devoted to the promotion of rationalism and secular education. Haldane speculated on the production of large, carbon-based molecules under the energetic influence of the sun's ultraviolet radiation. Much as in Oparin's book, Haldane proposed that the first living objects were self-replicating, specialized molecules that accumulated in the ocean, like a "hot, dilute soup."

Oparin and Haldane both offered original and intriguing ideas. More importantly, their simple chemical postulates were subject to straightforward experimental testing. And yet, for reasons that aren't at all obvious to me, Oparin, Haldane, and their contemporaries never tried to mimic the primordial soup in the lab. In fact, it was not until well after World War II that chemists began to design experiments specifically to investigate life's origin. We will learn more about those efforts in Lecture Nine. Suffice it to say, that even armed with these theoretical ideas, the experimental pursuit of life's origins must have seemed a daunting challenge.

Actually, it still does seem a daunting challenge. All of us who work in origins of life research have to deal with the seemingly vast gap between the lifeless geochemical world on the one hand, and the vibrant living Earth on the other. How did all of the mind-boggling biochemical complexity arise?

It is virtually impossible to conceive of this history as a single step that might be replicated in a single experiment. Rather, today's conventional wisdom is that living cells emerged from the barren, primordial geochemical world through a sequence of chemical steps. Therefore, I conclude that the only reasonable experimental approach is to reduce biological complexity to a series of simple chemistry experiments. Each of these steps must be performed within reasonable experimental time frames of a few weeks or months. But how do we begin? Even the simplest known living cell is complex beyond anything we can imagine. Even a single cell depends on the intricate interplay of millions of molecules engaged in literally hundreds of different, yet interdependent chemical reactions. I think it is very possible that human brains just aren't wired in a way to comprehend this kind of multidimensional complexity.

Over the past two hundred years, scientists have developed numerous chemical techniques, and their laboratories are filled with sophisticated analytical apparatus. Based on these advances, chemists continue to synthesize remarkable new materials: adhesives, paints, cleaners, drugs, cosmetics, and a host of other useful products. Yet, when confronted with the challenge of synthesizing life, it is easy to become hopelessly mired in confusion. So how can we hope to tackle the transition from geochemistry to the chemical complexity of life?

One logical approach to documenting life's origin lies in reducing the living cell to the much simpler level of life's basic chemical components, that is, the small carbon-based organic molecules, as well as the collective structures that they form. Therefore, we start by studying simple systems and then work our way up to systems of greater and greater complexity. It turns out that a fascinating theoretical field called "emergence" points to a most promising research strategy.

Let me set the stage by reviewing the natural laws that deal with how energy behaves—the laws of thermodynamics. If you would like more information, you might check out lectures seven through 10 in my other Teaching Company course, *The Joy of Science*. Two magnificent natural laws, both of which were discovered in the 19th century, quantify the behavior of energy in our lives. The first law of thermodynamics states a principle called "the conservation of energy." According to this law, energy, which is simply a measure of a system's ability to do work, can appear in diverse forms. Thus, heat is a form of energy; so is light. Energy can occur in the form of moving objects, that is, kinetic energy. It can also be stored in the form of the chemical potential energy of a battery, the gravitational potential energy of water behind a dam, and so forth. The first law of thermodynamics also tells us that energy can change from any one form to another repeatedly, but the total amount of energy does not change.

The second law of thermodynamics expands on the first by placing severe limitations on how energy can change from one form to another. In other words, there are constraints on how we can use energy. The second law of thermodynamics states that heat energy, for example, always flows from warmer to cooler regions. Heat never flows the other way. That is why the concentrated heat energy of a blazing campfire or your car's hot engine gradually radiates away. Note that all of the dissipated heat energy still exists somewhere, perhaps radiated out into space, but you can't use it to do anything useful. Another way of stating the second law is that all natural systems tend spontaneously to become disordered, or messier, if you will. Scientists measure this disorder with a characteristic called "entropy." It turns out that any collection of atoms, including your shiny new automobile, a new pair of shoes, or even your body, gradually deteriorates. For this reason, many people find the second law of thermodynamics more than a little depressing.

However, clearly that is not the whole story. Look around you. You see houses. You see holly trees, and hummingbirds; all of them remarkably ordered systems. So, in spite of the second law's pronouncement that entropy inexorably increases, it's obvious that disorder is not the only end point in the universe. Observations of all kinds of familiar phenomena, including sand ripples, slime mold, and star clusters reveal that the two laws of thermodynamics do not seem to provide a complete picture. Indeed, some of us who have studied these systems suspect that a fundamental law of nature, the law that describes the emergence of this kind of complex ordered system, including every living cell, has yet to be formulated. We suspect that a law of nature is missing.

So, what are natural laws? Each of these laws quantifies the behavior of the most fundamental aspects of nature: motion, forces, energy, and matter. These laws include Newton's three laws of motion, the law of gravity, the laws of thermodynamics, and laws related to electricity and magnetism. The formulation of these dozen or so natural laws represents a crowning scientific achievement of the past four centuries. Collectively, these laws describe the behavior of matter, energy, forces, and motions in almost every human experience. Each law is expressed as a simple mathematical equation that can be applied to an infinite number of events, at scales from atoms to stars. The tremendous importance of these laws lies in their universality. Armed with these laws, scientists and engineers are able to analyze the behavior of almost any physical system.

These natural laws are so powerful and inclusive, that a number of late-19th-century scholars suggested that all of the important scientific concepts had been deduced. It turns out this is not the case, since 20th-century physics research revealed a number of new, unexpected phenomena that occur at the extremes of the very small, the very fast, and the very massive. Nevertheless, the classic laws of nature are valid in our everyday lives.

But it is clear we do not know everything. In spite of the labor of countless thousands of scientists over the centuries, we do not understand one of nature's most transforming phenomena, the emergence of complexity. The second law of thermodynamics tells us that systems as a whole do become more disordered with time. Nevertheless, at the local scale of a single living cell, or an ant colony, or your amazing conscious brain, we do not see disorder; we see the complexity that emerges.

A true pioneer in this work was the chemist Ilya Prigogine, who was born in Russia in 1917, the year of the Russian Revolution. Prigogine was educated in Brussels, and lived until 2003. He recognized that all sorts of complex emergent systems arise when energy flows through a collection of many interacting particles. There are so many familiar examples—the arms of spiral galaxies form from hundreds of millions of stars; the rings of Saturn form from countless small pieces of ice; sand dunes form from myriad sand grains. Living cells, your conscious brain, bustling cities, and symphony orchestras are all ordered structures that emerge when many interacting particles are subjected to an energy flow. Why does this happen? One way of thinking about emergent systems is that the formation of patterns in these systems speeds up the dissipation of energy that is just as mandated by the second law of thermodynamics. Therefore, for example, a sand surface covered with ripples or dunes is more efficient at breaking up the energetic flow of water or wind than a flat surface.

Whatever the process by which emergent systems arise, scientists and nonscientists alike value the wonderful order and inherent novelty of such emergent systems. In a universe of increasing entropy, emergence provides interest and novelty.

This field of emergence plays a key role in my thinking about the origins of life. The study of a variety of emergent systems provides a solid foundation for my research, because life is the quintessential emergent phenomenon. Think about that amazing transition from lifeless molecules of only a few atoms to the first living cell with its millions of interacting molecules; how do we grasp that? If we can manage to elucidate the principles that govern emergent systems, then we should be able to apply those insights to designing our experimental program.

To summarize this lecture, the question of life's origins has intrigued scholars for thousands of years. For most of human history, however, it was assumed that life easily arises spontaneously through the action of a pervasive life force. The refutation of this idea by Pasteur and others forced scientists to accept the reality of a much more ancient episode of spontaneous generation: the chemical origin of life.

Only recently has the conceptual framework of emergence helped to frame the origin of life as a sequence of chemical steps. In this scenario, each emergent step in life's origin increases the complexity of the system. Small molecules become larger molecules, larger molecules clump together to form so-called macromolecules, and macromolecules begin to make copies of themselves.

However, before we explore the bottom-up chemical steps leading to life, we need to understand more about life after it emerged from the prebiotic soup. What exactly is life? That's the subject of the next lecture. What, exactly, is life?

Lecture Three

What Is Life?

Scope: There is no simple answer to the question "What is life?" Indeed, this question, with its implicit division of all objects into nonliving or living systems, may well represent a false dichotomy. Life probably arose as a sequence of steps: first came the synthesis of simple organic molecules, then the assembly of macromolecules. Eventually, an evolving, self-replicating collection of macromolecules emerged. Each of these stages added some degree of chemical and structural complexity. Each step is worthy of study, and each deserves its own taxonomic identity.

Outline

I. How did life arise? Because the answer is not easy, let us take a step back and explore an equally fundamental question: What is life?

A. Noam Lahav's 1999 book, *Biogenesis*, sets out scientific definitions of life written by no fewer than 48 different authorities. These definitions span a 150-year period, from the mid-19th to the late-20th century.

B. Remarkably, no two of these definitions are the same. This lack of agreement represents a problem in the search for signs of living organisms on other worlds and for anyone interested in origins-of-life research.

1. Many biologists rely on straightforward genetic analysis—that is, more-or-less automated tests for the presence of DNA or diagnostic proteins—to identify the presence of life-forms on Earth.
2. However, those of us who search for life's origins, or possible life on other worlds, prefer not to presume too much from the possibly idiosyncratic chemistry of Earth life. Perhaps life elsewhere doesn't rely on DNA and proteins.
3. A more general definition that's able to distinguish all imaginable living objects from the diversity of nonliving objects remains elusive.

C. In the quest for an unambiguous definition of life, scientists have adopted two complementary approaches to distinguish that which is alive from that which is not.

1. The "top-down" approach scrutinizes unambiguous modern living organisms and fossil entities to identify the most primitive forms that are, or were, alive. Primitive microbes and ancient microfossils might provide clues about life's early chemistry. This strategy may be inherently limited, however, because at least so far, all known life-forms, whether living or fossil, are based on biochemically sophisticated cells containing complex molecules, including DNA and proteins. Any definition of life based on top-down research is limited to what appears to be modern biochemistry.
2. By contrast, the "bottom-up" approach devises laboratory experiments to mimic the emergent chemical process of environments on the ancient Earth. Ultimately, the goal is to synthesize a self-reproducing chemical system in the laboratory to help clarify the ancient transition from nonlife to life.
3. Gerald Joyce, a biologist at the Scripps Research Institute and a member of the NASA Exobiology panel, has developed a widely accepted "working definition" for life. He concluded that: "Life is a self-sustained chemical system capable of undergoing Darwinian evolution." Thus, life incorporates three distinctive characteristics:
   a. All life-forms must be chemical systems.
   b. Life grows and sustains itself by gathering energy and atoms from its surroundings—that's the essence of metabolism.
   c. All living entities must display variation.
4. There is still much we don't know, but this definition is probably as general, useful, and concise as any we are likely to come up with—at least until we discover more about what is actually out there.

II. Even armed with a pragmatic definition of life, it is almost impossible to know what Earth's very first life-form was like.

A. Our planet's earliest life may have been vastly different from anything we know today. Many experts suspect that the first living entity was nothing like a single isolated cell as we know it, because even the simplest modern cell incorporates bewildering chemical complexity.

B. Most researchers assume that the first life-form did not use DNA, given the exceedingly intricate genetic mechanism of life on Earth today. It may not even have used proteins, which today act as the chemical workhorses of cellular life.

C. Whatever that first life-form looked like, it must have arisen from chemical reactions among the oceans, the atmosphere, and rocks.

III. The overarching problem with studying life's origins is that even the simplest known life-form is immensely more complex than any nonliving components that might have contributed to it.

A. As the French anthropologist Claude Lévi-Strauss recognized, human beings share a deep-seated tendency to reduce complex situations to oversimplified dichotomies.

B. Any attempt to formulate an absolute definition of life—any definition that makes a sharp distinction between life and nonlife—must represent a false dichotomy. Here's why:

1. The first living cell did not just appear, fully formed, with all its chemical complexity and genetic machinery intact. Rather, life must have arisen through a step-wise sequence of emergent events, as a process of increasing chemical complexity. That sequence began with diverse processes of molecular synthesis in a variety of environments.
2. Synthesis was followed by the selection, concentration, encapsulation, and organization of those molecules into diverse molecular structures.
3. Next came the emergence of self-replicating molecules of increasing complexity and mutability.
4. Then, self-replicating molecular systems evolved through the process of natural selection, driven by competition for limited raw materials. That sequential process provides the organizing theme for much of this lecture series.

C. What looks like a yawning divide between nonliving matter and living cells tends to obscure the fact that the chemical evolution of life occurred in a stepwise sequence of increasingly complex stages of emergence. Most of that history is lost, because when modern cells emerged, they quickly consumed virtually all traces of the earlier stages of chemical evolution.

D. Our challenge, then, is to use every available clue to establish a progressive hierarchy of emergent steps.

   1. These steps may lead from a prebiotic Earth enriched in organic molecules, to clusters of molecules perhaps arranged on a rock or mineral surface, to self-replicating molecular systems that had the ability to copy themselves using the resources in their immediate environment, to encapsulation in membranes—in other words, to cellular life.
   2. We may never know for sure the exact sequence (or sequences) of these steps that occurred on the ancient Earth. Yet many of us in origins research suspect that the inevitable direction of increasing chemical complexity is more or less the same on any habitable planet or moon.

IV. Any attempt to define the *exact* point at which a system of gradually increasing complexity becomes "alive" is intrinsically arbitrary and really more of a question of perceived value than of science.

A. "What is life?" becomes a semantic question—a subjective matter of taxonomy rather than of any absolute divide.

B. Scientists in the early 21st century are in no position to define life. We have yet to articulate many of the theoretical underpinnings of biology; we don't know if life's biochemistry is highly constrained or if there are many chemical solutions to life.

C. Given only one unambiguous example of a living system—cellular life on Earth—it is best to keep an open mind and just describe the chemical characteristics of whatever we find.

V. Any universal theory of life and its origins will rest, at least in part, on the idea of gradually emergent complexity.

A. If life did arise as a sequence of emergent steps, then each of those steps should represent a taxonomically distinct, fundamentally important stage in life's molecular synthesis and organization. Each step deserves its own taxonomic niche.

B. Ultimately, the key to defining the progressive stages between nonlife and life rests in experimental studies of plausible chemical systems in realistic geochemical environments.

C. The concept of emergence simplifies this experimental endeavor, because it reduces an immensely complex historical process to a much more comprehensible succession of chemical steps. Each of these emergent steps provides a tempting focus for laboratory experimentation and theoretical modeling.

VI. The view of life's origin as a stepwise transition from chemistry to biology is especially useful in the search for life elsewhere in the universe.

A. It is possible that certain primitive prebiotic isotopic, molecular, and structural forms are inevitably eaten by more advanced cells. It is also possible that these chemicals will survive as "fossils" *only* if cellular life never developed to eat them. In this way, certain distinctive prebiotic chemical features may provide us with extraterrestrial *abiomarkers*.

B. Abiomarkers are clear evidence that the organization and evolution of biomolecules never progressed beyond a certain pre-cellular stage. As scientists expand their search for life elsewhere in the universe, they may be able to characterize alien environments according to their degree of emergence along the multi-step path to life.

Essential Reading:

Hazen, *Genesis: The Scientific Quest for Life's Origin*, ch. 2.

Lahav, Biogenesis: Theories of Life's Origin, part I.

Supplemental Reading:

Margulis and Sagan, *What Is Life?*

Pályi, Zucchi, and Caglioti, *Fundamentals of Life*.

Questions to Consider:

1. Why is the distinction between living and non-living things a false dichotomy?
2. How would you define life?

Lecture Three—Transcript

What Is Life?

A decade ago, when my colleagues and I began to think about research on the origins of life, we knew that it was a huge challenge. There is so much to learn, and so much we do not know. At the time, however, we thought that we at least understood the meaning of the core question, "How did life arise?" Well, it turns out we were wrong. You might think that life is easy to recognize, that it would be obvious if NASA finds life on Mars, or if some clever researcher makes life in the lab. A decade ago, our impression of what constitutes life mirrored the famous 1964 statement of Supreme Court Justice Potter Stewart about what constitutes obscene material, "I know it when I see it."

Well, it turns out it is not that easy. So in this lecture I'm going to take a step back from the origin problem and explore an equally fundamental question, "What is life?"

One of my all-time favorite books on the origin of life is Noam Lahav's *Biogenesis*. It was published in 1999 by Oxford University Press. Lahav works in Israel at the Hebrew University of Jerusalem, and he has been involved in origins research for almost 40 years. His book is filled with insights, as well as countless technical details. Well, as part of his text he prepared an appendix with lots of different scientific definitions of life written by no fewer than 48 different authorities. Those definitions span a 150-year period, from the mid-19th century to the late-20th century. It is worthwhile thinking about a few of those definitions.

Aleksandr Oparin, who is the Russian theorist who we met in lecture two, reflects the view of many authorities that life can be defined by a combination of traits. He says:

Life may be recognized only in bodies which have particularly special characteristics. These characteristics are peculiar to living things and are not seen in the world of the dead. What are these characteristics? In the first place there is a definite structure or organization. Then there is the ability of the organisms to metabolize, to reproduce others like themselves and also their response to stimulation.

The trouble is that Oparin's characteristics are not unique to life. Many non-living systems have "definite structure and organization." Think about your car or your TV. Organisms obtain energy from their surroundings to grow and reproduce, but so does a fire. And many natural non-living systems, such as flowing water or drifting clouds, respond to stimulation.

John Desmond Bernal, an influential 20th-century biological theorist who died in 1971, provides an even longer list of characteristics: He says, "Life is a partial, continuous, progressive, multiform, and conditionally interactive, self-realization of the potentialities of atomic electron states." I don't know about you, but that definition seems to me to be hopelessly fuzzy and unhelpful in distinguishing life from non-life.

Information theorist Stuart Kauffman of the Santa Fe Institute in New Mexico offered a more promising definition of life in 1993. He claimed that, "Life is an expected, collectively self-organized property of catalytic polymers." Embedded in this statement are a couple of key ideas. Kauffman said life is self-organized; that is, life is a collective, emergent phenomenon. He also states that life relies on chemicals that promote the production of more copies of themselves. In Kauffman's view, life might be a relatively simple collection of self-replicating chemicals. That is a much more primitive entity than modern cellular life.

The distinguished evolutionary biologist John Maynard Smith proposed a short, but persuasive, definition in 1975. He describes life as, "Any population of entities which has the properties of multiplication, heredity and variation." Here, Smith introduces two key ideas and thus comes closer to a useful set of criteria. First, all life possesses information that is passed from one generation to the next. That key idea of heredity may not be unique to life, but it is certainly one of life's most important characteristics. Second, life forms display variations. In life, heredity isn't perfect like a Xerox copy. Variations lead to evolution by natural selection.

And there are many other diverse opinions, as well. One expert says, "Life is the ability to communicate." Another says, "Life is a flow of energy, matter and information." Yet another says that, "Life is a self-sustained chemical system capable of undergoing Darwinian evolution."

The definitions go on and on, and I've only cited a half-dozen of the 48 in Noam Lahav's list. Yet, remarkably, no two definitions are the same.

I think you can see why this lack of agreement might represent a problem for those of us who search for signs of living organisms on other worlds, as well as for anyone interested in origin of life research. After all, how can you be sure that you've discovered life, or that you've figured out the process of life's origins, for that matter, when you can't come close to defining what exactly life is? In spite of generations of work by literally hundreds of thousands of biologists, and in spite of countless investigations of living organisms at every scale from molecules to continents, we still have no universally accepted definition.

If you think about it, this frustrating lack of a unifying definition shouldn't be all that surprising. For one thing, the question, "What is life?" is asked in very different contexts by different groups of people. For centuries, theologians have hotly debated life's definition in relation to the beginning of human life. Does life start at the moment of conception? Or does it begin when the fetal brain first responds? Or when the unborn heart first beats? In some theological doctrines, life commences not with any physical process, but rather at the unknowable supposed instant known as "ensoulment."

Then, at the other end of our human journey, doctors, lawyers, and politicians require a definition of life in order to deal ethically with patients who are brain dead or otherwise terminally unresponsive. As we saw with the contentious case of Terri Schiavo, the Florida woman who spent more than a decade in a coma as different groups debated her right to live or to die; lots of people have intense and emotional views on this issue.

In sharp contrast to these ethically difficult and emotionally charged issues are the more abstract ongoing scientific efforts to define life. On the one hand, as a practical matter, many biologists rely on straightforward genetic analysis—that is, more-or-less automated tests for the presence of DNA or diagnostic proteins—to identify the presence of life forms on Earth today. But those of us who search for life's origins or possible life on other worlds don't want to presume too much from the possibly idiosyncratic chemistry of Earth life. Perhaps life elsewhere doesn't rely on DNA and proteins.

Still, a more general definition that is able to distinguish all imaginable living objects from the diversity of nonliving objects remains elusive. Even today, we know relatively little about the diversity of cellular life on Earth. It's been said that a shovel full of soil from any lawn in the United States will contain hundreds of microbial species that are unknown to science. And that's not to mention the vast range of plausible non-cellular life forms that might await discovery elsewhere in the universe, or even deep within the Earth's crust.

I have to conclude that endorsing any one sweeping definition of life based on such scanty knowledge is a lot like trying to define music after listening to a single recording, for example an Elvis Presley song or a Bach solo cello suite, over and over again. Elvis and Bach are great examples of music, to be sure, but they are hardly adequate to characterize the entire art form. So, what do we do?

As you can imagine, scientists crave an unambiguous definition of life. Such a definition remains elusive, but we've adopted two complementary approaches in our efforts to distinguish that which is alive from that which is not. On the one hand, many scientists adopt the so-called "top-down" approach. Top-down refers to the effort to scrutinize all manner of unambiguous modern living organisms and fossil entities to identify the most primitive forms that are, or were, alive. It turns out that primitive microbes and ancient microfossils have the potential to provide relevant clues about life's early chemistry, as we will see in Lectures Four through Eight. I must confess, though, that I find this top-down strategy inherently limited because, at least so far, all known life forms, whether living or fossil, are based on biochemically sophisticated cells containing complex molecules, including DNA and proteins. Any definition of life based on "top-down" research is correspondingly limited to what appears to be modern biochemistry.

By contrast, a small but determined army of investigators adopts the so-called "bottom-up" approach. A principal objective of bottom-up researchers is to devise laboratory experiments to mimic the emergent chemical process of environments on the ancient Earth. Ultimately, the bottom-up goal is to synthesize a self-reproducing chemical system in the laboratory from scratch. That is an effort that might help to clarify the ancient transition from non-life to life.

You might think that all bottom-up researchers hold a common view of what will constitute the first synthetic life form, but such research actually leads to an amusing range of passionate and diverging opinions regarding what is alive. Indeed, each scientist has a tendency to define life primarily in terms of his or her own chosen chemical or biological specialty. As a result, one notable group focuses on the origin of cell membranes. To them, naturally, life began when the first encapsulating membrane appeared. Other well-respected research teams study the emergence of metabolic cycles—that is the process by which cells gather and use atoms and energy. So, naturally, for them the origin of life coincided with the origin of metabolism. Lots of other groups investigate the origin and behavior of primordial RNA, which, as we'll see in Lecture Twenty-Two, many experts assume was DNA's precursor genetic material. For them, the origin of RNA is equivalent to the origin of life. Then there are many other workers who focus on viruses, or on minerals, or even on artificial intelligence, and each researcher advocates his or her definition of what constitutes life.

Into this mix, quite a few philosophers, theologians, and science fiction writers have injected a variety of more abstract views and speculations on the full range of possible phenomena that might be said to be alive. The possibilities seem endless—conscious clouds in space, high-temperature silicate minerals, robotic life, computer life, and even a self-aware Internet. Such proposals sound at times more like a fantasy convention than science, but there's so much we don't know. Many origin-of-life scientists are loath to draw too narrow a definition in our present state of ignorance. Consequently, the scientific community, with the full support of NASA and other governmental agencies, holds regular meetings to explore the definition of life. After all, if one of NASA's prime missions is to look for life on other worlds, then we better have a clear definition for planning future missions.

As you may have sensed, it is amazing how the what-is-life question sparks passionate arguments and fosters hard-line positions. Scientists may be good at many things, but it is safe to say that compromise is not always one of them. Nevertheless, Gerald Joyce, who is a biologist working at The Scripps Research Institute and who served on a NASA Exobiology panel, developed a widely accepted working definition for life, at least in the context of NASA's space exploration. He concluded that, "Life is a self-sustained chemical system capable of undergoing Darwinian evolution."

Think about the essence of this definition of life. According to this opinion, life incorporates three distinctive characteristics. First, all life forms must be chemical systems. That means that computer programs, or robots activated with microchips, or any other electronic entities are not alive according to this definition.

A second characteristic is that life grows and sustains itself by gathering energy and atoms from its surroundings. That's the essence of metabolism. So in recognizing life, it's also essential to describe the available environmental sources of atoms and energy.

Finally, all living entities must display some sort of variation. According to the concepts of Darwinian evolution, natural selection of more fit individuals inevitably leads to evolution and the emergence of more complex entities. A system that does not have the potential to evolve does not fit this definition of life. There is still so much we do not know, but this NASA-inspired definition is probably as general, useful, and concise as any we are likely to come up with, at least until we discover more about what is actually out there.

Even armed with this pragmatic definition of life, it is almost impossible to know what Earth's very first life form was like. One very real possibility is that our planet's earliest life may have been vastly different from anything we know today. Many experts, for example, suspect that the first living entity was nothing like a single isolated cell as we know it, because even the simplest modern cell incorporates bewildering chemical complexity. Most workers assume that the first life form did not use DNA, given the exceedingly intricate genetic mechanism of life on Earth today. It may not even have used proteins, which today act as the chemical workhorses of cellular life.

Naturally, experts in different fields propose very different ideas regarding Earth's first life form. As a geologist who has spent decades studying rocks and minerals, my favorite theory is that the earliest entity to fit NASA's provisional definition of life was an extremely thin molecular coating on a rock. It is easy to imagine the simple behavior of such "flat life"; it would have just spread across mineral grains in a layer only a few billionths of a meter thick. Flat life would have exploited energy-rich mineral surfaces; it slowly spread outward like lichen from rock to rock.

Well, whatever that first life form looked like—flat or round, free-floating or attached to a mineral surface—it must have arisen from chemical reactions among the oceans, the atmosphere, and rocks. Yet, we come back to the central dilemma of origins research. The overarching problem with studying life's origins is that even the simplest known life form is immensely more complex than any nonliving components that might have contributed to it. So how did such astonishing, intricate complexity arise from lifeless raw materials?

The French anthropologist Claude Lévi-Strauss, who was born in 1908, investigated the mythologies of many cultures. In the process, he recognized a deep-seated human tendency to reduce all sorts of complex situations to oversimplified dichotomies: we tend to divide people into friend and enemy, we divide the afterlife into heaven and hell, and we divide actions into good and evil. Now, we all know that most situations are much more subtle and complex, but it is a natural tendency to simplify.

The long history of science reveals that scientists are in no way immune to the trap of this kind of oversimplification. In the 18th century, for example, one group of naturalists called "Neptunists" favored a watery origin for rocks. They fought pitched battles with the "Plutonists," who favored heat to describe the origin of rocks. Well it turns out that both were right. Rocks sometimes form by the action of water, sometimes by the action of heat, and sometimes by a combination of the two.

A similar contentious and ultimately misleading dichotomy raged between 18th-century earth scientists, who fell into two camps, the catastrophists and the uniformitarians. The catastrophists espoused the view that brief and cataclysmic events, like earthquakes and floods, dominated geological history of the Earth. The uniformitarians countered that geological processes are, for the most part, gradual and ongoing. Again, both groups are correct. Geological change occurs gradually over millions of years, but discrete catastrophic events like the impact of big asteroids also influence Earth history.

Similarly, there was a time when sharp distinctions were seen between plants and animals—I learned that in high school—and between single-celled and multi-cellular organisms. However, those once-sharp distinctions have become blurred.

I believe that any attempt to formulate an absolute definition of life, any definition that makes a sharp distinction between life and non-life, must represent a similar false dichotomy. Here's why I say that. I think it's obvious that the first living cell did not just appear fully formed, with all its chemical complexity and genetic machinery intact. Rather, I suspect that life must have arisen through a step-wise sequence of emergent events. I see life's origin as a process of increasing chemical complexity. That process began with diverse processes of molecular synthesis in a variety of environments. Synthesis was followed by the selection, concentration, encapsulation, and organization of those molecules into diverse molecular structures. Next was the emergence of self-replicating molecules of increasing complexity and mutability. Then, self-replicating molecular systems evolved through the process of natural selection, driven by competition for limited raw materials. That sequential process provides the organizing theme for much of this lecture series.

What now looks to us as a yawning divide between non-living matter and living cells tends to obscure the fact that the chemical evolution of life occurred in this stepwise sequence of increasingly complex stages of emergence. Most of that history is lost, because when modern cells emerged, they quickly consumed virtually all traces of the earlier stages of chemical evolution. They ate the evidence! Whatever forms of "proto-life" might have first emerged, they became little more than a rich source of food. The early history was wiped clean by the all-consuming cellular life, sort of like a clever criminal leaving the scene of the crime.

Our challenge, then, is to become detectives, to use every available clue to establish a progressive hierarchy of emergent steps leading from a pre-biotic Earth enriched in organic molecules, to clusters of molecules perhaps arranged on a rock or mineral surface, to self-replicating molecular systems that had the ability to copy themselves using the resources in their immediate environment, to encapsulation in membranes; in other words, to cellular life. The exact nature and sequence of these steps may well have varied in different environments, and we may never know for sure the exact sequence, or sequences, that occurred on the ancient Earth. Yet many of us in origins research suspect that the inevitable direction of increasing chemical complexity is more or less the same on any habitable planet or moon.

So perhaps now you can see how this view of life as a stepwise sequence of emergent events also informs the central question of this lecture, "What is life?" You see, any attempt to define the exact point at which a system of gradually increasing complexity becomes "alive" is intrinsically arbitrary. Where you, or I, or anyone else chooses to draw such a line is really more of a question of perceived value than of science. So, for example, if you value the intrinsic isolation of each living thing, then for you life's origin probably corresponds to the stage when chemicals are encapsulated by a flexible cell-like membrane. Or perhaps you most value life's ability to reproduce, the extraordinary ability of one creature to become two and more. If so, then self-replication may well become the demarcation point for life. Many scientists today place special value on information as the key to life, so they argue that life began with a genetic mechanism that passed information from one generation to the next.

In this context, can you see how the question—"What is life?"—becomes fundamentally a semantic question? It is a subjective matter of taxonomy rather than any absolute divide. Nature supports a rich variety of complex, emergent chemical systems. Scientists are also learning to craft a wide variety of such systems in the laboratory as well. However, no matter how curious or novel the behavior of any of these systems may be, none of them comes with an unambiguous label, "life" or "nonlife."

Do not get me wrong. Labels are extremely important in science, and experts convene earnest conferences and appoint august committees to decide on all sorts of taxonomic issues. Valid taxonomy is vital for effective communication, and it certainly provides a foundation for effective communication in any scientific pursuit. The problem facing us with respect to defining life, however, is that valid taxonomies rely on some minimum level of understanding. For example, early attempts at classifying animals purely by their color, or their shape, or other superficial features ultimately failed. Similarly, early efforts to classify chemical elements according to their physical state—solid, liquid, or gas—were unhelpful in developing a predictive chemical theory.

Recently, philosopher Carol Cleland of the University of Colorado and planetary scientist Christopher Chyba of the SETI Institute in Mountain View, California, equated modern attempts to define life with similar ill-fated 18th-century efforts to characterize water. Before the discovery of molecules and atomic theory in the 19th century, water could be characterized only by a series of non-unique traits. So water is clear and wet, but so are many oils, and muddy water is not all that clear. Water sustains life, but so do many foods; and water that contains a few dangerous microbes can kill you. The list of water's properties is long—water freezes when it gets cold, it soaks into wood, it flows downhill, and so the list grows. Nevertheless, no one of these traits, or any combination of these traits, is both necessary *and* sufficient. Indeed, no definition devised in the 18th century could possibly have captured what we now recognize as the true essence of water—the molecule with two hydrogen atoms and one oxygen atom, H_2O.

I would argue that scientists in the early 21st century are in the same boat we are in—no position to define life. We have yet to articulate many of the theoretical underpinnings of biology; we do not know if life's biochemistry is highly constrained or if there are many chemical solutions to life. We have nothing analogous to the periodic table for living entities. So, given only one unambiguous example of a living system that is cellular life on Earth, we are in no position to lock ourselves into any precise definition of life. It is much better at this point, therefore, to keep an open mind and just describe the chemical characteristics of whatever we find.

I suspect that any universal theory of life and its origins will rest, at least in part, on the idea of gradual, emergent complexity. And if life did arise as a sequence of emergent steps, then each of those steps should represent a taxonomically distinct, fundamentally important stage in life's molecular synthesis and organization. Each step deserves its own taxonomic niche. So, for example, a collection of molecules that can make copies of itself, or a thin layer of molecules growing on a mineral surface, should be recognized as transitional states between the nonliving and living worlds.

Ultimately, the key to defining the progressive stages between non-life and life rests in experimental studies of plausible chemical systems under realistic geochemical environments. The concept of emergence simplifies this experimental endeavor, because it reduces an immensely complex historical process to a much more comprehensible succession of chemical steps. Each of these emergent steps provides a tempting focus for laboratory experimentation and for theoretical modeling.

This fuzzy view of life's definition is admittedly nontraditional. But the view of life's origin as a stepwise transition from chemistry to biology is especially useful in planning a strategy for searching for life elsewhere in the universe. It's quite plausible, for example, that Mars, or Europa, or other bodies in our solar system and elsewhere progressed only part of the way along the path to cellular life. If so, then that is crucial to know, at least from NASA's point of view. It is very possible that each step in life's origin produces its own distinctive and measurable isotopic, or molecular, or structural signatures. If such markers can be identified, then these chemical features become critical observational targets for planned space missions.

It is very possible, for example, that certain primitive, pre-biotic, isotopic, molecular, and structural forms are inevitably eaten by more advanced cells. It is also possible that these chemicals will survive only as fossils if cellular life never developed to eat them. In this way, certain distinctive pre-biotic chemical features may provide us with extraterrestrial "abiomarkers." Abiomarkers are clear evidence that the organization and evolution of biomolecules never progressed beyond a certain pre-cellular stage. As scientists expand their search for life elsewhere in the universe, they may be able to characterize alien environments according to their degree of emergence along the multi-step path to life.

Saturn's large moon Titan, which was recently visited by the Huygens probe, is a particularly intriguing example. Titan is enshrouded in clouds. It possesses an atmosphere that is about one-and-a-half times thicker than Earth's. Titan's atmosphere is rich in the gases methane—that's what's sold commercially as natural gas—and ammonia—that's a gas that causes the distinctive smell of ammonia cleaners. Organic molecules color the atmosphere a hazy orange, and they rain onto the surface to form thick accumulations of organic gunk. The Huygens probe found that lakes of methane and ethane occur side-by-side with frozen expanses of rock-hard water ice.

Under normal circumstances the conditions are much too cold for liquid water to form or for any significant chemical progress toward life. However, every once in a while the impact of a large comet or asteroid may have melted large regions of ice on Titan. Following such an event, for periods of hundreds or even thousands of years, gradually cooling ice-covered lakes might have supported the first chemical steps in the path toward life. But inexorable cooling of the surface would cause such chemical progress to slow and stop, only to become frozen again. Such primitive biochemistry has been lost forever on Earth's scavenged surface, but it might conceivably survive in the deep freeze of Titan.

To summarize this lecture, there is no simple answer to the question, "What is life?" Indeed this question, with its implicit division of all objects into nonliving or living systems, may well represent a false dichotomy. Rather, life probably arose as a sequence of steps. First was the synthesis of simple organic molecules, then the assembly of macromolecules. Eventually, an evolving, self-replicating collection of macromolecules emerged, and each of these stages added some degree of chemical and structural complexity. Each step is interesting and worthy of study. Each step deserves its own taxonomic identity.

So much for speculation and conjecture; science is built on observations of the living world, coupled with relevant experiments. In the next lecture we'll see how studies of actual specimens illuminate the emergence of life, both here on Earth and elsewhere in our solar system.

Lecture Four

Is There Life on Mars?

Scope: In the last lecture, we saw the profound difficulty we face in crafting an unambiguous definition of what is (or was) alive. This lecture looks at the fascinating, age-old question of the search for life on Mars. As we examine this question, we'll see both the opportunities and the challenges associated with the top-down approach to life's origins. We'll also see why some respected scientists suspect that life did not arise on Earth. Rather, they argue, we are all descended from Martians!

Outline

I. Is there life on Mars? Few questions, scientific or otherwise, so fire the public imagination.

- **A.** There is good reason for intense interest. Mars is our planetary next-door neighbor and the most accessible on which to look for alien life.
- **B.** A flood of new data from NASA reveals that Mars, like Earth, once had an abundance of surface water, especially during the planet's first billion years.
- **C.** Mars once had sunlit lakes, underwater volcanic systems, and a benign temperature and atmosphere that might have sparked life.
- **D.** Indeed, Mars was probably habitable long before Earth, though each is about the same age, namely, 4.5 million years old.
- **E.** Mars still has water beneath its cold, dry surface. It is possible that fossils, molecular and otherwise, litter the surface. It is even possible that Martian microbial life still survives in protected pockets underground.

II. Scientists tend to be wary about the prospects of finding Martian life, a quest with a long and checkered history of folly and fraud. Wariness underscores the profound difficulty of recognizing life on the basis of limited, fragmentary data.

- **A.** The first serious proposals regarding life on Mars were fueled by telescope observations of the red planet's surface in the 1880s–

1890s by such early astronomers as the Italian Giovanni Schiaparelli and the American Percival Lowell.

B. In the 1960s–1970s, NASA began its efforts to probe nearby planets with robotic missions. NASA's four *Mariner* spacecraft flew close to Mars and produced thousands of images of a hostile, dry, desolate world with not even a hint of life-sustaining water. On July 20, 1976, the first of NASA's two remarkable *Viking* landers carried an array of instruments to the Martian surface to look for organic compounds and search for evidence of cellular activity. The results of these experiments confounded the mission scientists.

1. An experiment to look for microbial activity yielded a positive signal. This so-called *labeled release* experiment involved scooping up a small amount of Martian soil, exposing it to a nutrient-rich solution labeled with radioactive carbon atoms, and watching for the telltale release of radioactive carbon dioxide, a sign of life-like metabolism. The nutrients reacted vigorously with the soil, releasing radioactive carbon dioxide in what appeared to be a metabolic reaction.
2. However, a second experiment to identify carbon-based molecules in the soil seemed to contradict the labeled-release results. *Viking*'s organic analyzer found no trace of carbon-based molecules. This result was a mystery, because Mars (like Earth) is subjected to a steady rain of microscopic, organic-rich particles from space. The experimental result seemed to rule out any possibility of living cells. How could there be microbes but no carbon?

III. Scientists have also used another method to investigate the possibility of life on Mars, namely, by examining its rocks.

A. Rocks from space constantly bombard our planet, and a few make it through the atmosphere without burning up to land on Earth as meteorites. About two dozen of these meteorites are known to have come from Mars.

B. In the 1980s, chemists deduced their origin by analyzing the composition of tiny pockets of gas trapped inside them. That gas matched perfectly the known idiosyncrasies of the Martian atmosphere, which holds a mix of carbon dioxide, nitrogen, argon, and a few other gases.

C. Most Martian meteorites that fall to Earth are never found. The best place to look for them is the barren ice deserts of Antarctica. These clean, dry regions persist for thousands of years, and dark-colored meteorites stand out starkly against their white surface.

D. With the discovery of Martian meteorites, scientists could, for the first time, study what are mostly pieces of once-molten rock that existed near the Martian surface a billion years ago. Conventional wisdom would suggest that such meteorites should be utterly devoid of life.

E. One Mars meteorite proved strikingly different from the others. Collected in 1984 from the Allan Hills region of Antarctica, ALH84001 is much older than other rocks from Mars—at least 4 billion years old. This meteorite also holds a suite of minerals that suggested to some scientists the possibility of ancient interactions with liquid water.

1. A team of biologists, planetary scientists, and meteorite experts led by NASA geologist David McKay subjected pieces of the four-pound, potato-sized rock to a battery of rigorous analytical tests. They never expected to find evidence for Martian life, but a hint of freely flowing water on Mars would constitute a major discovery.
2. McKay and his colleagues saw tantalizing anomalies that could not be explained easily by normal igneous processes. As evidence mounted, the group came to believe that they had found plausible signs of Martian life.
3. In 1996, the NASA team sent their results to the journal *Science*. The Allan Hills discovery was a bombshell and a boon to planetary science. A billion dollars in new funding was quickly approved for a new push in *astrobiology*, which NASA defines as "the search for the origin, distribution, and future of life in the universe."

IV. In setting out the claim of Martian life, the NASA team pointed to data presented with meticulous care.

A. The meteorite ALH84001 contained a suite of organic molecules, including carbon-based compounds called *PAHs* (polycyclic aromatic hydrocarbons). PAHs are sturdy, long-lasting molecules that feature interlocking rings of six carbon atoms.

B. The meteorite held microscopic globules of carbonate minerals, which are often deposited through the action of liquid water passing through a system of cracks and fissures. Liquid water is the presumptive medium of all cells and, thus, a necessary condition for life.

C. The NASA team used an electron microscope to discover and characterize two iron-bearing minerals: an iron sulfide mineral called pyrrhotite and an iron oxide called magnetite. Of particular interest were the curious chainlike arrays of tiny magnetite crystals. The perfect shape of these alien crystals and their unusual chemical purity, coupled with their distinctive linear arrangements, seemed unlike anything ever seen except in a few types of bacteria called *magnetotactic microbes*.

D. ALH84001 contained myriad tiny sausage-shaped objects reminiscent of some species of terrestrial bacteria. They looked like Martian fossils.

V. The main text of the six-page article in *Science* by McKay et al. conveyed a sober and reasoned discussion of their findings, and they acknowledged that no single line of evidence was enough to trumpet the discovery of alien life. But the article concluded by stating: "Although there are alternative explanations for each of these phenomena taken individually, when they are considered collectively, particularly in view of their spatial association, we conclude that they are evidence for primitive life on early Mars."

A. Predictably, controversy exploded around the NASA scientists' bold claim. Experts pored over the paper, which was aggressively challenged on every one of the key points.

1. PAHs and other carbon molecules litter the cosmos and are among the most common constituents of pollution on Earth, including "pristine" ice from Antarctica. Even if the PAHs are from the Martian rock, there is no reason to conclude that these molecules represent remains of living cells.
2. The carbonate minerals could have formed in many ways other than by circulating water. Indeed, researchers found evidence to suggest that the minerals had formed at temperatures well above the boiling point of water.
3. The chainlike arrays of pure magnetite crystals are unusual but insufficient to prove the existence of Martian life.

Magnetotactic bacteria, furthermore, would have required a moderately strong Martian magnetic field; geophysical evidence suggests a field far weaker than that of Earth.

4. Finally, the "fossils" are an order of magnitude smaller than any known Earthly bacteria. They could contain no more than a few hundred biomolecules, not nearly enough to support the chemical complexity of any known living cell. Inorganic processes, including sample processing in the lab, are known to produce similar structures.

B. The Allan Hills story became even less credible when scientists began examining other meteorites. Surprisingly, *all* meteorites reveal signs of life—but it is Earth life. Unless hermetically sealed almost on arrival, any meteorite will be compromised. In a matter of months, microbes are able to migrate deep into a meteorite's interior, searching for nutrients that are provided by the meteorite's minerals.

C. One of the most vocal critics of the NASA team's findings was UCLA paleontologist J. William Schopf, an expert on microfossils and Earth's most ancient fossil life. In his 1999 book, *Cradle of Life*, Schopf criticized the NASA, but as we'll see in the next lecture, Schopf's words have come back to haunt him.

VI. The bottom line? Though informed opinions vary, most scientists remain unconvinced by the evidence for Martian life in the Allan Hills meteorite, yet most are also enthusiastic about getting more data.

Essential Reading:

Goldsmith, *The Hunt for Life on Mars*.

Hazen, *Genesis: The Scientific Quest for Life's Origin*, ch. 3.

Supplemental Reading:

Boyce, *The Smithsonian Book of Mars*.

Schopf, *Cradle of Life: Discovery of Earth's Earliest Fossils*, ch. 12.

Ward and Brownlee, *Rare Earth: Why Complex Life Is Uncommon in the Universe*.

Questions to Consider:

1. Why do some scientists think that life on Earth originally came from Mars?
2. The NASA scientists who claimed to have found evidence for life in a Martian meteorite argued that, while no one of their several individual lines of evidence was sufficient to prove life, the evidence *collectively* was adequate. Is this a reasonable argument? Might a similar argument be effective in a courtroom trial?

Lecture Four—Transcript
Is There Life on Mars?

Origin-of-life researchers tackle their subject from two complementary directions. On the one hand, the bottom-up approach starts from the chemical simplicity of geochemistry before any life existed, and then attempts to build complexity one step at a time. On the other hand, the top-down approach starts from the modern living Earth, with its abundant cells and fragmentary fossil record, and then attempts to work backward from biochemical complexity to the simpler chemistry of early life. Eventually, we all hope, the bottom-up and top-down approaches will converge at some primitive life form.

In the last lecture we saw the profound difficulty we face in crafting an unambiguous definition of what is, or was, alive. That problem came into particularly dramatic focus in 1996, with the discovery of supposed cellular fossils in a meteorite from Mars. In this lecture I want to look at this fascinating, age-old question of the search for life on Mars. In the process, we'll grapple with the great opportunities, as well as the daunting challenges, associated with the top-down approach to life's origins. We'll also see why some respected scientists suspect that life did not arise on Earth; rather, they argue, we are all descended from Martians!

Is there life on Mars? Few questions, scientific or otherwise, so fire the public imagination—and I have to confess, I've always been a sucker for science fiction novels and stories on the subject. I vividly remember, as a boy, reading a tale about a band of intrepid astronauts battling Martian dinosaurs, of all things. From H. G. Wells to David Bowie, speculations about Martian life have been a pervasive part of popular culture.

There is a good reason for this intense interest. Mars is our planetary next-door neighbor. It's the most exciting, important and accessible field area on which to look for alien life. Mars, like Earth, formed some 4.5 billion years ago. A flood of new data from NASA reveals that Mars, like Earth, once had an abundance of surface water, especially during the planet's first billion years. In fact, planetary geologists have dubbed that early period on Mars the "Noachian epoch." Mars once had sunlit lakes, underwater volcanic systems, and a benign temperature and atmosphere that might have sparked life. Indeed, Mars was probably habitable long before Earth. Mars still has water beneath its cold, dry surface. It is possible that fossils, molecular and otherwise, litter the surface. It is even possible that Martian microbial life still survives in protected pockets underground.

In spite of these fascinating possibilities, today's scientists tend to be wary when asked about the prospects of finding Mars life. This question has a long and checkered history of folly and fraud, a history that underscores the profound difficulty of recognizing life on the basis of limited, fragmentary data.

The first serious proposals regarding life on Mars were fueled by telescope observations of the red planet's surface. The Italian astronomer, Giovanni Schiaparelli, who lived from 1835 to 1910, first reported in 1887 what appeared to be faint straight-line markings on the Martian surface. He called these features *canali,* which is the Italian word for "channels." It is a neutral word with no suggestion of their origins. As it turns out, the *canali* were just optical illusions, a consequence of irregular surface markings. It turns out that the human brain tends to "connect the dots" between dark patches on a light background.

Ironically, Schiaparelli's descriptions were mistranslated into English as "canals." That designation fired the imagination of the wealthy amateur American astronomer, Percival Lowell, who lived from 1855 to 1916. He was born and educated in Boston, but in 1894, he used part of the family fortune to build and operate the famed Lowell Observatory in Flagstaff, Arizona, principally to examine Mars. By 1895, he reported his first observations of a network of canals on the red planet that he interpreted as evidence of an advanced civilization. Three years later he founded a journal that promoted the idea of an ancient intelligent civilization on the red planet. Such speculation inspired the imaginations of science fiction writers. H. G. Wells' novel, *The War of the Worlds*, was published in that same year of 1898.

Nevertheless, the scientific community was not convinced. In fact, hyperbolic claims like those of Lowell hardened the scientific community for generations against any proposals regarding life elsewhere in the universe.

Fast forward to the 1960s, when NASA began its wildly successful effort to probe nearby planets with robotic missions. Efforts such as the *Mariner* and *Viking* missions had a variety of goals—geology, geophysics, meteorology and more—but the search for life was always a prime objective.

Beginning in 1965 and continuing through the early 1970s, NASA's four *Mariner* spacecraft flew close to Mars and produced thousands of remarkable images of a hostile, dry, desolate world. The surface of Mars is pockmarked by craters and is dotted with immense volcanoes, much bigger than any volcanoes on Earth; but they found no canals and no cities. And not even a hint of life-sustaining water was found.

Then, on July 20, 1976, the first of NASA's two remarkable *Viking* landers carried an array of experiments to the Martian surface. One key objective of this instrument package was to look for organic compounds and to search for evidence of cellular activity.

The results of these experiments confounded the mission scientists. On the one hand, an experiment specifically designed to look for microbial activity yielded a positive signal. This so-called "labeled release" experiment involved scooping up a small amount of Martian soil, exposing it to a nutrient-rich solution that had been labeled with radioactive carbon atoms, and watching for the telltale release of radioactive carbon dioxide that's a sign of life-like metabolism. Sure enough, the experiment produced a big radioactive signal. The nutrients had reacted vigorously with the soil, releasing radioactive carbon dioxide in what appeared to be a metabolic reaction.

However, a second experiment designed to identify carbon-based molecules in the soil seemed to contradict the "labeled release" results. *Viking*'s organic analyzer found nothing at all, not even a trace of carbon-based molecules. This result was a mystery because Mars, like Earth, is subjected to a steady rain of microscopic, organic-rich particles from space. There should at least be a little carbon near the Martian surface. Yet the experiment showed nothing. That result seems to rule out any possibility of living cells. How could there be microbes and no carbon?

These ambiguous results have led to years of controversy. The majority of scientists say that the lack of organic molecules proves that there's no life, at least not where *Viking* looked. They explain the strong "labeled release" as the result of chemicals in the soil: so-called strong oxidants, such as hydrogen peroxide. According to this view, these potent chemicals reacted with the nutrients and caused the release of radioactive carbon dioxide.

But others are equally convinced that *Viking* did find life. The organic analyzer wasn't sensitive enough, they say. There would have had to be at least a hundred million cells per cubic centimeter for the instrument to detect the carbon. But many harsh climates, including most of Earth's deserts, don't have nearly that many cells. What's more, recent discoveries show that the Martian soil contains significant amounts of water, and water reacts to destroy any strong oxidants. So the positive result from the "labeled release" experiment is much more difficult to explain.

The bottom line is that Viking didn't answer the question of whether or not there is life on Mars. We have to go back and do more experiments. But NASA is taking a very cautious and measured approach to avoid any more ambiguous results—no more quick-and-dirty life-detection missions. So all of the recent efforts, Mars *Pathfinder*, Mars *Odyssey*, the *Spirit* and *Opportunity* rovers, and more, have had specific geological and geochemical objectives—to establish the geological history, identify the minerals, and to find water. So we'll all have to stay tuned until at least 2010 for the next NASA life-detection mission.

In the mean time, there's another way to study Mars and to investigate the possibility of Mars' life. Rocks from space constantly bombard our planet, and a few of these visitors make it through the atmosphere without burning up, to land on Earth as meteorites.

Of the countless thousands of meteorites that have been collected on Earth's surface, only a precious two dozen or so are known to have come to us from Mars. These fist-sized chunks of rock are pretty nondescript, and they certainly don't look rare or valuable, but in the 1980s, clever chemists deduced the distant origins of these rocks by analyzing the composition of tiny pockets of gas trapped inside them. That gas matches perfectly the known idiosyncrasies of the Martian atmosphere, which holds a mix of carbon dioxide, nitrogen, argon, and a few other gases. No other known source could have mimicked the distinctive air of Mars.

You might well wonder how a chunk of Mars could find its way to Earth. Several theorists have tackled this problem, and they maintain that giant asteroid impacts on Mars, though infrequent, are inevitable. Any such collision will hurl rocky debris away from the planet and into orbit around the Sun. The Sun and Jupiter will sweep up most of that Martian debris. That is because the Sun and Jupiter are the two most massive objects in our solar system; that is, they have the strongest gravitational pull. Eventually, often after millions of years, a tiny fraction of the rubble will inevitably find its way to Earth. Thus one out of every thousand or so meteorites that hits Earth comes from the red planet.

It is amazing to think that a piece of Mars has been transferred to Earth, and just imagine the implications. If that piece of rock holds Martian microbes, and if those microbes could withstand the long journey through space, and many scientists think the hardiest microbes could, then Earth could have been infected by Martian life. What's more, Mars was probably habitable hundreds of millions of years earlier than Earth. So if life emerged on Mars first and was transferred to Earth, then that may be how life began here. We may all be Martians! I know this idea sounds odd, but highly respected scientists are giving this proposal very serious thought.

The great majority of Martian meteorites that fall to Earth are never found. For one thing, about three quarters of all meteorites land in the ocean and fall to the bottom. Even the ones that fall on land look so much like ordinary Earth rocks that you'd probably just kick one aside. So you need to find a place where meteorites fall onto a surface where they stand out as alien objects. For that, there's no better place than a flat, white sheet of ice.

The barren ice deserts of Antarctica are the world's most productive hunting grounds for meteorites. These clean, dry regions persist for thousands of years, and dark-colored meteorites stand out starkly against their white surface. Scientists swoop down in helicopters or they glide across the ice on snowmobiles. They can collect hundreds of pristine specimens in a short field season.

Antarctic ice fields are valuable not only because they yield so many meteorites, but also because they provide a relatively unbiased inventory of meteorite types. All meteorites, whether they look exotic or commonplace, stand out against the barren white surface. In more temperate regions, stony meteorites tend to blend into the surrounding rocky landscape where they just weather away; only the resistant iron meteorites stand out, so they've always constituted the majority of meteorite finds from Europe and North America. However, Antarctic collecting areas reveal that stony meteorites are much more common. By the way, if you want to see pictures and learn more about this international meteorite collecting effort, just go to the web site of ANSMET, the Antarctic Search for Meteorites program, which is cosponsored by NASA and the Smithsonian Institution.

With the discovery of Martian meteorites, scientists could, for the first time, investigate actual pieces of another planet. As you can imagine, these nondescript chunks of dark-colored rock are incredibly valuable, and they receive the closest possible examination by earthbound scientists. Most Mars meteorites turn out to be pieces of igneous formations as much as a billion years old. These objects formed from once-molten rock near the Martian surface. Conventional wisdom would suggest that such meteorites should be utterly devoid of life. After all, you wouldn't expect life inside volcanic magma.

However, one Mars meteorite proved strikingly different from the others, and it naturally attracted extra close scrutiny. That meteorite was collected in 1984 from the Allan Hills region of Antarctica, which is how it got its scientific designation, ALH84001. For one thing, that meteorite is much older than other Mars rocks; it's at least 4 billion years old. This meteorite also holds a suite of minerals that suggested to some scientists the possibility of ancient interactions with liquid water.

Proof that Mars was once wet would be a scientific coup, so a team of biologists, planetary scientists, and meteorite experts led by NASA geologist David McKay subjected pieces of the four-pound, potato-sized rock to a battery of rigorous analytical tests. They probed the meteorite with X-rays, lasers, gamma rays, and beams of electrons. In the process, they observed characteristics as small as a billionth of an inch across. I doubt if any rock in history had received such minute examination.

No one on the NASA team had ever expected to find evidence for Martian life in the Allan Hills meteorite. All they really hoped for was a hint of freely flowing water on Mars that would constitute a major discovery. Yet gradually, as more and more data piled up, David McKay and his colleagues began to see tantalizing anomalies that could not be explained easily by normal igneous processes but were plausible signs of life. As evidence mounted, the group came to believe that they had found the smoking gun for Martian life.

After much internal debate and cautious evaluation of their data, in 1996, they decided to go public. The NASA team wrote up their results and sent the paper to the premier journal *Science*. It was accepted in short order and scheduled for publication in mid August. The NASA scientists had to agree to hold off any public statement until the August 16th embargo date.

In his engaging book, *The Hunt for Life on Mars*, astronomer Donald Goldsmith describes the leak that caused NASA to call a hasty news conference several days early. Naturally, with a discovery of this magnitude, the highest levels of government, including the White House, were alerted. Well, it turns out that President Clinton's chief political advisor, who learned of the story, then bragged about the NASA discovery to a prostitute with whom he had weekly liaisons. Unbeknownst to him, the prostitute had been selling his secrets to a weekly tabloid. So, by early August, NASA's news was out.

That is why it was that on August 7, 1996 the Allan Hills team publicly claimed the discovery of tiny elongated objects that were once living Martian microbes. Naturally, the announcement was a bombshell. Headlines of austere newspapers screamed "Life on Mars!" while tabloids had a field day. The tabloid *Weekly World News* ran an enlarged photo of an insect with the headline, "New Photo of the Life on Mars: Frightening Picture NASA Didn't Want the World to See!" Meanwhile, *Science* published the NASA team's results in an article with a rather hyped title, at least for a scientific journal, "Search for Past Life on Mars—Possible Relic Biogenic Activity in Martian Meteorite ALH84001."

This discovery was a huge deal for NASA and its planetary exploration programs. They flooded the media with press releases, held news conferences, and convened scientific meetings to discuss the findings. President Clinton even got into the act by holding a national press conference, during which he basked in the reflected glory of NASA's triumph. In that speech, President Clinton said, "Like all discoveries, this one will and should continue to be reviewed, examined, and scrutinized. … I am determined that the American space program will put its full intellectual power and technological prowess behind the search for further evidence of life on Mars."

The Allan Hills discovery was thus a boon to planetary science. A billion dollars in new funding was quickly approved for a new push in "astrobiology," which NASA defines as "the search for the origin, distribution, and future of life in the universe." In fact, my own research is done under a grant from NASA's Astrobiology Institute, which is a direct outcome of this episode.

So what was the basis of McKay and coworker's claim? Did they find convincing evidence for life on Mars? In their *Science* paper, McKay and his eight coworkers pointed to five separate types of data, which they presented with meticulous care, point by point, like a prosecutor at a jury trial. Given their extraordinary claims, they really were on scientific trial by the rest of the research community.

Point number one, the meteorite was found to contain a suite of organic molecules, including carbon-based compounds called PAHs, which stands for poly-cyclic aromatic hydrocarbons. PAHs are sturdy, long-lasting molecules that feature interlocking rings of six carbon atoms. PAHs commonly form when once-living cells or other organic chemicals are subjected to high temperature. PAHs are not proof of life, by any means, but carbon is the key element of life as we know it. The presence of PAHs in ALH84001 distinguished that specimen from most of the other Martian meteorites, and was of extraordinary significance.

Point two, the meteorite held microscopic globules of carbonate minerals similar to those that make the graceful formations on the walls of caves on Earth. Such carbonates are often deposited through the action of liquid water passing through a system of cracks and fissures. Now, liquid water is the presumptive medium of all cells, and thus a necessary condition for life. What's more, their tiny structures, about a ten-thousandth of an inch in diameter, reminded some observers of minerals precipitated by microbes on Earth.

The third and fourth points relied on examining tiny minerals with sophisticated analytical tools. The NASA team used an electron microscope to discover and characterize two iron-bearing minerals. One was an iron sulfide mineral called "pyrrhotite," and the other was an iron oxide called "magnetite."

Of particular interest were the curious chainlike arrays of tiny magnetite crystals. "Magnetite" is a magnetic mineral that's found in abundance in rocks of all types. The perfect shape of these alien crystals and their unusual chemical purity, coupled with their distinctive linear arrangements, seemed unlike anything ever seen except in a few remarkable types of bacteria called "magnetotactic" microbes.

Look at this remarkable chain of black crystals. This is a modern bacterium. These microbes tend to live in thin layers of sediment where chemical conditions change rapidly with depth. They use their internal magnets to distinguish "up" from "down" by sensing the inclination of Earth's magnetic field. "Magnetotactic" bacteria are so sensitive to their vertical position that bacteria from the Northern Hemisphere tend to move in the wrong direction and die when placed in Southern Hemisphere soils, where magnetic "up" and "down" are reversed. Knowing all this, the NASA scientists enumerated all the unusual characteristics of their magnetite particles, and they claimed that no known inorganic process could have produced such an ordered crystalline array.

Finally, the fifth point, ALH84001 holds myriad tiny, sausage-shaped objects reminiscent of some species of terrestrial bacteria. Here's one of these features. Though much smaller than any known Earthly microbes, these suggestive forms provided the public with its most convincing evidence for Mars life. They actually looked like Martian fossils. Hundreds of newspapers and magazines reproduced the NASA electron microscope images with captions identifying them as "Martian microbes."

The main text of the six-page article in *Science* by McKay, et al, conveyed a sober and reasoned discussion of their findings, and they acknowledged that no single line of evidence was enough to trumpet the discovery of alien life. But the concluding sentence shifted tone and pushed the limits of many readers' credibility, "Although there are alternative explanations for each of these phenomena taken individually, when they are considered collectively, particularly in view of their spatial association, we conclude that they are evidence for primitive life on early Mars."

To paraphrase the late Carl Sagan, extraordinary claims require extraordinary proof. Predictably, controversy exploded around the NASA scientists' bold claim. Experts pored over the paper, which was aggressively challenged on every one of the key points. Opinions ranged from cautious skepticism to outright contempt. One meteorite expert, Edward Anders, dismissed it completely. He called the article, "half-baked work that should not have been published."

So let us look at each of the NASA team's claims. Point number one, the PAHs, those organic chemicals charred by high heat. It turns out that PAHs and other carbon molecules litter the cosmos, notably in the interstellar dust that forms comets and asteroids. Those are the raw materials that formed Mars. Furthermore, such molecules would have been synthesized in abundance by natural chemical processes at or near the primitive surface of Mars. What's more, PAHs are among the most common constituents of pollution on Earth. Analyze so-called "pristine" ice from Antarctica, and you'll invariably find PAHs. That's why many scientists have concluded that the meteorite could have become contaminated while sitting on the ice. And even if the PAHs are from the Martian rock, there's no reason to conclude that these molecules represent the remains of living cells.

Point two, the carbonate globules—mineralogists quickly pointed out that the carbonate minerals could have formed in many ways other than by circulating water. Carbonates can form in reactions of rock with carbon dioxide, which is the most common atmospheric gas on Mars. Carbonates commonly grow as alteration products, long after the host rock forms, or they can crystallize directly from melts by igneous processes. Indeed, a number of researchers reanalyzed the minerals and found evidence that they had formed at temperatures well above the boiling point of water, though that result is still a matter of debate.

Then, the magnetic minerals—skeptical experts also argued that the minute magnetite crystals prove nothing since they're cosmic constituents of meteorites that bear no possible signs of life. The chainlike arrays of exceptionally pure magnetite crystals are unusual, to be sure, but most observers feel that magnetite grains are insufficient by themselves to prove the existence of Martian life. "Magnetotactic" bacteria, furthermore, would have required that there was a moderately strong Martian magnetic field, whereas geophysical evidence suggests a field much weaker than that of Earth. After all, there's no reason for a microbe to go to all that trouble of making a compass needle if the compass doesn't work.

Finally, those purported fossil shapes that look like microbes—many biologists attacked this claim because the "fossils" are too small, an order of magnitude smaller than any known earthly bacteria. In fact, they are so small that they could contain no more than a few hundred biomolecules. That is not nearly enough to support the chemical complexity of any known living cell, and there's no reason to characterize these elongated shapes as fossils, since inorganic processes, including sample-processing in the lab, are known to produce similar structures.

The Allan Hills story became even less credible when scientists began examining other meteorites, Martian and otherwise, in the same meticulous detail afforded the Allan Hills specimen. Surprisingly, all meteorites reveal signs of life, but it is Earth life. Meteorites fall to the Earth's surface, where our planet's microbes are everywhere. Bacteria inevitably contaminate any rock on the surface. Almost every meteorite ever found has been on the contaminated ground for periods ranging from several days to many thousands of years. Once found, most meteorites are usually handled, breathed on, and otherwise exposed to more contamination. Even meteorites collected in the pristine Antarctic have been exposed to air for centuries. Unless hermetically sealed almost immediately on arrival, any meteorite will be compromised. In a matter of months, microbes are able to migrate deep into a meteorite's interior. They exploit every crack and crevice while they search for nutrients that are provided by the meteorite's minerals. Given such a contaminated Earth environment, how could anyone ever be sure about the microbial history of the Allan Hills meteorite?

Right from the start, one of the most vocal critics of the Martian claim was UCLA paleontologist J. William Schopf. Bill Schopf has been a leading expert on microfossils and an authority on Earth's most ancient fossil life for at least 40 years. He was outraged at what he regarded as the NASA team's shoddy analysis and unwarranted conclusions. I was surprised to learn, therefore, that Schopf was invited by NASA to participate as an objective, dissenting voice at the well-publicized August 7th press conference in 1996, which was the one at which the discovery was announced.

Schopf described the event in his popular book on micropaleontology, *Cradle of Life*, which was published in 1999. "I was like Daniel in the lion's den," he says. He didn't want to publicly humiliate the NASA crowd, at least at their own coming out party, so he was somewhat restrained in that public forum. He said, "I had tried to be reasonable, even gentle," but he underscored his criticisms of the NASA work in a sharp addendum to his book. There he attacked the NASA team with a blistering analysis. He amplified the critique by juxtaposing his analysis of ALH84001 with stories of the most egregious paleontological blunders of all time. Of the late famed meteorite, he wrote, "The minerals can't prove it. The PAHs can't either. The 'fossils' could, but they don't, and there are good reasons to question whether they are in any way related to life."

The bottom line is that while informed opinions vary, the majority of scientists at this point are unconvinced by the evidence for Martian life in the Allan Hills meteorite. But everyone I know is also wildly enthusiastic about getting more data.

From the canals of Percival Lowell, to the ambiguous results of the *Viking* lander, to ALH84001, the hunt for life on Mars has featured conflicts between our vivid imaginations and the cold, hard scientific facts. The heated and entertaining debate over purported fossils in the Allan Hills Martian meteorite is just another chapter in the effort to discover whether there is life on Mars. Given such a troubled context, NASA will certainly choose its next round of life detection experiments with the greatest of care.

Bill Schopf concluded his 1999 analysis on a philosophical note, "There are fine lines between what is known, guessed, and hoped for, and because science is done by real people these lines are sometimes crossed. But science is not guessing." This statement, which underscores the uncertainties in the search for life on Mars, applies equally well to the search for ancient life on Earth. This is a search that has been led by Schopf himself. As we'll see in the next lecture, Schopf's cautious words of wisdom have come back to haunt him.

Lecture Five

Earth's Oldest Fossils

Scope: In this lecture, we continue to examine the top-down approach to life's origins by considering the intensely controversial nature of black, carbon-bearing rock from the wilds of Australia, which may or may not entomb the oldest fossils on our planet.

Outline

I. As we have seen, the top-down approach to our inquiry attempts to tease out characteristics of ancient life-forms by examining the evidence of living and fossil organisms and working backwards.

A. Success depends on our ability to ferret out and characterize Earth's most ancient fossil life, which is difficult to find but may reveal key steps in life's emergence, evidence of the size and shape of ancient life, and the timing of life's origin.

B. Those fragile, fragmentary clues may help us bridge the vast gulf between geochemistry and biochemistry.

II. It is critical to know roughly *when* life occurred on Earth and how quickly. A fast emergence supports the idea that life is a cosmic imperative. Our best guess about the timing of life's origin is in the form of a bracket in time.

A. At the older end, we know that planet Earth formed about 4.5 billion years ago as a giant, molten, red-hot, glowing sphere, the result of the accumulation of countless comets, asteroids, and other cosmic debris.

1. As the planet grew, gravitational potential energy was converted to heat energy that would have turned the surface red hot.
2. Prior to maybe 4.4 billion years ago, that heat would have prevented the formation of any life.
3. For the next half-billion years, as recently as 4 billion years ago, meteorites pulverized every square inch of Earth's surface.

4. What's more, every few tens of millions of years, an epic impact of an object 100 kilometers or more across must have punctuated the steady rain of smaller boulders.
5. Such catastrophic events would have repeatedly vaporized any nascent oceans and blasted much of the primitive atmosphere into space. No imaginable life-form could have survived the hellish onslaught of that Hadean period, before about 4 billion years ago.

B. On the other hand, fossil evidence suggests that 3.5 billion years ago, life had established a firm foothold on Earth. And some geologists claim that Earth was a living planet perhaps even 3.85 billion years ago.

III. UCLA paleontologist J. William Schopf, a leading expert on microfossils whom we met in Lecture Four, made the news in April 1993, when he announced the discovery of Earth's oldest fossils.

A. Schopf claimed to have identified actual single cells of several different species. These cells may have been preserved for 3.465 billion years in the microcrystals of chert, a rock that is similar to flint and found in the sun-baked northwestern corner of Western Australia.

B. Even more surprising, these cells occurred in filament-like chains strongly reminiscent of those formed by modern microbes that are photosynthetic. The modern photosynthetic cells have the relatively advanced chemical capability to harvest sunlight. Schopf hinted that the fossil species might have also been chemically advanced.

C. Such studies deserve the closest scrutiny, as they help constrain the time interval during which life emerged. They also provide a tempting challenge in our efforts to tease out signs of life from the ancient world.

1. Schopf had spent decades establishing a reputation as one the world's leading experts in finding and describing ancient single-celled fossils. He and his students had cataloged dozens of new microbial species from rocks 1 to 2 billion years old.
2. In the process, they established a rigorous set of standards for the identification and reporting of new finds.
3. As a result of these well-documented guidelines and Schopf's own solid reputation for cautious, conservative science, the

work on the what was now called the Apex Chert was initially accepted.

D. One aspect of Schopf's 1993 study—the claim that some of the microbes were photosynthetic and, hence, oxygen-producing—remained deeply puzzling.

  1. Geochemical evidence from Earth's oldest rocks points to an oxygen-poor atmosphere prior to about 2.2 billion years ago. That's the time that most researchers identify with the rise of oxygen in the atmosphere.

  2. Many scientists wondered how oxygen-producing microbes could be present more than a billion years earlier.

IV. Controversy erupted in March 2002.

A. Oxford paleontologist Martin Brasier and a team of colleagues reexamined the original type specimens of the Apex Chert fossils.

  1. They found that the familiar images from Schopf's 1993 paper were visible if the focus of the microscope was set in just the right plane, but scan up or down and some of the most famous "fossils" appeared to be more extended shapes that seemed to have little to do with microbes. Rather than chains of cells, these researchers saw irregular branching streaks and sheet-like structures.

  2. To refine their view of the Apex fossils, Brasier and his team employed a microscopic technique called *image montage*, which allowed them to use sharp images of the original thin sections at many different levels within the rock slice to reveal three-dimensional details not previously obvious.

  3. These meticulous microscopic investigations cast the Apex fossils in a new light. Their three-dimensional structures seemed to differ sharply from those of any known cellular assemblages. In some cases, the "filaments" appeared to be more like irregular planes or sheets. In others, they branched, a feature never observed in cells.

  4. The thin sections with the most convincing cell-like objects contained far more black shapes that bore no resemblance to cells—forms that Schopf must have seen but failed to describe.

B. Further study by Brasier's geological colleagues in Western Australia called the interpretation of the Apex Chert into question.

1. Schopf had visited the site only briefly and, based on the linear character of the outcrop, he had described a standard layered sedimentary sequence with the black chert lying between other layers—a typical ocean-floor scenario that might support life.
2. After detailed field mapping of the site, Australian geologists Martin van Kranendonk and John Lindsay came to a different conclusion. The geological setting of the Apex Chert was much more complex than a simple layered formation.
3. According to the reinterpretation, the black chert formed as a result of hot volcanic fluids circulating through cracks and fissures in an active underground system as part of a cross-cutting vein.
4. This discovery caused problems for Schopf's interpretation. This relationship, with a vein of chert cutting across older rocks, means that the Apex Chert could be much younger than the surrounding rock. Further, if the chert was formed by hydrothermal activity, then the temperatures would have been far above the permissible limits for life.

C. In 2002, Brasier and his colleagues challenged Schopf's claims in a public attack. The man who had taken to task so many other scientists, including NASA's Allan Hills team, was now on the defensive.

1. The Brasier team report, and Schopf's rebuttal, appeared back-to-back in the March 7, 2002, issue of *Nature*. The debate came to a head on April 9, 2002, at the second biennial NASA Astrobiology Science Conference, where Schopf and Brasier squared off.
2. Schopf took the audience through his dozen or so necessary and sufficient criteria to establish the authenticity of ancient fossil cells. He reviewed data on the distinctive size and distribution of the Apex Chert fossil objects, he showed pictures of their elongated shapes, and he discussed the geological setting.
3. Schopf also offered some striking new analytical data. The smudgy black Apex Chert "fossils" are composed principally of carbon. The carbon remains of fossil cells, he claimed, are much less ordered than the crystalline carbon deposited as a lifeless mineral. Schopf argued that the degree of crystallinity

can be measured by the technique of *Raman spectroscopy*, and he presented a suite of new Raman spectra. Indeed, sharp, spiky peaks are characteristic of inorganic carbon. These peaks stood in contrast to what Schopf saw as the obviously biological broad humps in the Raman spectra from the Apex Chert.

4. Brasier made his rebuttal. The purported microscopic fossils were the most troubling evidence. Use the right lighting, trim the field of view, and adjust the level of focus, and the Apex features do look a lot like strings of cells, he said. However, if the focus is raised or lowered slightly, or shifted to another field of view, significant doubts arise. What are all those shapeless black blobs next to the "fossil"? How can a supposed straight chain of cells suddenly branch like a Y?
5. Schopf had the geology all wrong, Brasier said. His detailed geological map of the Apex area showed the black chert as a cross-cutting vein, which was evidence that the chert had formed later than the surrounding rocks.
6. Brasier also outlined new chemical experiments that had produced cell-like chains of precipitates in a purely inorganic setting. That meant that nonliving structures similar to the supposed Apex fossils might form with ease under the right chemical circumstances.
7. Brasier demonstrated how carbon-rich deposits might have formed nonbiologically through a familiar industrial process called the *Fischer-Tropsch*, or *F-T*, *synthesis*. In F-T synthesis, organic compounds are manufactured when hot gases pass over an iron metal surface. In nature, the same thing can happen in the presence of iron-rich minerals.
8. Brasier showed his own Raman spectroscopic data of inorganic carbon. In this case, several examples of nonbiological carbon clearly displayed the same broad features as the carbon of Schopf's fossils. Broad Raman spectra may be necessary to demonstrate ancient life, but they're not sufficient to prove the case.

V. Brasier and friends were successful in casting some doubt on Schopf's claims of ancient microbial fossils, but they did not disprove the theory that the carbon residues were biogenic in origin.

A. The debate outlined the problem and established what we know and what we don't know. Much of the Apex story is based on the interpretation of fuzzy black objects in a fuzzier geological context.

B. We have no way of knowing what 3.5 billion years of decay might have done to ancient microbes, and in many ways Brasier's arguments were just as subjective as Schopf's.

C. The knotty technical issues deserve further study. Armed with new techniques, paleontologists keep searching thin sections of ancient rocks in hopes of finding Earth's earliest fossils.

VI. Unambiguous identification of ancient life from microscopic structures is fraught with difficulty. Tiny rods and spheres are not always useful indicators of biology. If fossils are to provide any clues about life's emergence, then we have to look—as we will in the next lecture—beyond microscopic structures to the tiniest fossils of all. We have to look at atoms and their isotopes.

Essential Reading:

Hazen, *Genesis: The Scientific Quest for Life's Origin*, ch. 3.

Schopf, *Cradle of Life: The Discovery of Earth's Earliest Fossils*.

Supplemental Reading:

Dick and Strick, *The Living Universe: NASA and the Development of Astrobiology*.

Knoll, *Life on a Young Planet: The First Three Billion Years of Evolution on Earth*.

Rudwick, *The Meaning of Fossils: Episodes in the History of Paleontology*.

Schopf, *Earth's Earliest Biosphere: Its Origin and Evolution*.

Questions to Consider:

1. Which position do you think requires fewer assumptions: that evidence for life in a 3.5-billion-year-old rock is likely or that life in such a rock is unlikely?

2. In a scientific debate, can there be "winners" and "losers"? If so, who do you think won the Schopf-Brasier debate?

Lecture Five—Transcript
Earth's Oldest Fossils

In our quest to understand life's origins, we rely on two complementary strategies. Bottom-up research starts with chemical simplicity and builds complexity. I'll take that approach starting in Lecture Nine. By contrast, as we saw in the last lecture on Mars life, the top-down approach attempts to tease out characteristics of ancient life forms by examining the evidence of living and fossil organisms and working backwards.

Success in the top-down approach to life's origins depends on our ability to ferret out and characterize Earth's most ancient fossil life. Ancient microbial fossils are exceptionally rare. They're difficult to find and challenging to describe, but those fragile, fragmentary clues may help us bridge the vast gulf between geochemistry and biochemistry. Ancient fossils may reveal key steps in life's emergence.

You might think that it would be vastly easier to detect and analyze fossil microbial life in Earth's ancient rocks than in the handful of meteorite fragments from Mars. It's logical to assume that the difficulties associated with the interpretation of data from the Allan Hills meteorite shouldn't apply to rocks from Earth. After all, geologists can go into the field and collect literally tons of specimens. They can spend weeks on site scrutinizing their geological setting, and they can check and recheck any critical measurements by sending duplicate samples to lots of different laboratories. It's certainly true that some of these specimens are difficult to find and challenging to collect, but it doesn't really matter how remote the rocks or treacherous the journey. It's well worth the effort because Earth's earliest fossils provide us with unique information about life's beginnings. For one thing, they provide evidence regarding the size and shape of ancient life, the kinds of cells and how they clump together. In addition, ancient fossils reveal a lot about the timing of life's origin.

So, in this lecture, I'll continue our examination of the top-down approach by looking at the story of other intensely controversial rocks, including a black, carbon-bearing rock from the wilds of Australia that, according to paleontologist Bill Schopf, entombs the oldest fossils on our planet. In thinking about life's origins, it's critical to know roughly when it happened, and how quickly. After all, if life emerged relatively quickly, then the process may be relatively easy. A fast emergence of life supports the idea that life is a cosmic imperative.

Our best guess about the timing of life's origin is in the form of a bracket in time. At the older end, we know that planet Earth formed about 4.5 billion years ago as a giant, molten, red-hot, glowing sphere. This formation, which occurred at the same time as the formation of the Sun and the other planets, was the result of the accumulation of countless comets and asteroids and other cosmic debris. As the planet grew, gravitational potential energy was converted to heat energy that would have turned the surface red hot. Prior to maybe 4.4 billion years ago, that heat would have prevented the formation of any life.

For the next half-billion years, as recently as 4 billion years ago, a rain of meteorites pulverized every square inch of Earth's surface. The geological record of this so-called "Hadean Era" has been completely eroded away on Earth, but we have only to look at the pockmarked face of the Moon to chronicle that period of intense meteorite activity. What's more, as the Moon attests, every few tens of millions of years, an epic impact of an object a hundred kilometers or more across must have punctuated the steady rain of smaller boulders. Such catastrophic events would have repeatedly vaporized any nascent oceans and blasted much of the primitive atmosphere into space. No imaginable life form could have survived the hellish onslaught of that Hadean period, before about 4 billion years ago.

On the other hand, fossil evidence suggests that 3.5 billion years ago life had established a firm foothold on Earth. Some geologists claim that Earth was a living planet perhaps even 3.85 billion years ago. I'll discuss the evidence for those incredibly ancient fossils in this lecture and in the next. We'll see how chemical evidence for life in Earth's oldest known rocks—including formations 3.8 to 3.5 billion years old from Greenland, South Africa, and Australia—seems to establish this remarkably ancient beginning for life.

So we don't know exactly when cellular life arose, but the window of opportunity appears to have been surprisingly short. In any case, the short period of no more than a few hundred million years, and perhaps shorter than 150 million years, suggests that life's emergence was rapid, at least on a geological time scale.

I promised you at the beginning of this lecture series that we'd be taking an interdisciplinary scientific journey. Sure enough, we now have to turn to the fields of geology and paleontology. Paleontologists devote their lives to looking for fragmentary signs of life in rocks.

I've noticed that paleontologists, perhaps more than scientists in any other discipline, can generate attention-grabbing headlines. We live in an age of hyperbole, and the media loves to relate stories of Guinness-style records. We've all read stories about the discovery of history's biggest shark, or the most massive dinosaur, or the oldest human skull. These stories inspire the public imagination.

So it's not very surprising that Bill Schopf made the evening news and an entry in *Guinness World Records* in April 1993, when he announced in *Science* the discovery of Earth's oldest fossils. His article was entitled "Microfossils of the Early Archean Apex Chert—New Evidence of the Antiquity of Life." Schopf claimed to have identified actual single cells of several different species. These cells may have been preserved for 3.465 billion years in the microcrystals of chert—a rock very similar to flint, and found in the sun-baked northwestern corner of Western Australia. Even more surprising, these cells occurred in filament-like chains, strongly reminiscent of those formed by modern microbes that are photosynthetic. These modern photosynthetic cells have the relatively advanced chemical capability to harvest sunlight. So Schopf hinted that the fossil species might have also been chemically advanced.

I should pause here to make a couple of caveats. For one thing, this 1993 "discovery" wasn't really all that new. Fossil microbes from the same region of Australia and of similar age had been known since the late 1970s. What's more, Schopf and his graduate student, Bonnie Packer, had published a report on these same Apex Chert fossils several years earlier. It's not clear to me why the 1993 paper attracted so much publicity, compared to these earlier similar papers.

I should also state for the record that many scientists think studies of these ancient fossils tell us absolutely nothing about the origins of life. One such expert is Bruce Runnegar, a highly respected paleontologist and the Director of NASA's Astrobiology Institute. According to Runnegar, every one of these fossil microbes is "a highly derived organism. It tells us very little about Earth's earliest cells in the same way that living birds do not reveal the attributes of dinosaurs." In other words, these fossil cells represent organisms that are already so advanced that we can't deduce anything about the transition from non-life to life.

In spite of these cautions, I think that the work of Bill Schopf and others deserves the closest scrutiny. On the one hand, as I've said, they reveal the antiquity of life and they constrain the time interval during which life emerged, but they also provide a tempting challenge in our efforts to tease out signs of life from the ancient world. Basically, if we can't recognize fossils in old Earth rocks, then we're going to have trouble when it comes to fossil microbes from Mars or other alien worlds.

As in the subsequent Allan Hills meteorite incident, which you'll recall took place three years later in 1996, Bill Schopf's claims to have found Earth's oldest fossils were extraordinary, and consequently, they demanded extraordinary proof. In this case, however, the geological community was by and large ready to accept Schopf's claims. Bill Schopf had spent decades establishing a reputation as one the world's leading experts in finding and describing ancient single-celled fossils. Schopf and his many students had managed to catalog dozens of new microbial species from rocks 1 to 2 billion years old from all around the world. In the process, they established a rigorous set of standards for the cautious identification and conservative reporting of new finds. These latest fossils from the Apex Chert were thus a part of a much bigger and very well established story. The Apex fossils did little more than push back the record for the world's oldest life by a few hundred million years.

Schopf and his UCLA colleagues developed a straightforward set of protocols for documenting specimens in the maturing field of micropaleontology. First, you conduct field trips to Earth's many geological formations of the Archean Eon; that's the period from 4 billion to 2.5 billion years ago. Next, use the methods of geology to identify layers of sediment that were deposited in ocean environments; that's the logical place to look for ancient life. Then, scour the region for outcrops that might contain fossils, and you pay special attention to distinctive carbon-rich rocks called black chert, which are known to preserve cellular structures.

In this effort, graduate students and post-docs collect hundreds of pounds of Archean rocks. They break off hunks of the most promising rock specimens, carry them back to base camp, pack them up, and then ship them back in crates to the California laboratory. These rocks are then sliced into hundreds of transparent thin sections, each a few hundredths of an inch thick.

At this stage, the research protocol for finding ancient microbes is often exceptionally time consuming and tedious. Graduate students are coaxed and coerced into spending literally thousands of hours examining every detail of every slide, micron by eye-straining micron. They have to raster back and forth across each section, to examine, systematically, every bit of surface area. As it turns out, black chert isn't really black at all when viewed in thin section. If you illuminate a slide from beneath, and view it in a powerful microscope, you suddenly see a window on the ancient living world.

The typical cherty matrix is crammed full of little black, carbon-rich blobs and smudges, most of which have no discernable biological structures. Indeed, most black chert is seemingly barren of life. But once in a while a thin section from ancient rocks reveals a host of tiny spheres, disks, rods, or chains. These are fossil forms that appear to be just like modern bacteria. Schopf was fortunate that, in 1986, one especially sharp-eyed and conscientious graduate student named Bonnie Packer scrutinized the most promising Australian specimens from the Apex Chert. Most of the hundreds of thin sections yielded absolutely nothing at all of interest, but her discovery of unambiguous microfossils in several ancient rock units led to the first prominent publication on these purported microbes. The work of Bonnie Packer set the stage for the subsequent Apex controversy.

Appearances can be deceiving. All kinds of non-biological processes can produce round specks and enigmatic squiggles. It's all too tempting to see exactly what you want to see in an ancient rock. After all, you're not going to get a publication in *Science* for finding smudges. That's why Bill Schopf and his colleagues have developed a battery of critical tests. For one thing, size matters. Single-celled organisms come in a fairly narrow size range; for example, they can't be too small or too big.

Even more critical, all microbial populations tend to cluster pretty closely around one preferred size, as opposed to the more random sizes of structures produced by most non-biological processes. So a population of cells has a pretty distinctive size range. Consequently, Schopf always advocates a statistical analysis of size distributions of the fossils. The result should be a narrow, bell-shaped curve. Uniformity of shape is yet another key. Given enough black specks, some of them will most likely bear a resemblance to microbes. So, it's misleading to photograph just one or two suggestively contoured black bits while ignoring a multitude of shapeless blobs in the same thin section.

Added to these criteria, Schopf also demanded rigor in the description of local geological setting and in the proper dating of his samples. Valid fossils must come from sedimentary rocks that were deposited in an environment that might have supported life. As a result of all these well documented guidelines, Schopf's work on the Apex Chert was initially accepted. He had established a solid reputation for cautious, conservative science, so people naturally listened.

Nevertheless, one aspect of Schopf's 1993 studies—the claim that some of these microbes were photosynthetic and hence oxygen-producing—remained deeply puzzling. Lots of geochemical evidence from Earth's oldest rocks points to an oxygen-poor atmosphere prior to about 2.2 billion years ago, the time that most researchers identify with the rise of oxygen in the atmosphere. Many scientists wondered how oxygen-producing microbes could be present more than a billion years earlier. In spite of these questions, within a few years, Schopf's claims for the earliest fossils were standard textbook fare.

His pictures of Apex fossils, first published in *Science*, had become among the most frequently reproduced of all paleontological images. Bill Schopf himself highlighted the historic findings in his popular book, *Cradle of Life—The Discovery of Earth's Earliest Fossils*. For a time, it seemed as though the antiquity of life on Earth was as well established as any fact in the geological sciences.

But it was not to be. In science, you can be sure that, if you make an extraordinary claim, somebody is going to come along and check out your story. Schopf's announcement of having discovered Earth's oldest fossils was just such a claim, so naturally others took a closer look at the Apex fossils.

The controversy erupted in March 2002, several months after Oxford paleontologist Martin Brasier and a team of seven British and Australian colleagues began to conduct a careful reexamination of the original type specimens of the Apex Chert fossils. Most of those specimens had been deposited at the Natural History Museum in London. As Brasier peered at the thin sections, he first became puzzled, and then suspicious. Even in a thin section, microbial fossils are three-dimensional. You can change the microscope focus up and down and watch as different slices through the specimens come into focus. Brasier found that the familiar images from Schopf's 1993 paper were there if you set the focus in just the right plane, but scan up and down, and some of the most famous "fossils" appeared to be more extended shapes that seemed to have little to do with microbes.

Rather than chains of cells, he saw irregular branching streaks and sheet-like structures.

To refine his view of the Apex fossils, Brasier employed a microscopic technique called image montage, which allowed him to use sharp images of the original thin sections at many different levels within the rock slice to reveal three-dimensional details that were not previously obvious. These meticulous microscopic investigations cast the Apex fossils in a new light. Their 3-D structures seemed to differ sharply from those of any known cellular assemblages. In some cases, the "filaments" appeared to be more like irregular planes or sheets. In others, they branched, and that's a feature never observed with cells. Brasier, who by this point was convinced the Apex contained no fossils at all, gave some of the more curious shapes nicknames like "wrong trousers" and "Loch Ness monster." What's more, he found that the thin sections with the most convincing cell-like objects contained far more black shapes that bore no resemblance at all to cells—forms that Schopf must have seen but failed to detail in his *Science* paper.

Brasier brought other scientists into his investigation. The geological setting was especially important. Further study by Brasier's geological colleagues in Western Australia called the entire interpretation of the Apex Chert geology into question. Schopf had visited the site only briefly and, based on the linear character of the outcrop, he had described a standard layered sedimentary sequence, with the black chert lying between other layers. That's a typical ocean-floor scenario and one that supported the idea of life. But after detailed mapping of the site, Australian geologists Martin van Kranendonk and John Lindsay came to a very different conclusion. They realized that the geological setting of the Apex Chert was much more complex than the simple layered formation Schopf had described. Indeed, it now appears that the Apex Chert formed at the site of significant hydrothermal activity, at a zone where hot volcanic fluids circulated through cracks and fissures. According to their reinterpretation, the black chert formed as a consequence of fluids circulating through such an active underground system as part of a crosscutting vein. Well, this discovery causes a couple of problems for Schopf's interpretation. For one thing, this relationship—with a vein of chert cutting across older rocks—means we don't really know the exact age of the Apex Chert. It could be much younger than the surrounding rock. In addition, if the chert was formed by hydrothermal activity, then it formed at temperatures far above the permissible limits for life.

After gathering all this counterevidence together, Brasier and his colleagues challenged Schopf's claims in an article titled, "Questioning the Evidence for Earth's Oldest Fossils." This article was published in 2002 in the widely read journal *Nature*. Their bold conclusion summed up the paper, "Here we report new research on the type and re-collected material, involving mapping, optical and electron microscopy, digital image analysis, micro-Raman spectroscopy and other geochemical techniques. We reinterpret the purported microfossil-like structures as secondary artifacts formed from amorphous graphite."

The *Nature* article was a stunning and very public attack on Schopf's credibility. The man who had taken to task so many other scientists, including NASA's Allan Hills team was now on the defensive.

In an unusual move, the editors of *Nature* delayed the Brasier, et al. article for many months to allow Schopf time to prepare a rebuttal. The two conflicting articles appeared back-to-back in the dramatic March 7, 2002, issue. First was Schopf's article, "Laser-Raman Imagery of Earth's Earliest Fossils." Immediately following that was the Brasier, et al, article, "Questioning the Evidence." In addition, *Nature* ran an accompanying "News and Views" analysis by staffer Henry Gee, who emphasized the irony of Schopf's predicament.

Seldom has a scientific debate held such high drama. Bill Schopf had made his solid reputation in large part by staking claim to Earth's oldest life, while cutting no slack for the questionable claims of others. More than any other scientist, he had publicly thrown cold water on the NASA pronouncement of life on Mars. He seemed to revel in reminding the public of past paleontological follies and, by implication, stake a claim to his own higher standards. There was little wonder, then, that science journalists were quick to highlight the controversy. One clever headline even alluded back to Schopf's own book title. The article was called "Cradle of Life or Cauldron of Crud?"

This contentious debate came to a dramatic high on April 9, 2002, at the second biennial NASA Astrobiology Science Conference. I was there at that remarkable spectacle. Even the setting was bizarre. It took place inside the immense antique dirigible hanger at Moffett Field, which is an old Navy base 30 miles south of San Francisco. Moffett Field is now home to the NASA Ames Research Center, which is also headquarters of the Astrobiology Institute. That's where Schopf and Brasier squared off.

Imagine the scene. There was a raised stage with a sturdy lectern to the left side. It was embossed with the NASA logo. A big projection screen, perhaps 12-feet square, dominated the stage, and there was a rapt audience of several hundred scientists in row after row of folding chairs.

Schopf spoke first, and it certainly was a dynamic and intense presentation. Bill is a flamboyant and excitable presenter even under normal circumstances, but at this meeting he was literally trying to preserve his scientific credibility. Schopf's strategy was to take the audience step by step through his criteria for recognizing ancient life. He had long advocated a dozen or so necessary and sufficient criteria to establish the authenticity of ancient fossil cells. He reviewed data on the distinctive size distribution of the Apex Chert fossil objects. He showed pictures of their elongated shapes, and he discussed the geological setting,

Bill Schopf also had some striking new analytical data that he claimed would prove his case. It turns out that those smudgy black Apex Chert "fossils" are composed principally of carbon, the essential element of life. Such carbon concentrations may arise by both biological and non-biological processes, so the carbon is not itself diagnostic of life. However, Schopf pointed to an important difference, "The carbon remains of fossil cells," he claimed, "are much less ordered than the crystalline carbon deposited as a lifeless mineral." He argued that the degree of crystallinity could be measured by the well established technique of Raman spectroscopy. So, Bill Schopf presented a suite of new Raman spectra; indeed, we saw that sharp spiky peaks are characteristic of inorganic carbon. These peaks stood in sharp contrast to what he saw as the obviously biological broad humps in the Raman spectra from the Apex Chert. Schopf then briefly summed up all his data, and concluded with a rhetorical flourish, "If it fits with all other evidence for life, well folks, most likely it's life."

Martin Brasier took the podium and began his rebuttal with a dismissive putdown of Schopf's presentation. He said, "Well, thank you Bill for a truly hydrothermal performance. More heat than light, perhaps." He then proceeded to make his case by examining the evidence point by point to cast doubt on Schopf's claims. The purported microscopic fossils were the most troubling evidence. If you use the right lighting, trim the field of view, and adjust the level of focus, the Apex features do look a lot like a string of cells. The size of the cells is right, and the shapes are more than a little convincing; one can even make out regularly spaced dark divisions that look like cell walls. But if you raise or lower the focus slightly, or shift to another field of view, significant doubts arise. What are all those shapeless black blobs next to the "fossil"? How can a supposed straight chain of cells suddenly branch like a "Y"?

Brasier presented other new data. Schopf had the geology all wrong, he claimed. His new detailed geological map of the Apex area showed the black chert as a cross-cutting vein, which was evidence that the chert had formed later than the surrounding rocks. He also outlined surprising new chemical experiments that had produced cell-like chains of precipitates in a purely inorganic setting. That meant that non-living structures similar to the supposed Apex fossils might form with ease under the right chemical circumstances.

Brasier also demonstrated how carbon-rich deposits might have formed non-biologically through a familiar industrial process called the Fischer-Tropsch—or F-T synthesis. In F-T synthesis, organic compounds are manufactured when hot gases pass over an iron metal surface. In nature, the same thing can happen in the presence of iron-rich minerals. Brasier even showed his own Raman spectroscopic data of inorganic carbon. In this case, several examples of non-biological carbon clearly displayed the same broad features as the carbon of Schopf's fossils. So broad Raman spectra may be necessary to demonstrate ancient life, but they're not sufficient to prove the case.

Once the two presentations ended, there was an extended period for audience questions and comments, after which we sat around trying to decide who won. Now, we all know that science isn't about winning. Those black smudges in the Apex Chert are either fossil microbes or they aren't. We all assume that the truth will eventually be known. Any debate, such as the Schopf-Brasier confrontation, really does little more than outline the problem and establishes what we know and what we don't know. But we still talked about who won.

On the one hand, Schopf is a distinguished scientist with a stellar track record. His systematic point-by-point analysis of the fossils seemed to be both logical and persuasive. But Brasier raised numerous valid questions, which have not yet been adequately addressed. It turns out that much of the Apex story is based on the interpretation of fuzzy black objects in a fuzzier geological context. There may be some doubt regarding Schopf's claims, but there's just as much doubt regarding some of Brasier's interpretations. It's just as difficult to disprove any biological activity in Apex by pointing to irregular black shapes. We have no way of knowing what 3.5 billion years of decay might have done to ancient microbes, and in many ways Brasier's arguments were just as subjective as Schopf's. So, instead of showing the audience with the unambiguous smoking gun that would thoroughly discredit Schopf, Brasier simply raised a number of doubts. Lots of knotty technical issues now deserve further study.

I want to make a subtle point about scientific evidence here. Scientists are taught to evaluate alternative hypotheses based on what's known as Occam's Razor; we accept the hypothesis that requires the fewest assumptions. But how do you choose between evidence for and against the existence of ancient microbial life, or life in a Martian meteorite, for that matter?

If your philosophical bias is that life emerges rapidly, then life is widespread, or even ubiquitous, in the universe. That means that any tantalizing sign of life, carbon smudges or scrappy fossils, seems to provide encouraging supporting evidence. If, on the other hand, you think that life in the universe is rare or even unique, then you'll likely come to the opposite conclusion. Consequently, to prove either case, that an ancient rock does or does not hold evidence for life, is extraordinarily difficult. Brasier and friends were successful in casting some doubt on Schopf's claims of ancient microbial fossils, but they came nowhere close to disproving that the carbon residues were biogenic in origin.

Schopf and Brasier both have now taken the constructive approach of going back and looking harder. An important recent advance is the imaging of microscopic fossils in three dimensions. Traditional microscopy provides only a fuzzy, two-dimensional view of small fossils. But with the recent development of laser-scanning confocal microscopy, it's possible to produce sharp, focused images at many different levels. It's possible to create amazing 3D movies of fossils that reveal details never before seen. It's also possible to use a variety of analytical techniques to map chemical information from tiny fossils. So now we can obtain 3D images of both the shape and the chemistry of microscopic fossils. Armed with these techniques, paleontologists around the world, Bill Schopf and Martin Brasier included, keep searching thin sections of ancient rocks in hopes of finding Earth's earliest fossils.

If there is a moral to the Allan Hills meteorite and Apex Chert controversies, it's that unambiguous identification of ancient life from microscopic structures is fraught with difficulty. Tiny rods and spheres are not always useful indicators of biology. The older the rock, the more difficult the interpretation of such vague features becomes. So if fossils are to provide any clues about life's ancient emergence, then we have to look, as we will in the next lecture, beyond microscopic structures, to the tiniest fossils of all. We have to look at atoms and their isotopes.

Lecture Six
Fossil Isotopes

Scope: Other fossil evidence besides microscopic cell-like spheres or segmented filaments contributes to our understanding of early life. Some paleontologists look for the very atoms and molecules of once-living organisms. The electron microprobe can map the distribution of carbon atoms in a fossil and reveal new cellular details, and mass spectrometry can measure carbon isotope ratios with exquisite accuracy. However, questions still remain about the meaning of a light carbon isotope signature. Is light carbon necessary and sufficient to prove life?

Outline

I. Paleontologists look for the very atoms and molecules of once-living organisms. How is that possible?

A. Usually when an organism dies, it decays and breaks down into small chemical bits that disperse in the environment. The essential atoms of biochemistry, including carbon, oxygen, and nitrogen, are recycled.

B. Every so often, a dying organism becomes entombed in a rock that is impermeable, and it is possible that the original atoms and molecules of that organism will be sealed off from the outside world. It is also possible that those atoms and molecules will persist for hundreds of millions of years in more or less original form.

C. Even if the organism decays into a shapeless blob, the hermetically sealed atoms and molecules represent remains of ancient life. That's the definition of a fossil—remains or evidence of past life.

II. Not many Earth rocks are as old as 3 billion years, and even fewer hold fossil evidence. Rare samples of such rocks require new approaches to the study of fossils using analytical chemistry.

A. Bones and shells are most likely to survive burial and decay. Soft tissues of animals, plants, and microbes, by contrast, almost always rot away. Any fossil of a single-celled organism is significant.

B. The very earliest fossil cells, like those from Bill Schopf's Apex Chert, are nondescript and difficult to identify. Nevertheless, paleontologists have discovered many localities with unambiguous microfossils dating back to about 3 billion years. These microfossils take the form of rods, spheres, chainlike filaments, and corkscrew spirals, which predate the appearance of multicellular organisms about a billion years ago.

C. In the summer of 1997, I began a collaboration with Andrew Knoll, a professor of paleontology at Harvard University, to study such fossils using equipment to measure atoms and isotopes.

III. At the Carnegie Institution, we quickly confirmed that we could provide three useful types of microanalytical data: chemical composition, including the presence of carbon; the isotopes of carbon and other elements; and measurements of molecules—small collections of atoms bonded together.

A. To measure chemical composition, we used an electron microprobe, which focuses a narrow beam of electrons onto a polished piece of rock.

1. The electron beam causes the rock's atoms to send out a spray of X-rays of different wavelengths. Each element has its own characteristic wavelengths.
2. The microprobe captures these X-rays and determines their diagnostic wavelengths, enabling researchers to see whether a sample contains an element such as carbon and its distribution.

B. The microprobe produces a map of the fossil's composition. Regions with a lot of carbon are colored red and yellow, and cell walls can be resolved to less than $1/10{,}000^{th}$ of an inch across. Each map consists of an array of spot analyses, similar to the pixels on a computer screen.

AN ELECTRON MICROPROBE

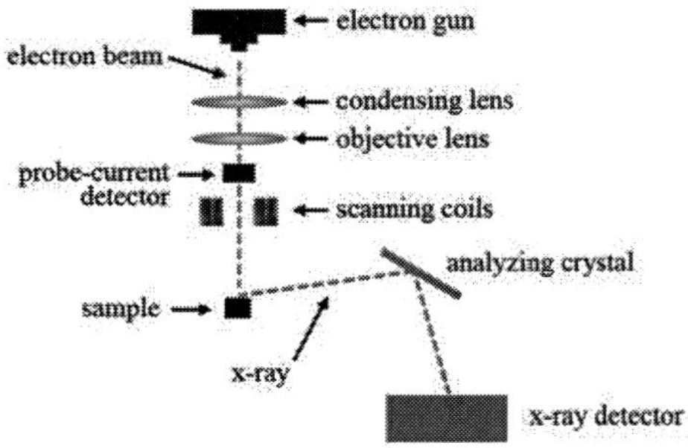

C. These maps of fossil carbon reveal subtle, cellular-scale details not previously recognized, yet they are not much more than a logical extension of the traditional focus on the morphology of fossils.

IV. Because carbon-rich fossils have the potential to preserve more information than just chemical composition, more paleontologists are turning to the study of fossil isotopes.

A. Paleontologists have zeroed in on individual elements, because every living cell has the ability to transform any collection of atoms in certain ways. This is especially true of carbon atoms, which come in two varieties. These isotopes are called carbon-12 and carbon-13.

1. Every carbon atom needs exactly 6 positively-charged protons in its nucleus. The number 6 is the atomic number of carbon, and it provides a complete chemical definition of *carbon*. Carbon is the element with 6 protons.
2. The distinction between these two common carbon isotopes, carbon-12 and carbon-13, lies in the number of neutrons, a second kind of particle in the nucleus. Carbon-12 has 6 protons and 6 neutrons in the nucleus, for a total of 12 particles. And carbon-13 has 6 protons and 7 neutrons, for a total of 13.
3. The number of neutrons, whether it is 6 or 7, has no bearing at all on carbon's chemical behavior, yet there is one physical difference: Because an atom of carbon-13 has an extra neutron in its nucleus, it is about 8 percent heavier than an atom of carbon-12.

4. Atoms of carbon-13 move a little slower and are more sluggish than carbon-12 atoms in some biological chemical reactions. Over time, as living cells digest carbon-rich foods, they tend to become slightly enriched in the lighter isotope carbon-12. If a cell dies and the cell's atoms are encased in rock, then that characteristic light isotopic signature can be preserved for billions of years.

B. With regard to ancient rocks that contain carbon atoms, there are two distinct trends. Most of the carbon in Earth's crust is held in thick rock deposits, notably of limestone, which is composed of carbon, oxygen, and other elements. Almost everywhere worldwide, the carbon atoms in limestone have a characteristic ratio of about 99 carbon-12 atoms for every 1 carbon-13 atom. Geologists take that ratio as the standard reference value, designated as 0 on the scale of carbon isotopes.

1. Recall, however, that living cells preferentially take in carbon-12, making living things isotopically lighter, with a slightly higher percentage of carbon-12 than is found in limestone.
2. Geochemists have developed a scale in which a 1 percent excess of carbon-12 relative to ordinary limestone is called *–10 per mil*, a 2 percent excess is *–20 per mil*, and so forth.
3. If the carbon in a rock is "light," with a value of –20 or –30, then there is a good chance the carbon was once processed by living cells. Countless ancient fossil specimens bear out this relationship.

C. Much attention is focused on the oldest specimens of black chert and black shale from billion-year-old rocks. These specimens all have negative carbon isotope values, but there is a lot of scatter in the data.

1. Photosynthetic microbes, which get their energy from sunlight, tend to be in the –20 to –30 range. During the last 2 billion years, when Earth's atmosphere became oxygen-rich, these organisms have dominated the fossil record—a typical "modern" carbon isotope signature.
2. Older specimens are more varied. Some primitive single-celled organisms are more efficient at concentrating carbon-12. These microbes display unusually light carbon isotopes, with values as low as –50. Thus, carbon isotopes have the

potential to reveal not just the presence of microbial life but also the microbial lifestyle.

V. In July 2002, I was asked to analyze some fragments of Strelley Pool Chert that were nearly as old as Apex Chert. I enlisted the help of Marilyn Fogel, a biologist and an expert in the way that living cells process carbon and other isotopes.

A. Isotopes of carbon and nitrogen, in particular, provide an excellent approach to tracking the way elements move from one organism to another. Carbon and nitrogen isotopes both get progressively lighter in moving through the food chain.

B. Fogel measured the ratio of isotopes in the samples using a mass spectrometer. The point was to separate carbon atoms according to their masses.

A MASS SPECTROMETER

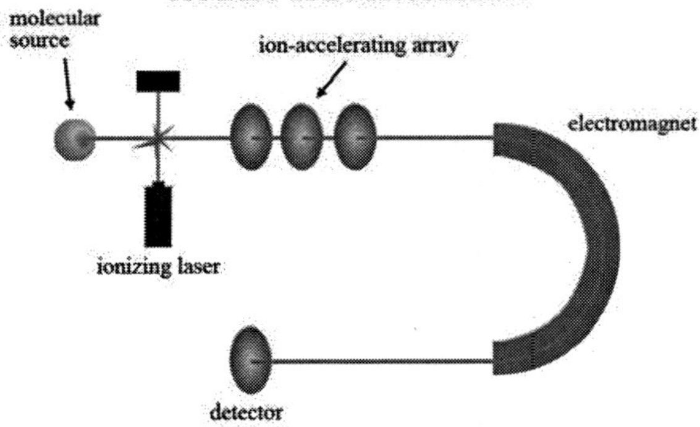

C. If an equal force is applied to two different masses, then the lighter object will accelerate more. Because carbon-12 and carbon-13 differ by about 8 percent in their masses, the carbon-12 atom will accelerate about 8 percent faster than the carbon-13 atom.

D. A single number near the end of the analysis was the key: –25.7 ± 0.5. The carbon was light, just the sort of number we've learned to expect from ancient microbes. The result for the duplicate sample, –25.9, was consistent with the first value.

E. Despite these typical values for carbon isotopes—close to –26—we still had nagging doubts about the biological origin of the carbon.

1. For one thing, the black specks in the chert looked nothing like fossil forms. And there has always been a concern that there might be nonbiological ways to produce a light carbon signature.
2. All known life-forms have a negative isotopic signature, but that doesn't mean that a negative isotope value is enough to prove that the sample was once alive.

VI. This episode raises an important question. We can measure the ratio of carbon-12 to carbon-13 in an ancient rock to a thousandth of a percent. So what does a negative carbon-isotope value tell us about that ancient rock?

A. This question came into focus in November 1996, when a group of scientists claimed to have found the most ancient fossil carbon on Earth.

1. The oldest known rock outcrops occur on the island of Akilia, off the coast of Greenland. These formations have been dated at 3.85 billion years old. The rocks don't contain any fossils, but they do hold a tiny amount of carbon that is isotopically light.
2. The Earth's oldest rocks have been heated, squeezed, and chemically altered. Any fossil forms have been thoroughly erased, and fluids that flowed through cracks and fissures would have long since removed any original carbon.
3. Nevertheless, according to geochemist Stephen J. Mojzsis at the University of Colorado and his colleagues at the Scripps Institution of Oceanography, some of the original carbon atoms were protected in crystals of the phosphorus-rich mineral apatite.
4. In analyzing the carbon isotopes of these rocks, Mojzsis found that the carbon was extremely light—on average, 3.7 percent lighter than reference limestone. That's –37 on our scale.
5. Because we don't know any nonbiological process that produces such a low value, many geologists were convinced that life had been well established on Earth by 3.85 billion years ago.

B. If true, the data held enormous significance and would drastically narrow the time window during which life originated. If there are fossil carbon atoms from 3.85-billion-year-old rocks, then life must have emerged very quickly—in less than 150 million years.

C. The interpretation of Akilia rocks turns out to be problematic.
   1. Mojzsis saw intensely folded layers and interpreted them as metamorphosed oceanic sediments, which is a logical kind of rock in which to find early cellular life.
   2. Other geologists said that the Akilia rocks are actually an ancient molten igneous mass that gradually solidified deep underground, cooling from temperatures close to 1,000° C. In this view, any carbon atoms in the rocks must have collected under extreme metamorphic conditions of high temperature and pressure and could not contain any trace of life from the time of their formation.

VII. Questions remain about the meaning of a light carbon isotope signature.

A. Is light carbon necessary to prove life? Probably so.

B. Is light carbon sufficient proof of ancient life? It is possible that some natural *non*biological processes also generate light carbon.
   1. Today, living organisms regulate the carbon cycle, and carbon compounds derived from living organisms are almost always isotopically light.
   2. Still, even before the appearance of microbes, there might have been other geochemical processes that efficiently separated carbon-12 from carbon-13.
   3. If that's true, then carbon isotopes cannot be used as unambiguous evidence for ancient life.

VIII. In the next lecture, we'll see that in their quest for signs of ancient life, paleontologists are turning to even more elusive fossils—fragments of life's oldest biomolecules.

Essential Reading:

Hazen, *Genesis: The Scientific Quest for Life's Origin*, ch. 4.

Schopf, Cradle of Life: The Discovery of Earth's Earliest Fossils.

Supplemental Reading:

Knoll, *Life on a Young Planet: The First Three Billion Years of Evolution on Earth.*

Rudwick, *The Meaning of Fossils: Episodes in the History of Paleontology.*

Schopf, *Earth's Earliest Biosphere: Its Origin and Evolution.*

Questions to Consider:

1. What is a fossil? Is it reasonable to call an ancient atom or isotope a "fossil"?
2. Why is the potential discovery of life in rocks 3.85 billion years old important? What would such a discovery reveal about life's origin?

Lecture Six—Transcript

Fossil Isotopes

In the last two lectures we came face to face with the exceptional difficulty in proving that an ancient microscopic object or a black smudge is evidence for life. The bitter controversies over the Allan Hills Martian meteorite and the Australian Apex Chert are typical of this kind of debate.

As it turns out, there are other kinds of fossil evidence that contribute to our understanding of early life. Rather than resorting to microscopic cell-like spheres or segmented filaments, some paleontologists actually look for the very atoms and molecules of once-living organisms.

How is that possible? Usually, when an organism dies, it quickly decays and breaks down into small chemical bits that disperse in the environment. The essential atoms of biochemistry, including carbon, oxygen, nitrogen, and more are recycled. These atoms are used over and over again. It is amazing to think that the majority of atoms in your body were once part of trilobites, or dinosaurs, or even those earliest living cells. Those creatures died, their atoms became part of the soil or the air, and now you eat food and breathe air made of those same atoms. Take a minute to look at your hand. A carbon atom in your little finger was once part of the stem of a jungle fern in an ancient forest; an oxygen atom in your thumbnail was once breathed in by a wooly mammoth. Imagine the amazing histories of every one of the trillions of atoms. It is all because our planet is a fabulous recycling machine.

I have said that the atoms in cells almost always recycle when the cells die, but every so often, under just the right circumstances, a dying organism becomes entombed in a rock that is impermeable. So, if a worm becomes buried under a pile of fine-grained mud, or if an insect gets trapped in sticky tree sap, then it is possible that the original atoms and molecules of that organism will be sealed off from the outside world. It is possible that those atoms and molecules will persist for hundreds of millions of years in more or less their original form. And even if the organism decays into a shapeless blob, the hermetically sealed atoms and molecules represent the remains of ancient life, and that is the definition of a fossil—remains or evidence of past life.

Not many Earth rocks are as old as 3 billion years, and even fewer have any chance to hold fossil evidence. I can tell you, it feels amazing to hold a piece of rock that old that once teemed with living organisms. I've had the chance to work on a few ancient formations, including 3.4 billion-year-old black chert from Australia and 3.1 billion-year-old sediments from South Africa. These rocks contain the very atoms and molecules of cells from the dawn of life. You can imagine that such samples, as rare and precious as they are, require us to develop new approaches to the study of fossils. You can't just rely on descriptive paleontology of sizes and shapes. We have to rely on analytical chemistry.

What do you think of when I use the word "fossil"? Most people would say dinosaur bones and teeth, or perhaps the spiny shell of a trilobite; and, it is true that bones and shells are most likely to survive burial and decay. Soft tissues of animals, plants, and microbes, by contrast, almost always rot away. When a rare fossil does preserve soft parts, it is a cause for celebration and headlines. So any fossil of a single-celled organism is significant.

The very earliest fossil cells, like those from Bill Schopf's Apex Chert, are nondescript and difficult to identify. Nevertheless, paleontologists have discovered many localities with unambiguous microfossils dating back to about 3 billion years. These microfossils take the form of rods, spheres, chainlike filaments, and curious corkscrew spirals, which predate the appearance of multi-cellular organisms about a billion years ago.

I first got into this field of study after what seemed to be a casual conversation over a beer. It was during the summer of 1997, late at night, and I was sitting with a longtime friend, Andrew Knoll. We went to school together, and he's now professor of paleontology at Harvard University. Andy and I were attending the Gordon Research Conference on Origins of Life. It was being held at New England College in Henniker, New Hampshire.

While Andy Knoll and I talked at the Gordon conference, I was struck by how many of the earliest fossils are preserved in black chert or shale. Those rock types tend to be pretty impermeable to the flow of water, so they're the most likely rocks to preserve chemical traces of original bacteria. I asked Andy if paleontologists ever study these fossils with the kind of machines that geochemists and mineralogists use to measure the atoms and isotopes of our samples. Well, apparently they didn't, so he asked me if I wanted to collaborate, and I readily agreed.

It was easy to think up a whole battery of experiments that might be brought to bear on the problem. We spent the evening talking about technical details, such as how many samples, their sizes and shapes, and more.

At the Carnegie lab, using a suite of 400-million-year-old plants from Canada, along with several different pieces of fossil wood, we quickly confirmed that these fossils could provide three useful types of micro analytical data. First, we could measure the chemical composition, including life's essential element, carbon. Second, we could measure the isotopes of carbon and other elements. I'll talk about that later in this lecture. And third, we could measure molecules, which are small collections of atoms bounded together. Of these three opportunities, the chemical composition is easiest to measure.

To measure composition we use a fancy piece of hardware called an electron microprobe. These machines are expensive, on the order of half million dollars, but they are indispensable. The electron microprobe works by focusing a narrow beam of electrons onto a polished piece of rock. The electron beam causes the rock's atoms to send out a spray of X-rays of different wavelengths. Each element has its own characteristic wavelengths. The microprobe captures these X-rays and determines their diagnostic wavelengths. So you can identify if a sample has an element like carbon, and you can see how that element is distributed.

These days most microprobe analyses are pretty easy and automated. The machine at my lab often operates 24 hours a day analyzing all kinds of rocks and minerals. So analyzing fossils in this way was pretty straightforward. The first step was to take a thin section of a fossil-bearing rock and coat it with an ultra thin layer of aluminum. That's because the surface of the fossil has to conduct electricity for this technique to work. Otherwise, electrons would pile up on the rock surface and the resulting electric charge would disrupt the data.

Once the sample was coated, it was loaded into the probe. The Geophysical Lab electron microprobe sits in its own small, dedicated room. I inserted one of the thin fossil sections into a sample holder, closed the sample port and waited a couple of minutes for the machine to achieve a high vacuum. That's necessary in order to stabilize the beam of electrons. We hadn't done much work on carbon analyses before. Most rocks are made of heavier elements, and light elements like carbon, number 6 in the periodic table, are a little bit trickier. But it worked quite well. I selected half a dozen promising areas on the thin section, each area about a fiftieth of an inch square. Then the machine automatically produced maps of the fossil's composition. It takes about three hours to produce a single map, so we usually just set it on automatic and come back after a while.

Within a few weeks we had figured out the best parameters for making maps of carbon in ancient fossils. The maps are richly colored according to the amount of carbon present. Regions with a lot of carbon are colored red and yellow, so they stand out against the dark, carbon-poor fossil matrix. We can resolve features like cell walls that are less than a ten-thousandth of an inch across. We made maps of 400-million-year-old wood that actually has all the cell structure preserved. Harvard graduate student Kevin Boyce made the most of these maps. With his sophisticated botanical eye, he was able to identify cellular details that had never been seen before.

It's really fun making these carbon maps and watching all the fine fossil details emerge. Each map consists of an array of spot analyses; it's a lot like the pixels on your computer screen. Most of our preliminary maps use a 400- x 400-point array; that's fast, and can reveal where we want to focus our attention. Then we make slower 500- x 500-pixel maps to achieve a higher resolution.

On the one hand, these maps of fossil carbon are quite wonderful because they reveal all kinds of subtle cellular-scale details not previously recognized. On the other hand, this analytical work is not much more than a logical extension of traditional paleontology. We still focus on the morphology of fossils; it's a study of size and shape. It is a slightly more technical way to make images of the specimens.

But these carbon-rich fossils have the potential to preserve much more information than just the chemical composition. That's why more and more paleontologists are turning to the study of fossil isotopes.

So let's shift gears and talk about fossil carbon isotopes. Paleontologists have zeroed in on individual elements, because every living cell has the ability to transform any collection of atoms in rather subtle and surprising ways. This is especially true of carbon atoms, which come in two distinct common varieties. These isotopes are called "carbon-12" and "carbon-13."

Every carbon atom has to have exactly six positively charged protons in its nucleus; that number six is the atomic number of carbon, and it provides a complete chemical definition of "carbon." Carbon is the element with six protons.

However, there's a second kind of particle in the nucleus, called the neutron, so the distinction between these two common carbon isotopes, carbon-12 and carbon-13, lies in that number of neutrons. Carbon-12 has six protons and six neutrons in the nucleus, for a total of 12 particles. Carbon-13 has six protons and seven neutrons, for a total of 13. You might recall that there's yet another kind of carbon isotope as well, carbon-14, with six protons and eight neutrons. That relatively rare radioactive isotope is used in age dating. There is a lot more on this topic in Lecture Twenty-Seven of my Teaching Company course, *The Joy of Science*.

So we have carbon-12 and carbon-13. Well it turns out that the number of neutrons, whether it is six or seven, has no bearing at all on carbon's chemical behavior. That means you could survive just as well on a diet of food made with pure carbon-12 or with pure carbon-13. Yet there is one physical difference. Because an atom of carbon-13 has one extra neutron in its nucleus, it is about 8 per cent heavier than an atom of carbon-12. That means that a typical person who weighs about 150 pounds would weigh about two pounds more if he or she were made of carbon-13, compared with the exact same person made entirely of carbon-12. It is a small difference, but it winds up having real and measurable effects.

In particular, heavier atoms of carbon-13 move a little slower than lighter atoms of carbon-12. As a consequence, carbon-13 atoms are more sluggish than carbon-12 atoms when taking part in some biological chemical reactions. Over time, as living cells digest carbon-rich foods, they tend to become slightly enriched in the lighter isotope carbon-12. If a cell dies and the cell's atoms are encased in rock, then that characteristic light isotopic signature can be preserved for billions of years.

So what do we find when we look at ancient rocks that contain carbon atoms? It turns out that there are two distinct trends. Most of the carbon in Earth's crust is held in thick rock deposits, notably of limestone—that is, a rock composed of carbon, oxygen, and other elements. You often see massive limestone cliffs, as in the peaks of the Canadian Rockies. Limestone also dissolves to open up most of the big caves you've seen, including Mammoth Cave and Carlsbad Caverns. It turns out that almost everywhere you find limestone worldwide, the carbon atoms have a characteristic ratio of about 99 carbon-12 atoms for every one carbon-13 atom. Geologists take that ratio as the standard reference value, which is designated as zero on the scale of carbon isotopes.

Recall, however, that living cells preferentially take in carbon-12 compared to carbon-13. That means that living things are invariably isotopically lighter, with a slightly higher percentage of carbon-12 than is found in limestone. Geochemists have developed a peculiar scale in which a one percent excess of carbon-12 relative to ordinary limestone is called "minus 10 per mil," a two percent excess is "minus 20 per mil," and so forth. Conventional wisdom is that if the carbon in a rock is "light," with a value of minus 20 or minus 30, then there is a good chance the carbon was once processed by living cells. Countless ancient fossil specimens bear out this relationship. At our lab we have measured all kinds of specimens. A 10,000-year-old fossil mammoth bone was minus 21. A 3-billion-year-old microbial mat in South African sandstones was about minus 27. Black coal represents the remains of 300-million-year-old swamp life—it typically has the carbon isotope value of minus 24 to minus 25. Perhaps my favorite carbon-rich fossils are from the Burgess Shale, a 530-million-year-old rock formation from British Columbia. It was made famous by Stephen Jay Gould's popular book, *Wonderful Life*. I found values between minus 25 and minus 27 for these specimens. So it's pretty easy to conclude that, if a rock has preserved carbon atoms from ancient life, then that carbon will be light. This would be true even if all visible traces of life were gone.

As you might expect, following the big debate between Bill Schopf and Martin Brasier, the most attention is focused on the oldest specimens of black chert and black shale from billion-year-old rocks from Africa, Australia, Europe, and North America. These specimens all have negative carbon isotope values, but there's a lot of scatter in the data. Photosynthetic microbes, which get their energy from sunlight, tend to range from minus 20 to minus 30. During the last 2 billion years, when Earth's atmosphere became oxygen-rich, these organisms have dominated the fossil record. So that's a typical "modern" carbon isotope signature.

Older specimens are much more varied. Some very primitive single-celled organisms, which get their energy from unstable minerals, are often more efficient at concentrating carbon-12. These microbes display unusually light carbon isotopes, with values as low as minus 50. So carbon isotopes have the potential to reveal not just the presence of microbial life, but also the microbial lifestyle.

I'd like to give you a sense of what it's like to make a measurement of carbon isotopes. Part of my work with Andy Knoll and his Harvard students was to help with these measurements. However, the most memorable specimen I worked on came from Australia. I was given that carbon-rich sample in July of 2002. I was giving a lecture in Australia and had made a side trip to the Australian Centre for Astrobiology. That facility is on the campus of Macquarie University in the northwestern suburbs of Sydney.

I was especially eager to meet the Centre's director, paleontologist Malcolm Walter, who's famous for his work on ancient Australian microfossils. He's a world authority on dome-shaped structures called stromatolites, which are among the oldest known signs of life on our planet. So I showed up at his office, and he suddenly perked up when I mentioned Andy Knoll and the work with his students. It's really useful to be associated with the big names in the field. So I outlined some of our procedures for using the electron microprobe to map carbon atoms. I also described our recent carbon-isotope results.

Then he did something really unexpected. He handed me a small cloth bag containing a bunch of black rock fragments. "You might be interested in these," he said. "It is Strelley Pool Chert. It is almost as old as the Apex." So I was actually holding one of the oldest rocks on Earth.

So naturally I offered to study it to measure the carbon. I told him I'd do it as soon as I returned to Washington the following week. It turns out that few labs in Australia were equipped to measure carbon isotopes, so he handed me this sample, which is just a small fraction of his collection. "Perhaps you could have a look at this," he had said. Throughout the rest of my stay in Australia this little rock was in the back of my mind, and I started working on it as soon as I got back to Washington, DC.

The most important measurement was the carbon isotope value. Would the carbon be light, like other black cherts? Might it be as low as minus 50, which would be especially convincing evidence for life? So when I got back to the Carnegie Institution, I went to the office of Marilyn Fogel. She's a biologist who's an expert in the way that living cells process carbon and other isotopes. Fogel and her many students investigate how elements cycle through ecosystems. It turns out that isotopes of carbon and nitrogen in particular provide an excellent way to track the way that elements move from one organism to another. Carbon and nitrogen isotopes both get progressively lighter as you move through the food chain. So the carbon atoms of herbivores are lighter than the carbon atoms of the plants they eat. Likewise, the carbon atoms of carnivores are lighter than the carbon atoms of the herbivores that they eat. So I used Fogel's impressive array of analytical equipment to measure these isotopes.

I needed to examine the sample and prepare it for isotopic analysis. The first thing I did was just to break off a small chip and crush it to a fine powder in order to look at it under the microscope. Recall that I said that black cherts aren't really black? Well when I looked at the tiny rock fragments, no more than a hundredth of an inch across, they appeared to be transparent, but with countless little black dots. That's the carbon. Nothing looked like a microfossil. It was unlike any rock I'd ever seen before.

My principal task before the meeting in her lab was to grind and weigh samples for measurement. I used a mortar and pestle to grind up a small chip into as fine a powder as I could. Then I measured out a few milligrams of the powdered chert and put it into a little tin foil cup. I weighed that sample and crumpled up the foil to make a specimen about the size of a BB. That sample contained the carbon atoms that we wanted to measure.

When I showed up at Marilyn Fogel's lab, I gave her two samples, each wrapped in tin foil. Fogel then proceeded to put the samples into her mass spectrometer, which is the best experimental technique for measuring the ratio of isotopes.

Let me explain how a mass spectrometer works. It's really a beautifully designed device. The whole point of the machine is to separate carbon atoms according to their mass. Carbon-13 weighs more than carbon-12, so the two isotopes respond differently to a force. Newton's second law of motion states that, "Force equals mass times acceleration." In other words, if you apply an equal force to two different masses, then the lighter object will accelerate more than the heavier one. So, since carbon-12 and carbon-13 differ by about 8 per cent in their mass, the carbon-12 atom will accelerate about 8 per cent faster than the carbon-13 atom.

But how do you accelerate atoms? That trick is accomplished by relying on another great principle of science, one related to electricity and magnetism. It turns out that magnetic fields can exert forces on electrically charged particles. So, in the mass spectrometer, the trick is to put an electrical charge on the carbon atoms; that is, to ionize the carbon atoms. That's done by zapping the carbon with a powerful laser that strips off a negatively charged electron from each atom. The result is every carbon atom winds up with a positive electric charge. Then we turn on a strong magnetic field that causes the carbon atoms to accelerate. In some mass spectrometers you can see a giant horseshoe-shaped magnet that bends the stream of carbon atoms. In that case, the less-massive carbon-12 atoms curve in a tighter arc than the carbon-13 atoms. That means that the stream of ionized carbon atoms separates into two. We place two ion detectors side-by-side. One measures the amount of carbon-12 and the other measures the amount of carbon-13. That gives us the isotope ratio.

So Marilyn took my two samples and inserted them into an autosampler along with a set of standards. Those standards are samples with well documented carbon-isotope values. So there's a circular sample holder with little wells, each holding a crumpled metal sphere about the size of a BB. One by one those samples drop into the machine. A computer control system takes over and all we have to do is wait for the numbers to appear.

I'm always amazed at how nerve wracking it can be to wait for a machine to produce just one number, one carbon isotope value. It takes just a few minutes per sample, but you have to run standards first and it always seems like a long time. But, if everything works properly, the machine just churns out data, and it did work. Finally, I got the one critical value that I had been waiting for. Each analysis appeared on a single printed sheet of white paper with lots columns of numbers. The single number near the bottom was the key, minus 25.7 plus or minus 0.5. So the carbon was light, and just the sort of number we've learned to expect from ancient microbes. We also got the results of the duplicate sample. It was minus 25.9, which was very consistent with the first value. That's what you want to see.

I have to confess, in spite of these typical values for carbon isotopes—close to minus 26—we still had nagging doubts about the biological origin of the carbon. For one thing, the black specks in the chert looked nothing like fossil forms, and there's always been a concern that there might be non-biological ways to produce a light carbon signature. Just because all known life forms have a negative isotopic signature, that doesn't mean that a negative isotope value is enough to prove that the sample was once alive.

In any case, I sent an e-mail to Malcolm Walter that afternoon with the data. I also suggested a few more experiments that could lead to a quick publication on the chert. Well, his reply came as something of a disappointment. In fact, he asked us to stop working on the samples. It turns out that he had also given Strelley Pool chert samples to another paleontologist, who had been promised the chance to analyze the new find. We worked a lot faster, but these specimens were to be the basis of a Ph.D. thesis, and it would not do for us to undermine that study. Well, there were plenty of other samples to study, so we put the chert aside, but I think you can sense how much fun it was while it lasted.

This episode with the Australian black chert raises an important question. We can measure the ratio of carbon-12 to carbon-13 in an ancient rock to a thousandth of a per cent, but what exactly does a negative carbon-isotope value tell us about that ancient rock?

This question came into especially sharp focus in November of 1996, when a group of scientists claimed to have discovered the most ancient fossil carbon on Earth. The oldest known rock outcrops occur on the remote island of Akilia, which lies off the southwest coast of Greenland. These formations have been dated at 3.85 billion years old. The rocks do not contain anything that looks even remotely like a fossil. They do, however, hold a tiny amount of carbon, and that carbon turns out to be isotopically light.

The Earth's oldest rocks have been through a lot. They have been heated and squeezed and chemically altered. Any fossil forms would have been thoroughly erased, and fluids that flowed through the cracks and fissures would have long since removed any original carbon.

Nevertheless, according to geochemist Stephen J. Mojzsis at the University of Colorado and his colleagues at the Scripps Institution of Oceanography, some of the original carbon atoms were protected in crystals of the phosphorus rich mineral, "apatite." Mojzsis collected these rocks and he analyzed their carbon isotopes. Lo and behold, the carbon was extremely light, on average a whopping 3.7 per cent lighter than referenced limestone. That is minus 37 on our scale. We don't know of any non-biological process that produces such a low value, so many geologists were convinced that life had been well established on Earth by 3.85 billion years ago.

This is very significant result. It does much more than establish a world record for the oldest fossil. Indeed, this data of Stephen Mojzsis would drastically narrow the time window during which life originated. Life presumably couldn't have appeared until after the last giant asteroid impact that sterilized the planet, roughly 4 billion years ago. If there are fossil carbon atoms from 3.85-billion-year-old rocks, then life must have emerged very fast—in less than 150 million years.

Well, it is a great story, but there is a difficulty. The interpretation of Akilia rocks turns out to be problematic. They've been through a lot in 3.85 billion years, so it's not easy to document their history. Mojzsis saw intensely folded layers and interpreted them as metamorphosed oceanic sediments, which is a logical kind of rock in which to find early cellular life, but other geologists who came along and did a more detailed analysis of the geology arrived at a completely different conclusion. They say the Akilia rocks are actually an ancient molten igneous mass that gradually solidified deep underground, cooling from temperatures close to 1,000°C. In their view, any carbon atoms in those rocks must have collected under extreme metamorphic conditions of high temperature and high pressure deep in the crust. Under no circumstance is it possible that those rocks contain any trace of life from the time of their formation.

To make matters worse, other geologists have gone back to Akilia to collect more "apatite" crystals, which are pretty common in some rocks. However, they can't even find any traces of that mineral in these old formations. So it's even possible that specimens got mixed up and the carbon with the minus 37-isotope value is from somewhere else!

To summarize this lecture, paleontologists are more and more turning from suggestive shapes—things that actually look like fossil cells—to fossil elements and isotopes. We can use the electron microprobe to map the distribution of carbon atoms in a fossil and reveal new cellular details. We can also use mass spectrometry to measure carbon isotope ratios with exquisite accuracy and precision.

Questions still remain about the meaning of a light carbon isotope signature. Is light carbon necessary to prove life? I think probably so. However, is light carbon sufficient to prove ancient life? There, the jury is still out, because it's very possible that some natural *non*-biological processes also generate light carbon. Today, living organisms regulate the carbon cycle, and carbon compounds derived from living organisms are almost always isotopically light. However, before the appearance of microbes, there might have been other geochemical processes that efficiently separated carbon-12 from carbon-13. If that's true, then carbon isotopes can't be used as unambiguous evidence for ancient life.

In the next lecture, we'll see that, in their quest for signs of ancient life, paleontologists are turning to even more elusive fossils, which are fragments of life's oldest biomolecules.

Lecture Seven

Molecular Biosignatures

Scope: Fossils can provide important clues to understand life's emergence. Even when morphological evidence, such as bones or shells, is lacking, fossil elements, isotopes, and biosignature molecules point us to the nature of primitive biochemical processes and give us the best hope for narrowing the time window for life's emergence. What's more, if we're lucky, fossils from Mars or some other extraterrestrial body may someday provide us with the best evidence that life has emerged more than once in the universe.

Outline

I. The top-down study of life's origins relies on interpretation of ancient fossil scraps. By looking far back in history, we might catch a glimpse of that enigmatic gap between the nonliving and living worlds.

A. Is there *any* aspect of old fossils that can give us unambiguous evidence? Biomolecules may be the key. A few types of life's most hardy molecules can be preserved in rocks for billions of years.

B. What makes life's molecules so useful to paleontologists is that they are astonishingly idiosyncratic. Carbon can combine with other common elements to form literally millions of different small molecules, but life uses only a small fraction of a percent of these.

C. Let's focus on hydrocarbons, which serve many cellular functions: They form flexible cell membranes, they provide an efficient means of storing energy, they act as internal support structures, and much more.

1. Cells have an amazing tool kit of protein catalysts that work together to make chains of hydrocarbons. This collection of chemical steps is called the *fatty acid synthase*. The process begins by adding a molecule with 3 carbon atoms to a growing chain.
2. Then, the fatty acid synthase removes 1 of these carbon atoms. The net result is that carbon is added in pairs.

3. This procedure leads to a distinctive pattern in the hydrocarbons in living cells. They are characterized by a preference for chains with even numbers of carbon atoms: 10 or 12, instead of 11 or 13.
4. Given a collection of molecules from an unknown source and an analytical machine, such as a mass spectrometer that measures the size distribution of those molecules, it is easy to tell if the molecules came from living cells or if they're not biological in origin.

D. Even more remarkable examples of life's chemical idiosyncrasies are the polycyclic molecules.

1. These chemicals include a rich variety of carbon-based compounds that have two or more interlocking rings, each with 5 or 6 carbon atoms.
2. These molecules are found just about everywhere in the natural environment, because they're easily formed in many processes. They're manufactured abundantly in deep interstellar space and during star formation; thus, they litter the galaxy and are often found in meteorites.
3. Our planet was assembled from interstellar gas and dust, so polycyclic molecules must have been present in large amounts on the primitive Earth. The PAHs (polycyclic aromatic hydrocarbons) from the Allan Hills Martian meteorite are typical examples of these ubiquitous compounds.
4. Polycyclic molecules also are continuously synthesized on Earth today as a consequence of burning. Thus, the most remote places on Earth contain PAHs.

E. Cells make a variety of polycyclic carbon compounds that are essential to life, but just as we saw with hydrocarbon chains, these polycyclic compounds are highly idiosyncratic. They're much less varied than the ring structures produced in space, for example, or by burning.

1. The 4-ring molecules called *sterols*, including cholesterol, steroids, and other important biomolecules, are essential to life. Literally dozens of different possible 4-ring molecules exist, and the relatively random processes of combustion or interstellar synthesis produce lots of them.

2. Yet life zeroes in almost exclusively on sterols and a few of their byproducts. Once again, the selection of sterols is a consequence of a remarkably quirky synthesis pathway.
3. Despite hundreds of different simple sugar molecules, the key molecules of genetics, DNA and RNA, rely on ribose and deoxyribose.

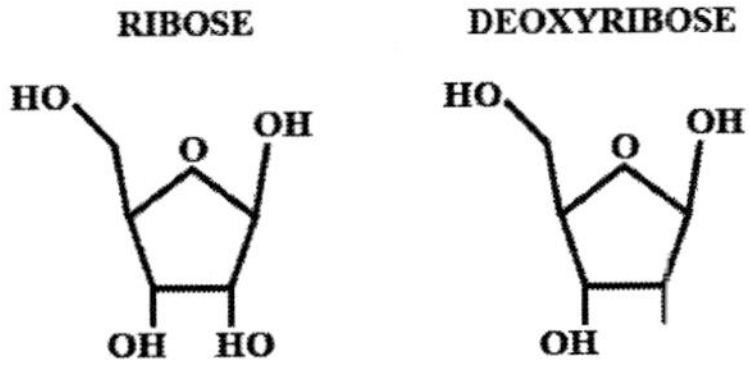

F. Life is exceedingly choosy about its chemical building blocks. Of the countless millions of known small organic molecules, cells typically employ just a few hundred different ones.
   1. This remarkable molecular selectivity is the most diagnostic chemical characteristic of living versus nonliving systems.
   2. And this conclusion is tremendously important for paleontologists. If an old rock has a diverse and nondescript suite of organic molecules, then there is little we can say about its origins. By contrast, if an ancient rock contains a highly selective suite of molecules, such as sterols or even-numbered hydrocarbon chains, then that's strong evidence that life was involved at some point in the rock's history.

II. To search for life here or on other worlds, the molecules in question must be stable for very long periods of time—typically billions of years. Very few of life's molecules survive anywhere near that long. Fortunately, some polycyclic compounds, including sterols, may decompose more slowly.

   A. Occasionally, under exceptional circumstances, a very old rock is found to hold small amounts of a black residue.
      1. These carbon-rich accumulations are typically hydrocarbons that may represent the remains of ancient marine algae.
      2. When such residues were discovered in Australia decades ago, most scientists concluded that the organic remains were relatively recent.

3. However, careful study and improved analytical techniques are convincing many in the geological community that sterol-related compounds can survive in ancient rock, provided that those rocks never experienced high temperatures.
4. What we want to find is one or more unambiguous molecular biosignatures.

B. An influential group of Australian scientists has now focused attention on polycyclic molecules called *hopanes*, which look like sterols except that they have a fifth ring; another hexagon ring. Hopanes in an ancient rock are likely to be molecular fossils, and they may reveal the specific lifestyles of those cells as well.

A FOUR-RING STEROL BACKBONE

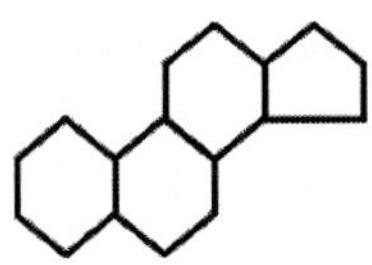

HOPANE

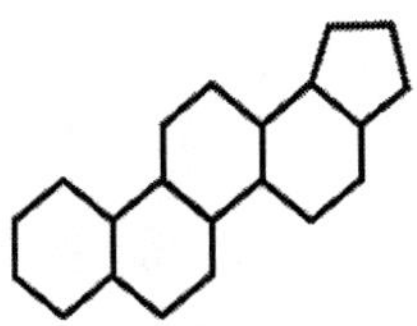

1. In 1999, geologist Roger Summons and his colleagues published evidence for the presence of hopanes in 2.7-billion-year-old sedimentary rocks from western Australia. The molecules were recovered from black, carbon-rich shale layers that had been extracted from a drill core at a depth of more than 2,000 feet. Minerals in those rocks reveal that they never experienced temperatures above about 300°C.
2. The team was also able to rule out modern sources of contamination because all the molecules preserved in the shale were "mature"; that is, they did not include any of the less stable organic species that would point to recent microbial activity.

3. The Summons group published its findings in two remarkable papers. In *Science*, the group described extraction of hopanes that shattered the previous record for the oldest molecular biosignature by about a billion years. In *Nature*, it reported on hopanes from a 2.5-billion-year-old Australian black shale. That's a younger formation, but there is a twist.
4. In the younger formation, the group found what are known as 2-methylhopanoids, a variant of hopane known *only* in cyanobacteria, which are the primitive photosynthetic microbes thought to be responsible for generating Earth's oxygen-rich atmosphere. The Australian team had found tantalizing evidence that cyanobacteria were thriving long before 2 billion years ago, when Earth's atmosphere is thought to have reached today's level of oxygen.

C. As paleontologists search for unambiguous molecular biosignatures, such as hopanes, it is equally important to search for unambiguous *abiosignatures*—that is, chemical evidence that life was *never* present in a particular environment. Abiosignatures might prove especially valuable in shaping our picture of life's emergence.

1. If life emerges through a sequence of chemical steps, then each step might produce distinctive molecules or isotopic ratios. Subsequent steps would tend to erase the evidence of the previous steps.
2. Thus, the discovery of odd suites of molecules might actually point to a setting where life never got past an early emergent step.

D. Hopanes provide a good model for the type of evidence that we're looking for on Earth, Mars, and elsewhere. We're not likely to find unambiguous Martian fossils, but we may find local concentrations of carbon-based molecules. From those samples, we can determine carbon isotope ratios, but a simple isotopic ratio is probably not sufficient to distinguish living from nonliving chemical systems.

E. A suite of carbon-based molecules holds much greater promise. As we've seen, molecular fragments from cells will probably be fundamentally different from molecules synthesized abiotically.

F. Not all molecules are useful in looking for life. Biosignatures and abiosignatures possess three characteristics.

1. They must consist, respectively, of molecules that are essential and unique either to cellular processes or to exclusively nonbiological processes.
2. Signature molecules should be stable. Both biosignatures and abiosignatures must be able to survive through geological time.
3. The molecules must be reasonably abundant to allow researchers to detect the compound by mass spectrometry or some other standard analytical techniques.

G. Hopanes and related polycyclic molecules make excellent biosignatures but are probably not the ultimate solution to looking for life on Mars: They seldom occur in abundance, and their absence cannot be taken as a reliable abiosignature.

III. An alternative search strategy—one that's appropriate for both biosignatures and abiosignatures—would be to identify distinctive ratios of molecular fragments. Which pairs of molecules should be studied to determine these ratios?

A. George Cody, a geochemist I work with at the Carnegie Institution, has found that virtually all carbon-bearing meteorites have measurable quantities of two of the most common polycyclic molecules, anthracene and phenanthrene.

1. Both of these 3-ring molecules have 14 carbon atoms and 10 hydrogen atoms, but they differ in the arrangement of those rings. In anthracene, the 3 rings line up, but in phenanthrene, they form a dogleg. Because both molecules are distinctive, relatively stable, and easy to detect in trace amounts, they might provide an excellent biomarker.

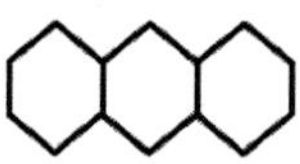

Anthracene

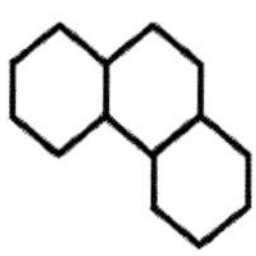

Phenanthrene

2. Phenanthrene and anthracene both form in abundance in nonbiological processes, such as any burning process that produces soot. Both molecules are also found in deep space, which is why they contribute to the molecular inventory of the

carbon-rich meteorites. Cody found that such meteorites often have close to a 1-to-1 ratio of phenanthrene to anthracene.

B. Preliminary work suggests that biochemical processes produce a different ratio, because sterols, hopanes, and other common biomolecules incorporate a 3-ring dogleg.

1. Phenanthrene, the dogleg, is a common and expected fragment from decayed cells, and it should persist in rocks and soils.
2. However, life almost never seems to use the linear arrangement of 3 rings found in anthracene. Thus, anthracene should be correspondingly rare as a biomolecular fragment, and Cody found that some coals hold 10 times more phenanthrene than anthracene.
3. Might the ratio of phenanthrene to anthracene be a useful biomarker? To test this idea, the ratios of cyclic compounds were measured in dozens of different coals, black shales, carbon-rich fossils, soot, and more. At first, the data seemed to support the hypothesis. With more samples, the results began to scatter.
4. Cody and I missed what coal experts have known for a long time: Anthracene is slightly less stable than phenanthrene. With exposure to prolonged high temperatures and pressures, the ratio rises, topping 20-to-1 in some specimens.
5. Our only useful conclusion was that specimens of a biological origin *never* seem to display ratios of less than about 2 to 1. A low ratio of phenanthrene to anthracene may be a valid abiomarker.

IV. All of this work has implications for the search for life on Mars. NASA's robotic rovers have uncovered amazing results, but they have not yet looked specifically for signs of life. Working toward that goal is astrobiologist Andrew Steele at the Carnegie Institution.

A. How to detect life on Mars with a robot?

1. One strategy is to look for life's distinctive molecules. It is essential to develop a procedure to find tiny amounts of any target molecule.
2. Trained in clinical medical technology, Steele's colleague Jake Maule has been developing molecular antibodies—proteins with specialized shapes that lock onto only one type of target molecule.

3. In recent years, Maule has sought to develop a rapid field test for microbes, both on Earth and Mars. Though his process for identifying hopanes is now fast and automated, it is a long way from an instrument that can withstand a trip to Mars.

B. Steele's proposed instrument is called the Microarray Assay for Solar System Exploration (MASSE). His team is in the process of building both flight-ready and hand-held devices that use antibodies to detect trace amounts of dozens of diagnostic biomolecules: hopanes, sterols, DNA, amino acids, a variety of proteins, and even rocket exhaust.

V. Though promising, the study of fossil chemistry raises questions about the chemical processes underlying the earliest forms of life.

Essential Reading:

Hazen, *Genesis: The Scientific Quest for Life's Origin*, ch. 5.

Supplemental Reading:

Knoll, *Life on a Young Planet: The First Three Billion Years of Evolution on Earth.*

Schopf, *Earth's Earliest Biosphere: Its Origin and Evolution.*

Questions to Consider:

1. Given the vast number of different carbon-based molecules, why do cells use such a small number?

2. The ratio of carbon-12 to carbon-13 is a single number, such as –25. Is it reasonable to expect a single number (as opposed to a combination of many factors) to distinguish between living and nonliving systems? Where else in our lives are single numbers used to characterize complex situations?

Lecture Seven—Transcript
Molecular Biosignatures

The top-down study of life's origins relies on the interpretation of ancient fossil scraps. We hope that by looking far back in the history of life, we might be lucky enough to catch a glimpse of that enigmatic gap between the non-living and the living worlds. But, as we have seen in the last three lectures, that is not always an easy task.

In Lecture Four we learned that world-class scientists have been unable to agree on whether the Allan Hills meteorite contains evidence for Martian life. Then, in Lecture Five, we saw that paleontologists Bill Schopf and Martin Brasier still argue about the origins of black smudges in the Apex Chert. And in the last lecture, I told you why no one is absolutely sure that a negative carbon isotope value in an ancient rock is proof of life.

So, is there any aspect of old fossils that can give us unambiguous evidence? It turns out that biomolecules may be the key. In this lecture, the seventh of our course, we'll look at how a few types of life's most hardy molecules can be preserved in rocks for billions of years.

As we have seen, carbon is the element of life. It combines with a variety of other atoms in every living cell to form the essential molecules of life. In the last lecture, we saw that ancient rocks can preserve the original carbon isotopes from cells. It turns out that under just the right circumstances, rocks can also entomb larger fragments of biomolecules. These molecular remnants have tremendous potential for identifying ancient cells, because life is remarkably idiosyncratic in the chemical building blocks that it uses.

So, what are these distinctive molecules of life? You can get a pretty good sense by looking at the nutrition labels on the side of your favorite packaged foods. You'll typically see data on fats, on carbohydrates, and on proteins, all of which help to satisfy life's energy needs. That is, these molecules all have calories. These three kinds of nutrition molecules also provide some of life's most basic molecular building blocks. Cells use these molecules to make their physical structures. After all, you are what you eat.

What makes life's molecules so useful to paleontologists is that they are astonishingly idiosyncratic. Israeli chemist Noam Lahav puts it this way, "The ability of the major atomic components of the cell to combine into molecules of considerable complexity is enormous. However, the actual number of compounds that are used in biology is relatively small, comprising only hundreds of compounds." In other words, carbon can combine with other common elements to form literally millions of different small molecules, but life winds up using only a small fraction of a percent of all these possible types.

I could review all of the major types of molecules, including amino acids, sugars, and more. But let us focus on the molecular family that includes fats and oils, which are made primarily of long, chainlike segments of carbon and hydrogen atoms. These so-called "hydrocarbons" serve many cellular functions. They form flexible cell membranes; they provide an efficient means of storing energy; they act as internal support structures, and lots more.

The easiest way to make this kind of long hydrocarbon molecule is to extend the chain by adding carbon atoms one at a time. So we go stepwise from methane with one carbon, to ethane, to propane, to butane, and so forth. Indeed, industrial chemists have perfected this synthesis process in the manufacture of waxes, fuels and many other useful products. All are formed by the same stepwise mechanism.

Life builds hydrocarbons in a different way, indeed, an unexpected way. Cells have an amazing tool kit of protein catalysts that work together to make chains of hydrocarbons. This collection of chemical steps is called the "fatty acid synthase."

The process begins by adding a molecule with three carbon atoms to a growing chain. Then the fatty acid synthase removes one of these carbon atoms. The net result is that carbon is added in pairs. Well, this odd procedure leads to a distinctive pattern in the hydrocarbons in living cells. They are typically characterized by a preference for chains with even numbers of carbon atom—10 or 12 or 14, instead of 11 or 13 or 15.

So think what that means to a paleontologist looking for ancient molecules. If you're given a collection of molecules from an unknown source, and if you have an analytical machine such as a mass spectrometer that measures the size distribution of those molecules, then it's easy to tell if those molecules came from living cells or if they're not biological in origin.

To me, an even more remarkable example of life's chemical idiosyncrasies is the polycyclic molecules. These chemicals include a rich variety of carbon-based compounds that have two or more interlocking rings, each with 5- or 6-carbon atoms. These molecules are found just about everywhere in the natural environment, because they're easily formed in all kinds of processes. They are manufactured abundantly in deep interstellar space and during star formation, so they litter the galaxy and are often found in meteorites. Our planet was assembled from interstellar gas and dust, so polycyclic molecules must have been present in large amounts on the primitive Earth. Recall the PAHs, or polycyclic aromatic hydrocarbons from the Allan Hills Martian meteorite? Those are typical examples of these ubiquitous compounds.

Polycyclic molecules also are continuously synthesized on Earth today as a consequence of burning. They are the principal component of the soot from fireplaces, the smoke from incinerators and forest fires, and the exhaust of diesel engines. As a result, you can travel to the most remote places on Earth, including the Sahara desert, the deepest parts of the ocean, or the ice fields of Antarctica, and you'll be sure to find PAHs.

Cells make a variety of polycyclic carbon compounds essential to life, but just as we saw with hydrocarbon chains, these polycyclic compounds are highly idiosyncratic. They're much less varied than the ring structures produced in space, for example, or by burning.

Let us look at the group of 4-ring molecules called "sterols" that include cholesterol, and steroids, and other important biomolecules. You have probably heard stories about these molecules in the news a lot because of the questionable use of performance-enhancing steroids by a few athletes, but sterols in general are essential to life. Now, there are literally dozens of different possible 4-ring molecules, and the relatively random processes of combustion or interstellar synthesis produce lots of these. But life zeroes in almost exclusively on sterols and a few of their by-products.

Once again, the selection of sterols is a consequence of a remarkably quirky synthesis pathway. The first step in forming a sterol is to make lots of molecules of isoprene, which is a 5-carbon branching molecule, shaped like the letter Y, but with a kink at the bottom. The cell makes isoprene by linking three smaller molecules together.

To make a sterol you first have to link six isoprene molecules end to end, and that forms a 30-carbon molecule called "squalene." Twenty-four of those carbon atoms in squalene are in a long chain, while six-carbon atoms stick off to the sides at regular intervals, like twigs off a larger branch.

Once the cell has made this long squalene molecule, it then folds precisely into a particular 4-ring structure. That structure has three hexagon rings and one pentagon ring side by side. This 4-ring carbon backbone is common to all sterols. There are dozens of other ways to fit four carbon rings together, but life uses this one.

I could go on and on for an entire course about life's oddly choosy approaches to making molecules. Any biochemistry textbook will have dozens of descriptions of similarly elaborate and exacting pathways. We have the synthesis of the sugar glucose by photosynthesis, or the splitting of glucose to make the energy-rich molecule ATP, or the manufacture of urea in the liver; on and on the list goes. It is amazing how picky life can be.

Here is another example. There are hundreds of different simple sugar molecules, but the key molecules of genetics, DNA and RNA, rely on ribose and deoxyribose. What's more, many of life's most basic molecules, including amino acids and sugars, come in mirror-image pairs—what you might call "right-handed" and "left-handed" variants—but life uses only left-handed amino acids and only right-handed sugars. In Lecture Eighteen we will examine some of the possible explanations for handedness, which remains one of the most perplexing questions in origins research.

Well, the point here is obvious, but surprising. Life is exceedingly choosy about its chemical building blocks. Of the countless millions of known small organic molecules, your cells typically employ just a few hundred different ones. I think it's safe to say that this remarkable molecular selectivity is the most diagnostic chemical characteristic of living versus nonliving systems. This conclusion is tremendously important for paleontologists. If an old rock has a diverse and nondescript suite of organic molecules, then there's very little we can say about its origins. The rock might have once contained cells, or it might just be an accumulation of organic junk. But if an ancient rock contains a highly selective suite of molecules, such as sterols or even-numbered hydrocarbon chains, then that's pretty strong evidence that life was involved at some point in the rock's history.

Of course, it is not quite that easy. If this logic is to be applied to the search for life here or on other worlds, then the molecules in question must be stable for very long periods of time—typically billions of years. Very few of life's molecules survive anywhere near that long. Neither protein molecules, nor the 20 amino acids that make up proteins last that long. Most carbohydrates and hydrocarbon chains also decay, because water attacks the bonds of these biomolecules and breaks them into smaller fragments of no diagnostic use. Fortunately for us, some polycyclic compounds, including sterols, may decompose more slowly, and so they sometimes survive over geological spans of time. So there is a possible molecular way to discover life that is distant in both space and time.

Occasionally, under exceptional circumstances, a very old rock is found to hold small amounts of a black residue. These carbon-rich accumulations are typically hydrocarbons that may represent the remains of ancient marine algae. When such residues were discovered in Australia decades ago, most scientists concluded that the organic remains were relatively recent. They assumed that no molecules could survive billions of years of geological processing. However, careful study and improved analytical techniques are convincing many of us in the geological community that sterol-related compounds can survive in ancient rock, provided that those rocks never experienced high temperatures. What we want to find is one or more unambiguous molecular biosignature.

Of special note is an influential group of Australian scientists who now focus on polycyclic molecules, called "hopanes." Hopanes look like sterols, except they have a fifth ring, another hexagon ring. Hopanes are elegant 5-ring molecules that are found in nature only in cells, where they concentrate in the cell membranes. What's more, different varieties of hopanes are found in specific groups of microbes. So if you can find hopanes in an ancient rock, you can be pretty sure they are molecular fossils. In addition, they might just reveal the specific lifestyles of those cells.

Australian geologist Roger Summons and his colleagues announced the most dramatic discovery of hopanes in 1999. They published evidence for hopanes in 2.7 billion-year-old sedimentary rocks from Western Australia. The molecules were recovered from black, carbon-rich shale layers that had been extracted from a drill core at a depth of more than 2,000 feet. Minerals in those rocks reveal that they never experienced temperatures above about 300°C. That's an unusually tranquil history for such an old deposit.

Of course, there are always analytical concerns. Hopanes are present in all modern microbes, so the Australians had to rule out contamination from more recent life. It is also possible that their rocks might have been contaminated hundreds of millions of years ago by subsurface microbes, or long ago by groundwater carrying molecules from the surface; 2.7 billion years is a long time and a lot can happen. The team was able to rule out modern sources of contamination because all the molecules preserved in the shale were "mature"; that is, they don't include any of the less stable organic species that would point to recent microbial activity.

Summons and his colleagues developed exacting procedures to expose only fresh, uncontaminated rock surfaces. First, they'd break the rock, wash the broken surface, and measure the wash for contamination. They also used small rock fragments to avoid the impurities that inevitably seep in along cracks. After all these precautions, they could recover suites of presumably ancient hydrocarbon molecules, although the amount of hopanes is minuscule. Still, finding any hopanes is amazing, and they provided evidence for ancient microbial life.

The Summons group published their findings in August of 1999, in not one, but two remarkable papers. An article in *Science* described extraction of hopanes from the 2.7 billion-year-old shale. That result shattered the previous record for the oldest molecular biosignature by about a billion years. A second article from *Nature* reported on hopanes from 2.5 billion-year-old Australian black shale. That's a younger formation, but there's a twist.

In that formation, they found what are known as "2-methylhopanoids." That is a variant of hopane that is known only in "cyanobacteria." Cyanobacteria are the primitive photosynthetic microbes that are thought to be responsible for generating Earth's oxygen-rich atmosphere. So the Australian team had found tantalizing evidence that cyanobacteria were thriving long before two billion years ago, when Earth's atmosphere is thought to have reached today's level of oxygen.

I think you can see why scientists are excited by these discoveries. Ancient hopanes not only provide pretty convincing evidence for microbial life, but they also can point to specific lifestyles of those microbes. In this way, fossil biomolecules help to close the gap in our ignorance of life's emergence. They enhance the top-down story and push it just a little bit farther back in time.

Lots of paleontologists are now involved in the search for unambiguous molecular biosignatures, such as hopanes. However, there is a flip side to this story. I think it is equally important to find unambiguous "abiosignatures;" that is, chemical evidence that life was never present in a particular environment. Abiosignatures might prove especially valuable in shaping our picture of life's emergence.

Think of it this way. If life emerges through a sequence of chemical steps, then each step might produce a distinctive molecule or isotopic ratio. Subsequent steps would tend to erase the evidence of the previous steps. In particular, living cells would devour all previous molecules as food. So the discovery of odd suites of molecules might actually point to a setting where life never got past a very early emergent step.

Abiosignatures are especially important in the search for life in Martian meteorites and other exotic specimens. Is there any physical or chemical test that might preclude the presence of past life in those specimens? I realize that a headline like, "No Life Found on Mars!" would be sort of a bummer, but it is a possibility. And that discovery would have tremendous scientific and philosophical implications. Such a discovery would help to answer the question of whether life is a cosmic imperative that arises very quickly in any plausible environment.

Hopanes provide a really good model for the type of evidence that we're looking for on Earth, Mars and elsewhere. We're not likely to find unambiguous Martian fossils, at least not any time soon. Even if Martian fossils are common, there won't be a Mars sample return mission until well after 2010, and human exploration of Mars is many decades away. Even if we do get our hands on Mars rocks, we'd be incredibly lucky to find a convincing fossil. So we're more likely to find local concentrations of carbon-based molecules. From those samples, we can determine carbon isotope ratios; but a simple isotopic ratio is probably not sufficient to distinguish living from nonliving chemical systems.

On the other hand, a suite of carbon-based molecules holds much greater promise. We've seen that molecular fragments from cells will probably be fundamentally different from molecules synthesized abiotically. So that's now a major focus for exploration.

Not all molecules are useful in looking for life. Ideal signs of life, which are known as biosignatures—or signs of no life, which are known as abiosignatures—have to possess three characteristics. First of all, biosignature molecules should be compounds that are essential and unique to cellular processes. By the same token, abiosignatures should consist of molecules from exclusively nonbiological processes.

Second, signature molecules should be stable. Both biosignatures and abiosignatures have to be able to survive through geological time. Even the least altered Archean rocks have spent billions of years buried deep in the crust, where temperatures are higher than that of boiling water. Most molecules can't survive that. So we have to focus on relatively stable molecules.

Finally, the molecules have to be reasonably abundant, because you have to be able to detect the compound by mass spectrometry or some other standard analytical techniques.

We've seen that hopanes and related polycyclic molecules make excellent biosignatures. They're certainly distinctive and stable, but hopanes are probably not the ultimate solution to looking for life on Mars. They seldom occur in abundance, and their absence cannot be taken as a reliable abiosignature.

At the Geophysical Lab, we've been thinking about an alternative search strategy, one that's appropriate for both biosignatures and abiosignatures. We'd like to identify distinctive ratios of molecular fragments, something like measuring the ratio of carbon-12 to carbon-13. The immediate problem is that we're faced with a huge number of possible molecule pairs. Which pair of molecules should we study?

My colleague George Cody, a geochemist at the Carnegie Institution, suggested one possibility. Cody is an expert on analyzing organic molecules, and he spends a lot of his time studying carbon-rich meteorites. He found that virtually all carbon-bearing meteorites have measurable quantities of two of the commonest polycyclic molecules, called "anthracene" and "phenanthrene".

Both of these 3-ring molecules have 14-carbon atoms and 10-hydrogen atoms, but they differ in the arrangement of those rings. In anthracene, the three rings line up, but in phenanthrene they form a dogleg.

We realized that the relative amounts of these two molecules might provide a great biomarker. Both molecules are distinctive, they're both relatively stable, and both are easy to detect in trace amounts.

Now phenanthrene and anthracene both form in abundance in nonbiological processes such as any burning process that produces soot. You also find both molecules in deep space, which is why they contribute to the molecular inventory of the carbon-rich meteorites. Cody found that such meteorites often have close to a one-to-one ratio of phenanthrene to anthracene.

But our preliminary work suggests that biochemical processes produce a different ratio. That's because sterols, hopanes, and other common biomolecules incorporate a 3-ring dogleg. So phenanthrene, the dogleg, is a common and expected fragment from decayed cells, and it should persist in rocks and soils. However, life almost never seems to use the linear arrangement of three rings found in anthracene. So anthracene should be correspondingly rare as a biomolecular fragment. Sure enough, George Cody found that some coals, which are fossil fuels derived from ancient life, typically hold 10 times more phenanthrene than anthracene.

So, might the ratio of phenanthrene to anthracene be a useful biomarker? To test this idea we measured the ratios of cyclic compounds in lots of samples—dozens of different coals, black shales, carbon-rich fossils, soot, and more. At first, the data seemed to support our hypothesis. Meteorites showed phenanthrene-to-anthracene ratios close to two to one, while the biogenic Burgess Shale and a mature coal had higher ratios close to fifteen to one. But with more samples, the results began to scatter. Some coals had ratios less than five to one, while a fossil-rich black shale from Germany had a two-to-one ratio.

It turns out that we had missed what coal experts have known for a long time—anthracene is slightly less stable than phenanthrene. Upon exposure to prolonged high temperatures and pressures, the ratio rises, topping twenty to one in some specimens. Such a high ratio of phenanthrene to anthracene may result simply from prolonged heating, and have nothing at all to do with biology. Our promising hypothesis crumbled before our eyes, at least for this pair of molecules.

Our only useful conclusion was that specimens of a biological origin *never* seem to display ratios less than about two to one. So maybe a low ratio of phenanthrene to anthracene is a valid abiomarker.

That is not the end of this research. We can move on to other pairs of molecules. For example, we can try another ratio of polycyclic molecules, such as that of phenanthrene to pyrene. Pyrene is a very stable diamond-shaped molecule with four interlocking rings. Whatever we do, you can be sure that we'll never run out of experiments to try.

All this work has implications for the search for life on Mars. Researchers have to resign themselves to the fact that humans are not going to set foot on Mars, at least not any time soon. Instead, research must rely on robotic exploration, and that's what NASA's Mars rovers are for. We've seen the amazing results returned by the most recent rovers, *Sojourner*, *Spirit* and *Opportunity*. They carried a variety of instruments across the surface of Mars to probe rocks and analyze soils and to look for water. However, they didn't specifically look for signs of life. So how do we design a chemical analyzer for life?

One of the people working on chemical analyzers is astrobiologist Andrew Steele, a dynamic and fun-loving Brit. Everyone calls him "Steelie." His life's work is to fly a life-detection instrument to Mars. Just two weeks after Steelie defended his Ph.D. thesis in England in 1996, the Allan Hills meteorite story broke. He got in touch with Dave McKay right away and landed a job at NASA to study the meteorite for signs of life.

Well, it seems that Steelie may have done his job too well. He did high-resolution microscope studies on the Allan Hills meteorite and several other meteorites and discovered that they are all riddled with contaminating Earth microbes. This discovery threw any conclusions about evidence for Mars life in the Allan Hills specimen into greater doubt. After his findings, it was time for Andrew Steele to look for another job, and he was quickly hired at the Carnegie Institution, which is where he's building his Mars analyzer. I have been lucky enough to have an office right next door to him.

So how do you detect life on Mars with a robot? One obvious strategy is to look for life's distinctive molecules. It is essential to develop a procedure to find tiny amounts of any target molecule. That's where Steelie's post-octoral colleague, Jake Maule, came in. Maule is an aspiring astronaut who has been trained in clinical medical technology. His job has been to develop molecular antibodies, which are proteins with specialized shapes that lock onto only one type of target molecule.

In recent years, Maule has focused on producing antibodies for hopanes. He has sought to develop a rapid field test for microbes, both on Earth and Mars. He has injected dozens of mice with hopanes and then let the immune systems of the mice do most of the work. In about three weeks, the mice manufacture hopane-sensitive antibodies. He can retrieve these molecules and use them to analyze ancient rocks.

Basically, Maule crushes a promising rock sample and washes it in a solvent to extract any hopane residues. He then applies a small amount of hopane-sensitive antibody to the solution. If any hopanes are present, then the antibody sticks to the hopane. Finally, Maule adds a fluorescent chemical that binds to the antibody. That causes the hopane-bound antibodies to fluoresce, so there's a colorful signal if hopanes are present in the rock.

Jake Maule's process for identifying hopanes is now fast and automated. Still, it's a long way from that procedure to an instrument that can withstand a trip to Mars. The design and engineering of a flight-ready instrument is the next step. To go to Mars, the analyzer has to be totally reliable, shock resistant and very, very small. If Andrew Steele, Jake Maule and their colleagues can solve all the technical problems, their device will probably fly on NASA's 2013 mission to Mars.

Steelie's proposed instrument is called MASSE—the Microarray Assay for Solar System Exploration. His team is in the process of building both flight-ready and hand-held devices that use antibodies to detect trace amounts of dozens of all sorts of diagnostic biomolecules—hopanes, sterols, DNA, amino acids, a variety of proteins and even rocket exhaust. It's a neat concept, and I'm sure it will work, but the Carnegie team is not alone in developing such an instrument. Discovering life on Mars is a seductive challenge, and there's intense competition from other dedicated design teams, including groups in France, Spain, and Britain.

At this point there is no guarantee that MASSE will ever fly anywhere. Only one instrument package can fit on the lander. Even if MASSE does join the flight, there's no guarantee that it will arrive safely. If it arrives safely, we can't be sure that it will find anything of interest. Of course, that's not the only reason to build instruments like MASSE. Steelie and his group are learning lots of fascinating stuff along the way. In fact, thanks to Andrew Steele and colleagues, Earth-bound scientists may soon have a hand-held device that analyzes for a host of environmental chemicals, and even microbial life.

In summary, I hope that you have gotten a glimpse of how fossils can provide important clues in our age-old efforts to understand life's emergence. Even when morphological evidence such as bones or shells is lacking, fossil elements, isotopes and biosignature molecules are pointing us to the nature of primitive biochemical processes. They give us the best hope for narrowing the time window for life's emergence. What's more, if we are very lucky, fossils from Mars or some other extraterrestrial body may some day provide us with the best evidence that life has emerged more than once in the universe.

While promising, the study of fossil chemistry, in turn, also raises questions about the chemical processes underlying the earliest forms of life. Moreover, as valuable as these insights from fossil chemistry may be, all known fossils represent remains of extremely advanced cellular organisms, in most respects similar to cells that are alive today. There are few, if any, clues left about the much simpler biochemical steps that must have emerged before cells.

To understand how life arose, therefore, we have to go back to the very beginning, to the geochemical world of rocks, oceans, and air. So in the next lecture, we'll begin our exploration of the "bottom-up" approach to the origins of life.

Lecture Eight

Emergence

Scope: A precise definition of emergence and a useful mathematical formulation of this phenomenon remain elusive. The key to a definition may lie in the fact that all emergent systems display *complexity*, which seems to depend on four factors: (1) the concentration of interacting particles, (2) the nature of the interactions, (3) the flow of energy through the system, and (4) the cycling of that energy flow. Eventually, a law of emergence will provide us with a deeper understanding of every system of interacting agents, including the origins of life itself.

Outline

I. What is an emergent system? Think of it as a process in which many interactions among lots of simple objects give rise to behaviors that are much more complex. Lots of sand grains interact to form sand dunes, and lots of nerve cells interact to form our conscious brains.

- **A.** The first step in developing a law of emergent systems is to examine everyday examples in countless systems all around us, all the time.
  1. New patterns arise as individual objects, anything from atoms to entire galaxies, come into proximity with each other.
  2. This universal tendency for systems to become ordered when lots of objects interact is consistent with the first and second laws of thermodynamics, but neither of those laws addresses emergence explicitly.
- **B.** All emergent systems occur when energy flows through a collection of many interacting particles.
  1. Such systems tend spontaneously to become more patterned and to display new, often surprising behaviors.
  2. We find that, as new patterns arise, the energy is dissipated more efficiently. The resulting behavior appears to be much more than the sum of the parts.
- **C.** Even though emergent systems abound in our surroundings, a precise definition of emergence and a useful mathematical formulation of this phenomenon remain elusive.

D. The key to a definition may lie in the fact that all emergent systems display *complexity*.
   1. All complex systems, by definition, have an intricate or patterned structure.
   2. This property of complexity may also be applied to information content.

II. Complexity appears to be the hallmark of every emergent system, especially life. Indeed, that's one way that we recognize emergent systems. A simple mathematical equation would be helpful, one that relates the physical state of a system (the temperature or pressure, for example, expressed in numbers), to the resultant complexity of the system (also expressed as a number).
   A. If we could formulate such an equation, it would constitute that elusive missing "law of emergence." That law would serve as a guide for modeling the origin of life—the ultimate complex system.
   B. To formulate a law of complexity requires a quantitative definition of the complexity of a physical system. No one has such a definition, though scientists have thought about complex systems and ways to model them mathematically.
      1. Revered as a founder of the field of emergence, Santa Fe Institute guru John Holland has modeled emergent systems as if they were computer programs with a fixed set of operating instructions.
      2. California programmer Craig Reynolds has managed to duplicate the complex collective behavior of flocking birds, schooling fish, swarming insects, and other animals that move as a group with a relatively simple program called BOIDS.
      3. The mathematical prodigy Stephen Wolfram also sees complex behavior arising from systems with simple rules. He suggests that the complex evolution of the physical universe and all it contains is perhaps best modeled as a set of sequential instructions.
      4. The late Danish physicist Per Bak described complex systems in terms of a mathematical characteristic called *self-criticality*. These systems evolve by repeatedly reaching a critical point at which they collapse and regroup. The mathematical

description of this process is mimicked in many natural events, but it is hard to see the link to life's origins.

5. Santa Fe theorist Stuart Kauffman focuses on the emergence of chemical complexity by way of what he calls competitive *autocatalytic networks*, that is, large collections of chemical compounds that have the ability to catalyze their own formation. We'll see how this idea comes into play in Lecture Nineteen.

C. All these approaches inform the search for a universal law of emergence. Yet each model seems too abstract to apply directly to the design of bench-top chemical experiments on the origin of life.

III. The study of simple emergent systems may eventually point us to general physical factors that also lead to patterning in much more complex systems, including life. As a first factor, the concentration of interacting particles plays a central role in the complexity of a system.

A. Sand, for example, provides common and comprehensible examples of the kind of structures that arise when an energetic flow—wind or water—interacts with lots of agents.

1. Below a critical concentration of particles, we see no patterns of interest.
2. As the particle concentration increases, so does the complexity.
3. Yet complexity only increases to a point: Above some critical density of particles, no new patterns emerge.

B. Similar observations have been made about other emergent systems.

1. The army ant remains close to home as long as the colony has fewer than about 80,000 individuals. If an ant colony exceeds that number, however, it tends to exhibit a new emergent behavior called a *swarm raid*, in which the ants pour out to attack adjacent colonies.
2. At a galactic scale, a minimum number of about 100 million stars is required to trigger the development of the familiar, majestic spiral arm structure.
3. Human consciousness and self-awareness also appear to emerge from the complex interactions of trillions of neurons. Consciousness depends on the interactions of some large

number of brain cells. When a critical number of cells and their connections are lost, then self-awareness fades away.

C. All these observations suggest a mathematical relationship between the concentration of agents on the one hand and complexity on the other.

1. A qualitative picture of this relationship is given by the so-called *error function.*
2. This function starts at zero complexity for low concentrations of particles.
3. As the concentration of particles increases, it reaches some critical value where patterns begin to emerge, and complexity begins to rise.
4. Ultimately, at some higher concentration of agents, a system's complexity may achieve a maximum value.

D. The emergence of life depended on achieving some minimal concentration of biomolecules, the essential particles that lead to cellular life. If the concentration of life-forming molecules is too low, even if the environment is conducive to life, then life could not arise.

IV. A second factor in fostering emergent complexity is the interconnectivity of agents.

A. Quantifying emergent complexity will require incorporating the varied ways that agents can be interconnected, that is, quantifying the differences between the interactions of adjacent touching sand grains, versus the persistent chemical markers of moving ants, versus the elaborate variable impulses of neurons.

B. It is not obvious that any of these examples is analogous to the interactions of molecules that are necessary for life's origin.

C. The origins of life must have depended on many chemical interactions. A good experimental strategy is to optimize the number and types of molecular contacts to induce emergent behaviors of interest.

V. A third factor in fostering emergent complexity is the energy flow through the system.

A. No matter how many interacting particles exist in a system, if there isn't a reliable flow of energy to drive change, no patterns arise.

B. There is a limit to effective energy flow. Every complex patterned system has a critical magnitude of energy flow, above which it can no longer tolerate additional energy.

   1. During an energetic storm, the crashing waves will obliterate sand ripples; all signs of emergent patterning will disappear.

   2. During waking hours, the human brain maintains an intermediate level of electrical impulses. In contrast, deepest sleep corresponds to a sharp drop in electrical activity and a loss of consciousness. The intense electrical activity of an epileptic seizure halts conscious action by scrambling the usual patterned electrical flow.

C. In all these examples, the relationship between energy flow and emergent complexity has a qualitative similarity to the familiar bell-shaped curve.

   1. At the lowest energy flux, no patterning occurs and the complexity is zero.

   2. At a critical minimum energy flux, pattern formation begins, and complexity rises to a maximum value.

   3. Above some critical value, the energy flux becomes too large and emergent patterns disperse.

D. Emergence of complex patterns depends on an energy flow within rather narrow constraints.

   1. If energy flow is too low, then nothing happens.

   2. Too much energy flow and the patterns are randomized—entropy triumphs.

E. This conclusion carries an important lesson for the study of life's origins. A reliable energy source is essential, to be sure, but too much energy is unwanted.

   1. Intense energy sources, such as lightning and ultraviolet radiation, can fragment molecules and may be too extreme to foster life.

   2. It may be prudent to search for more benign chemical energy sources—perhaps the natural equivalent of a flashlight battery—to sustain the metabolism of primitive life.

VI. The fourth factor that appears to foster complexity is the cycling of energy flow.

A. Many natural systems are subjected to cycles of energy, such as day and night, high tide and low tide, or wetting and drying.

B. Such inevitable cycles may have played a fundamental role in the evolution of emergent systems, including life, though it is often difficult to recognize the effects of these subtle cycles in nature.

C. The behavior of sand and of water again provides insight.

1. Cycling energy flow across a sand surface generates new complex sand patterns. Physicist Jonas Hansen at the Niels Bohr Institute and a group of his colleagues discovered this result in 2001.
2. Most previous experimenters had used artificial waves that had a fixed wave height and frequency. If well designed, such studies typically produce perfectly spaced, straight ripples.
3. Hansen and his colleagues increased and decreased the height or frequency of their water waves, and the results were striking. Rather than simple parallel sand ripples, they found elegant intertwined and branching sand structures.
4. Similar sand features occur along the Chesapeake Bay when the water is only a few inches deep. Those shallow conditions may favor periodic fluctuations in wave amplitude.
5. Doctoral student Mark Kessler and Professor Brad Werner of the University of California, San Diego, in 2003 published photos of amazing stone circles and other so-called *patterned grounds* in the Arctic terrain of Alaska, where cyclical freezing and thawing causes these boulders to shift position slightly. The natural circles of stone emerge from interactions over many years.

D. The role of cycling in the emergence of patterns remains a frontier area of research. We know that the early Earth was subject to many different cycles—day/night, high tide/low tide, wet/dry, hot/cold, and more. Such cycles may have contributed to the emergence of life itself.

VII. In summary, four factors appear to contribute to the emergence of complexity: (1) the concentration of interacting particles, (2) the nature of the interactions, (3) the flow of energy through the system, and (4) the cycling of that energy flow.

A. These results suggest that a mathematical expression of complexity will take the form of a mathematical inequality.

B. The complexity of an emergent system is a number. That number will be less than or equal to some value that is a mathematical function of at least four factors: the concentration of interacting particles, the degree of those particles' interconnectivity, the energy flow through the system, and the time-variation of that energy flow, and perhaps other variables as well.

$$C \leq f[n, i, \nabla E(t)]$$

C stands for "complexity"
f indicates a mathematical function
n is the number of interacting particles
i is the degree of those particles' interconnectivity
$\nabla E(t)$ is the time-varying energy flow through the system

C. Two impediments stand in the way of the completion of this formulation.

1. We don't know how to assign *numbers* to different degrees of complexity. What is the complexity of an ant hill or the human brain? And what are the units?
2. We have no clear idea of the mathematical relationships between complexity and the four key factors.

VIII. The effort to quantify emergent complexity is not a mathematical abstraction.

A. Emergent systems surround us: our environment, our bodies, our minds, our culture. All these systems display emergent complexity.

B. Even if we don't have a complete definition and mathematical formulation of emergence, the principle can guide our quest to understand life's origins. In the next lecture, we'll explore the earliest emergent step in the path to life—the emergence of simple biomolecules.

Essential Reading:

Hazen, *Genesis: The Scientific Quest for Life's Origin*, ch. 1.

Morowitz, *The Emergence of Everything*.

Supplemental Reading:

Bak, *How Nature Works: The Science of Self-Organized Criticality*.

Chaisson, *Cosmic Evolution: The Rise of Complexity in Nature*.

Gell-Mann, *The Quark and the Jaguar*.

Holland, *Emergence: From Chaos to Order.*

Kauffman, *At Home in the Universe: The Search for Laws of Self-Order and Complexity.*

Prigogene, *Order out of Chaos: Man's New Dialogue with Nature.*

Wolfram, *A New Kind of Science.*

Questions to Consider:

1. Why is the field of emergence important to the study of life's origin?
2. How do studies of complex systems inform the design of experiments that mimic life's origin?

Lecture Eight—Transcript
Emergence

The last four lectures have looked at four "top-down" approaches to the origin of life—the possibility of life on Mars, ancient fossil microbes from Earth, carbon isotopes and biochemical markers. However, scientists are rarely content with what they find in the field. Scientists also want to do experiments.

Moreover, the process that transformed the lifeless geochemical world to a vibrant living Earth was much too complicated to occur all at once. That history is much too intricate to mimic in a single experiment. What we need is a framework that treats the origins of life as a sequence of simpler emergent events.

In this lecture, I'll explore the idea of emergence and show how it might help us to comprehend how life arose. The idea of emergence informs and pulls together much of the laboratory work, theory-building and speculation that I will be discussing in subsequent lectures.

So what exactly is an emergent system? John Holland, a founder of the scientific field of emergence, said in his 1998 book, *Emergence: From Chaos to Order*, "It is unlikely that a topic as complicated as emergence will submit meekly to a concise definition, and I have no such definition." Well, if he does not have a definition, then I am not going to be able to do any better. Perhaps you can get a rough idea of what emergence is all about if you think of it as a process in which many interactions among lots of simple objects give rise to behaviors that are much more complex. So, for example, lots of sand grains interact to form sand dunes, and lots of nerve cells interact to form our conscious brains.

The first step in developing a law of emergent systems is to examine everyday examples. You observe emergent behavior in countless systems all around us, all the time. Just watch the interactions of automobiles or ants. New patterns arise as individual objects—anything from atoms to entire galaxies—come into proximity with each other. This universal tendency for systems to become ordered when lots of objects interact is consistent with the first and second laws of thermodynamics, but neither of those laws address emergence explicitly.

If we are to understand emergence, then a central question is whether all emergent systems possess a unifying mathematical behavior. The fact that we do not have such a formula right now should not seem too unsettling. It took at least 50 years for the first two laws of thermodynamics to develop from qualitative ideas into quantitative laws. My guess is that we will have a mathematical formulation of emergence in the next couple of decades.

My optimism is based on the fact that we have come a long way. Scientists have already identified several key aspects of the problem. For example, we know that many familiar natural systems that lie close to equilibrium are stable and unchanging, and so do not display emergent behavior. If you cool water slowly to below the freezing point, it will equilibrate to become a clear piece of ice. By the same token, if you heat water gradually above its boiling point, it will similarly equilibrate as it converts to steam. Scientists have spent centuries documenting equilibrium processes in all kinds of natural and artificial systems. Something very different happens in systems that are away from equilibrium. If you bring water rapidly to boil, superheating it in the process, it will display complex, turbulent convection. Similarly, water flowing downhill in the steep gravitational gradient of a river valley will interact with sediments to produce all kinds of emergent landform patterns, including braided streams, meandering rivers, sandbars and deltas. These emergent patterns arise because energetic water moves across lots of particles.

All emergent systems share this characteristic. They occur when energy flows through a collection of many interacting particles. Such systems tend spontaneously to become more patterned and to display new, often surprising behaviors. We find that, as new patterns arise, the energy is dissipated more efficiently, just as dictated by the second law of thermodynamics. The resulting behavior appears to be much more than the sum of the parts.

I must admit that emergent patterns in a simple system, like water flowing over sand, may seem a far cry from the complexity of living organisms. But for those of us who study life's origins, there's potentially a big benefit in understanding simple systems. Life is the most dramatic of all known emergent phenomena. Perhaps studies of much simpler emergent systems can provide us with guidance for origins-of-life research.

So, how to proceed? Even though emergent systems abound in our surroundings, a precise definition of emergence and a useful mathematical formulation of this phenomenon remain elusive. The first step in discovering a natural law of emergent systems is to identify the essential properties of such systems; but that's a problem. What characteristics distinguish emergent systems from other less interesting collections of interacting objects?

The key, I think, is that all emergent systems display a rather subjective characteristic that we might call "complexity." That is a property without a precise quantitative definition, but you understand what I mean in a general sense. All complex systems, by definition, have an intricate or patterned structure, as in a complex machine, a complex argument, or a Bach fugue, for that matter. This property of complexity may also be applied to information content. So an advanced textbook in physics contains more detailed information, and is thus more complex than an elementary textbook. In this context, the varied interactions of ants in an ant colony or the intricate connections of neurons in the human brain are vastly more complex than the behavior of a pile of sand grains or a bowl of cereal.

Complexity appears to be the hallmark of every emergent system, especially life. Indeed, that's one way that we recognize emergent systems. What we'd like to know, therefore, is a simple mathematical equation that relates the physical state of a system—the temperature or pressure, for example, expressed in numbers—to the resultant complexity of the system, also expressed as a number. If we could formulate such an equation, it would constitute that elusive missing "law of emergence", and that law would serve as a guide for modeling the origin of life—the ultimate complex system. But there's a big problem. If we're going to formulate a law of complexity, then we must have a quantitative definition of the complexity of a physical system. No one has such a definition.

How to define complexity? A small band of scientists, several of them based at the Santa Fe Institute in New Mexico, have tried to come to grips with this problem. A number of brilliant people have thought long and hard about complex systems and ways to model them mathematically. But it turns out that their efforts yield quite different, even contradictory, views on how to approach the subject.

Santa Fe Institute guru, John Holland, is an ace at computer algorithms, and he's revered as a founder of the field of emergence. His approach is to model emergent systems as if they are computer programs with a fixed set of operating instructions. He suggests that other emergent phenomenon, including both simple systems like sand ripples and much more complex systems like ant colonies and the conscious brain, can be modeled as a set of selection rules. Indeed, Holland and his many followers have made great advances in duplicating natural phenomena with a few lines of computer code. In this context, Holland equates the complexity of a system to the minimum number of lines of computer code that are required to mimic that system's behavior.

Let me tell you about a wonderful example, called BOIDS, which you can investigate at your leisure on the Internet. BOIDS is a relatively simple computer program that was written by California programmer Craig Reynolds. Reynolds has managed to duplicate the complex collective behavior of flocking birds, schooling fish, swarming insects and other animals that move as a group, with astonishing accuracy. For more of this, just Google BOIDS, and follow the links. Reynolds relies on just three rules of motion—each individual moves based on the directions and spacing of its neighbors. Lest you dismiss this effort as some trivial exercise, just think of BOIDS when you watch such Hollywood blockbuster movies as *Jurassic Park* with its dinosaur herds on the run, or *Finding Nemo* with its realistically animated schools of fish.

The brilliant mathematical prodigy Stephen Wolfram has taken a similar approach. He also sees complex behavior arising from systems with simple rules. Like Holland, Wolfram has been fascinated by the ability of a set of simple instructions to produce elaborate visual patterns. In this behavior, he sensed a new paradigm for the description and analysis of all aspects of the natural world, and he has spent the past 20 years developing what he calls "a new kind of science," or simply NKS. He wrote a massive book in 2002, and he also maintains an elaborate web site, which you can explore for hours at *www.wolframscience.com.* At the site, you will find some of the truly mind-boggling ways by which geometric complexity arises from simple rules. Wolfram suggests that the complex evolution of the physical universe and all it contains is perhaps best modeled as a set of sequential instructions.

Well, there are lots of other approaches to understanding complexity as well. The late Danish physicist Per Bak, who described complex systems in terms of a mathematical characteristic called "self-criticality," took one influential tack. These systems evolve by repeatedly reaching a critical point at which they collapse and regroup. His classic example is a growing pile of sand that builds a little higher and then avalanches over and over again as new grains are added. The mathematical description of this process is mimicked in many natural events, from the shape of coastlines to the destruction of earthquakes. I must confess, though, that I don't see a direct link to life's origins in this approach.

Santa Fe theorist Stuart Kauffman proposes yet another model that's much more relevant to life. He focuses on the emergence of chemical complexity by way of what he calls competitive "autocatalytic networks"—that is, large collections of chemical compounds that have the ability to catalyze their own formation. We'll see how this idea comes into play in Lecture Nineteen.

All of these approaches and more are informing the search for a universal law of emergence. All of these ideas provide a glimpse of the answer, yet each model seems to be too abstract to apply directly to the design of bench-top chemical experiments on the origin of life. As an experimentalist, I need to control all sorts of details: what chemicals should I use at what starting concentrations? Should I make the solution acidic or basic? What run temperatures, pressures and times? I'd love to know if there's a way that the ideas of emergence can help.

The classic scientific approach to deducing general principles and laws is to study the behavior of specific simple systems. I suspect that the study of simple emergent systems may eventually point us to general physical factors that also lead to patterning in much more complex systems, including life. We can hope that observations of specific systems will eventually point to more general rules.

It does not take a fancy laboratory to study emergent phenomena. In fact, you can not go outside without seeing all sorts of emergent systems. Perhaps my favorite emergent phenomena are related to interactions of water and sand, which I've mentioned several times already. Sand provides common and comprehensible examples of the kinds of structures that arise when an energetic flow of wind or water interacts with lots of agents. Studying sand and water provides a great excuse to spend the day at the beach as well. Emergent patterns arise when moving water or wind disturbs a flat layer of sand. You commonly see the appearance of regular sand ripples, while sand grains become more sorted by their size, shape, and density. As the sand system becomes more orderly and patterned, the energy of the system is more efficiently dissipated.

By far, my favorite sandy system to study occurs about an hour's drive east of Washington, DC, along the western shore in Calvert County, Maryland. There you'll find fossil-rich, 100-foot-tall cliffs that border the Chesapeake Bay. The cliffs are loaded with fifteen-million-year-old whalebones, big sharks' teeth, and all sorts of coral reef shells that abound in the sandy wash zone.

The best time to walk is during an unusually low tide, when receding waters expose a slightly sloping, smooth surface of ancient sediments below the base of the cliff. This formation, called blue marl, commonly accumulates a thin layer of sand. These particles often display emergent patterns when the sand is subjected to the back-and-forth washing of shallow water.

Over the years I have observed four distinct factors that appear to contribute to the emergence of complex patterns in sand. The first factor is simply the concentration of sand grains on the surface. That is pretty obvious, because you can not have a pattern in sand if there's not enough sand to make a pattern. So I measure the density of sand grains, which is the number of sand particles per square centimeter of the hard surface. That is an easy number to estimate in the following way: I simply scrape up almost every grain of sand from an area 10 centimeters square, which is about the size of a small paper napkin. I take the sand back to the lab, dry it, and then weigh the whole sample. Then, using a microscope, I count out exactly 100 grains from the total sample and I weigh that batch. It turns out that the density of grains per square centimeter is approximately equal to the total weight of all the sand from the 10 x 10-centimeter area—that is, 100 square-centimeters, divided by the weight of 100 sand grains.

So what are the results? I have found out that when there are fewer than about 100 sand grains per square centimeter, then the density of particles is just too sparse for any patterns to emerge. The sand grains are so small that 100 grains-per-square centimeter provides little coverage, certainly less than 10 percent of the surface.

If a surface has a higher sand concentration of about 1,000 grains per square centimeter, then an intriguing pattern arises; I observe regularly spaced little sand piles, each pile perhaps a centimeter or two across. Furthermore, it is common to see a small circle of darker sand grains crowning each little tan sand pile. So it appears that a minimum concentration of at least several hundred grains per square centimeter is necessary to initiate sand patterning.

Now, if the sand concentration increases to a few thousand grains per square centimeter, we see discontinuous short bands of sand, at more or less right angles to the gentle back-and-forth wave motion of the water. Once again, each of the tan bands is topped off by darker grains.

Then, if the sand concentration exceeds 10,000 grains per square centimeter, we observe a continuous pattern of evenly spaced, black-capped sand ripples. On occasion, I've seen such a regular rippled surface over hundreds of square meters of shallow water.

That is the end of the story. If a surface has a higher concentration of sand grains, you simply have a deeper base for the regular sand ripples. That's because buried sand grains can't participate in the interaction with water, and consequently no new structures arise beyond the familiar wavelike forms on the surface.

So what do these observations tell us about the emergence of complexity? I conclude that the concentration of interacting particles plays a central role in the complexity of a system. Below a critical concentration of particles, we see no patterns of interest. Then as the particle concentration increases, so does the complexity. Complexity only increases to a point however; above some critical density of particles, no new patterns emerge.

Strikingly similar observations have been made on other emergent systems. For example, the army ant remains close to home as long as the colony has fewer than about 80,000 individuals. If an ant colony exceeds that number, however, then the colony tends to exhibit a new emergent behavior called a "swarm raid." Like a dam bursting, the army ants pour out to attack adjacent colonies. At even higher populations, roughly half of the ants may just leave to form a new colony. And recent studies of termite colonies reveal a similar kind of behavior. Above a critical density of individuals the colony begins construction of distinctive pillar-type mounds.

At a galactic scale, it requires a minimum number of about 100 million stars to trigger the development of the familiar majestic spiral arm structures. According to the calculations of astrophysicists, galactic arms form as a result of gravitational instabilities, which may depend on having a large central mass of stars.

At a more immediate scale, human consciousness and self-awareness also appear to emerge from the complex interactions of trillions of neurons. Consciousness depends on the interactions of some large number of brain cells. It's so hard to watch friends and relatives who are afflicted with Alzheimer's disease, which gradually destroys brain cells and their connections. When a critical number of cells and their connections are lost, then self-awareness fades away. All of these observations suggest a mathematical relationship between the concentration of agents on the one hand, and complexity on the other.

A qualitative picture of this relationship is given by the so-called "error function." This function starts at zero complexity for low concentrations of particles. So, if you plot concentration along the horizontal axis, and complexity on the vertical axis, then you start with a horizontal line at the lower left. As the concentration of particles increases, it reaches some critical value where patterns begin to emerge, and the complexity begins to rise. So the curve rises sharply in the middle of the graph. Ultimately, at some higher concentration of agents, I suggest that a system's complexity will achieve a maximum value. That means the line levels off at the upper right.

Well, how do these results inform studies of life's origins? I think it is pretty obvious that the emergence of life depends on achieving some minimal concentration of biomolecules, which are the essential particles that lead to cellular life. If the concentration of life-forming molecules is too low, even if the environment is conducive to life, then life could not arise. It turns out that is an extremely useful idea to bear in mind when planning origin-of-life experiments.

The second factor in fostering emergent complexity is the interconnectivity of agents. Particles can interact in so many different ways. At the simplest scale, sand grains interact with each other just by touching. A typical rounded grain at the surface of a sand pile typically contacts six to eight adjacent grains. These stabilizing contacts and gravity, on the one hand, are countered by the disruptive flow of water. The net result is a controlled shuffling of grains, which ultimately produces the rippled patterning of sand.

Ants in an ant colony typically interact over much greater distances because ants mark the ground with a variety of pheromones. These chemical signals, which alert other ants to food, danger and other information, decorate the ant's environment with an array of chemical gradients. As a consequence, each given ant has the ability to interact with potentially thousands of colony mates in a variety of ways. That is one reason why ant colonies display much more complex behavior than sand grains.

Then we come to the conscious brain, which is by far the most complex system we know. The brain is also the most complexly interconnected system we know. Every one of the trillions of neurons in the human brain interacts with hundreds of other nearby cells by way of a branched network of dendrites. What's more, the electrical signals that pass between neurons may be stronger or weaker, much like the variability of a dimmer switch on your lights. As a result, the interconnections of brain cells are vastly more intricate than those of ants or sand.

I suspect that one of the biggest challenges in quantifying emergent complexity will be to incorporate the varied ways that agents can be interconnected. How is it possible to quantify the differences between the interactions of adjacent touching sand grains, versus the persistent chemical markers of moving ants, versus the elaborate variable impulses of neurons? Nor is it obvious that any of these examples is analogous to the interactions of molecules that are necessary for life's origin.

In any event, it is easy to see that the origins of life must have depended on lots of different chemical interactions. So a good experimental strategy will be to figure out a way that optimizes the number and types of molecular contacts. That way, we're more likely to induce emergent behaviors of interest.

The third factor in fostering emergent complexity is the energy flow through the system. It doesn't matter how many interacting particles you have; if there isn't a reliable flow of energy to drive change, no patterns can arise. So, for example, sand grains can't start moving without a minimum water-wave speed. More energetic waves with greater speed and amplitude may move grains more easily and generate sand patterns more quickly, but the resulting patterns aren't appreciably different in their shapes.

There is a limit to effective energy flow, however. Every complex patterned system has a critical magnitude of energy flow, above which it can no longer tolerate. So, for example, during an energetic storm, the crashing waves obliterate sand ripples. All the sand grains will become jumbled, and all signs of emergent patterning will disappear.

The human brain exhibits strikingly similar behavior. During typical waking hours, your brain maintains an intermediate level of electrical impulses. Deepest sleep, on the other hand, corresponds to a sharp drop in electrical activity and a loss of consciousness. At the other extreme, the intense electrical activity of an epileptic seizure halts conscious action by scrambling the usual patterned electrical flow.

In all of these examples, the relationship between energy flow and emergent complexity has a qualitative similarity to the familiar bell-shaped curve. At the lowest energy flux, no patterning occurs and the complexity is zero. At a critical minimum energy flux, pattern formation begins and complexity rises to a maximum value. Then, above some critical value, the energy flux becomes too large and emergent patterns disperse.

I conclude that the emergence of complex patterns depends on an energy flow within rather narrow constraints. If energy flow is too low, then nothing happens, but too much energy flow and the patterns are randomized. Entropy triumphs. This conclusion carries an important lesson for the study of life's origins. A reliable energy source is essential, but you don't want too much energy. Intense energy sources, such as lightning and ultraviolet radiation, can fragment molecules, and may be too extreme to foster life. It may be prudent to search for more benign chemical energy sources, perhaps the natural equivalent of a flashlight battery, to sustain the metabolism of primitive life.

Finally, the fourth factor that appears to foster complexity is the cycling of energy flow. Lots of natural systems are subjected to cycles of energy; such as day and night, high tide and low tide, or wetting and drying. Such inevitable cycles may have played a fundamental role in the evolution of emergent systems, including life; though it is often quite difficult to recognize the effects of these subtle cycles in nature.

Once again, the behavior of sand and water provides valuable insight. In this case, it's best to use laboratory wave tanks, which permit carefully controlled experimental conditions. It turns out that cycling of energy flow across a sand surface generates new complex sand patterns. Physicist Jonas Hansen at the Niels Bohr Institute and a group of his Danish colleagues discovered this surprising result in 2001. Most previous experimenters had used artificial waves that had a fixed wave height and frequency. If well designed, such studies typically produce perfectly spaced, straight ripples.

Instead, Hansen and his colleagues increased and then decreased the height or the frequency of their water waves, and the results were striking. Look at these neat patterns. Rather than the simple parallel sand ripples, they found elegantly intertwined and branching sand structures. I've seen remarkably similar sand features along the Chesapeake Bay when the water is only a few inches deep. I suspect that those shallow conditions favor periodic fluctuations in wave amplitude.

There is another example of patterns formed by cycling. Mark Kessler, a Ph.D. student, and Professor Brad Werner of the University of California, San Diego, in 2003, published photos of amazing stone circles and other so-called "patterned grounds" that are found in the Arctic terrain of Alaska. There, the ground is subject to cyclical freezing and thawing. These boulders shift position slightly each time the ground freezes. As they interact with one another over many years, the natural circles of stone emerge.

At this point, the role of cycling in the emergence of patterns remains a frontier area of research. We know that the early Earth was subject to many different cycles—day and night, high tide and low tide, wet/dry, hot/cold, and more. We have to bear in mind that such cycles may have contributed to the emergence of life itself.

In summary, four factors appear to contribute to the emergence of complexity—the concentration of interacting particles, the nature of the interactions, the flow of energy through the system, and the cycling of that energy flow.

These results suggest that a mathematical expression of complexity will take the form of a mathematical inequality. For those of you who are listening on audio, you can see the expression in the course booklet.

That expression looks pretty daunting, but it's just a shorthand way of saying that the complexity of an emergent system, denoted here by the letter C for "complexity", is a number. That number will be less than or equal to some value that is a mathematical function of at least four factors—the concentration of interacting particles, the degree of those particles' interconnectivity, the energy flow through the system, and the time-variation of that energy flow, and perhaps other variables as well.

Two daunting impediments stand in the way of the completion of this formulation. First, we do not know how to assign numbers to different degrees of complexity. What is the complexity of an anthill or the human brain? What are the units? Every scientific measurement needs to have units, like kilograms or meters per second. We need to be able to say that this system or that system has a complexity of so-many hundreds or thousands of some complexity unit. Nobody knows how to do that.

In addition, we have no clear idea of the mathematical relationships between complexity and the four key factors. Simple patterned systems like sand and water may provide some tantalizing clues, but we're a long way from a useful formula.

The effort to quantify emergent complexity is not a mathematical abstraction. Emergent systems surround us: our environment, our bodies, our minds, and our culture. All of these systems display emergent complexity. Consequently, a comprehensive theory of emergence will foster myriad applications—long-range weather prediction, computer network design, traffic control and much more. A law of emergence will provide us with a deeper understanding of every system of interacting agents, including the origins of life itself.

Even if we do not have a complete definition and mathematical formulation of emergence, the principle can guide our quest to understand life's origins. So, in the next lecture we will explore the earliest emergent step in the path to life—the emergence of simple biomolecules.

Lecture Nine

The Miller-Urey Experiment

Scope: In 1953, the landscape of research on the origins of life changed forever with the Miller-Urey experiment. For the first time, an experimental protocol mimicked plausible life-forming processes. As we'll see, the emergence of simple biomolecules is arguably the best understood aspect of the origin of life. Decades of research have led many scientists to accept that by 4 billion years ago, Earth's oceans must have become a dilute soup of life's building blocks. This chemical system was not alive, but it was poised to undergo a sequence of increasingly complex emergent steps of molecular organization and evolution.

Outline

I. The experimental pursuit of life's chemical origin began at the University of Chicago in the early 1950s. Graduate student Stanley Miller and his mentor, Harold Urey, designed what is now known as the Miller-Urey experiment and together made scientific history.

A. Miller-Urey demonstrated for the first time that organic synthesis could have occurred quite readily, as Earth's primitive atmosphere and ocean were subjected to bolts of lightning and the Sun's intense radiation.

B. A Nobel laureate chemist, Harold Urey set the stage for the classic experiment when he proposed that biomolecules might have been synthesized in abundance in a plausible primitive atmosphere of hydrogen, methane, and other relatively reactive gases.

C. As Urey's student, Miller carried out the simple, elegant experiment, which simulated the primitive Earth on a tabletop in the lab.

THE MILLER-UREY EXPERIMENT

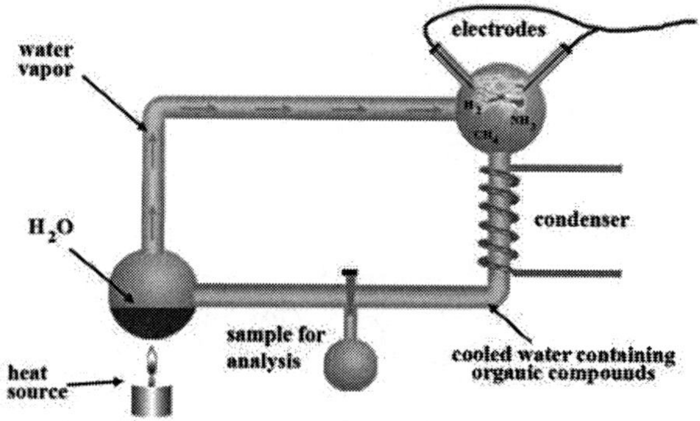

1. Miller filled a 5-inch-diameter spherical flask about two-thirds full with water to represent the primitive ocean. He then linked the water flask through glass tubing to a 10-inch-diameter gas-filled flask, situated off to the side and above.
2. His primitive atmosphere consisted of three relatively reactive gases: methane or CH_4, ammonia or NH_3, and hydrogen or H_2.
3. Into this plausible chemical environment, Miller introduced energy, using two simultaneous inputs. First, two pointed tungsten electrodes penetrated the upper flask to simulate lightning by sparking. Second, a flame gently heated the water-filled flask to mimic the ocean's continuous evaporation.
4. Miller pumped out the system with a vacuum pump to remove atmospheric oxygen. Then he filled his glassware with water and a 2:2:1 gas mixture of methane, ammonia, and hydrogen. He used a burner to heat the water, and then he set off a continuous sequence of tiny electric sparks in the gases to simulate lightning.
5. Within a couple of days, the solution had undergone dramatic chemical changes. It had begun to turn yellowish, and a black residue had begun to accumulate near the electrodes.
6. Miller wanted to find out if he had produced amino acids. He tried to separate and identify the molecules he made that turned his solution yellow, using an old technique called paper chromatography to perform the organic chemical analysis.

7. After performing a series of analyses, Miller found that the molecules he had made contained the amino acid glycine. Glycine is the smallest and simplest of life's amino acids, and the one most likely to be seem in any prebiotic synthesis experiment.

D. In just a couple of days, the graduate student and his advisor had transformed simple gases into one of the most basic building blocks of life. The tabletop experiment had worked much faster than anyone would have expected.

1. Miller repeated the entire experiment, this time for a full week's duration. He also cranked up the burner to bring the water to a slow boil. The results were amazing.
2. As before, the water quickly turned yellow, but it kept changing. It gradually turned intriguing shades of pink and ultimately to a deep red color. At the same time, a gooey black residue formed near the electrodes and oozed down the sides of the larger flask.
3. Clearly, the added time and temperature had led to an even richer accumulation of organic molecules.
4. Miller analyzed the contents of the flask as before, and this time, he found a complex mixture of potential biomolecules, including at least a half-dozen different essential amino acids.
5. Once again, the message was clear. Reactions of water and air, a plausible model of the primitive earth, had produced organic molecules in abundance.

II. The Miller-Urey experiment transformed the science of life's chemical origins and prompted thousands of similar experiments during the next 50 years. This work has shown both the promise and the limitations of the idea that life arose as a chemical process at the surface of the Earth.

A. The Miller-Urey experiment worked, even when scientists tried variations of the process. Over the years, almost every kind of biomolecule has been made, including most of life's amino acids, and membrane-forming hydrocarbons, energy-rich sugars and other carbohydrates, and metabolic acids.

B. The early inventories from Miller-Urey experiments even included some of the building blocks of the genetic molecules DNA and RNA. About the only important molecules *not* made were the 5-carbon sugar ribose—the *R* of RNA—and two polycyclic molecules, adenine and guanine. Those molecules are essential to both DNA and RNA.

C. A good bit of the molecular variety occurs because electric sparks and ultraviolet radiation fragment the gases to form reactive chemical species—molecules such as hydrogen cyanide (HCN) and formaldehyde (CH_2O). These reactive molecules easily bond to other molecules. Miller suspected that the amino acids he made arose when hydrogen cyanide reacts with formaldehyde and ammonia in a chemical process known as the *Strecker synthesis*.

D. In 1960, chemist John Oró of the University of Houston discovered that heating a concentrated hydrogen cyanide solution results in a significant quantity of adenine, one of the essential components of DNA and RNA that hadn't been made in Miller-Urey type experiments. Miller-Urey experiments had produced hydrogen cyanide.

E. Other chemists conducted similar experiments, starting with relatively concentrated solutions of formaldehyde. That's also a molecule produced in Miller-Urey experiments, and it is thought to have been common in some prebiotic environments. These experiments produced a variety of sugars, including a modest yield of ribose. Gradually, chemists filled in the gaps in the prebiotic inventory of life's molecules.

F. The biggest problem with the experiments of Oró and others was that they used relatively concentrated solutions of the reactive organic molecules, such as hydrogen, cyanide, and formaldehyde. The Miller-Urey experiments never yielded these molecules in concentrations high enough to make adenine or ribose. However, there is a possible way around this problem.

1. In the mid-1960s a discovery in the lab of Salk Institute pioneer Leslie Orgel pointed to an intriguing solution. Geologists suspect that the early Earth was not uniformly hot, just like today's Earth is colder at the poles than at the equator. The early Earth may have had ice caps, and it may even have periodically experienced more extensive "ice ages." Orgel realized that such icy conditions might promote intriguing organic reactions.
2. When a solution of water and organic molecules is frozen, only the pure water crystallizes into ice. As the ice crystals

grow, the dwindling reservoir of liquid becomes an increasingly concentrated organic brine.

3. Orgel and coworkers exploited this idea by gradually freezing containers of a dilute hydrogen cyanide solution to about –20°C. As the water froze, tiny volumes of extremely concentrated hydrogen cyanide were produced. Over a period of weeks to months, this concentrated HCN reacted to produce larger compounds that linked up to four HCN molecules together.
4. Sometime in the mid-1970s, Stanley Miller, by then a professor at the University of California, San Diego, repeated the Orgel experiment. He put a frozen flask of hydrogen cyanide solution in the back of a freezer and forgot about it.
5. In the late 1990s, more than two decades later, he and his colleagues found the old container and saw that the frozen solution had formed intriguing dark regions that were rich in organics.
6. When they analyzed the products, they discovered an abundance of adenine, which forms when six HCN molecules are linked together. Clearly, it takes time for reactions to work at extremely cold temperatures, but the primitive Earth had millions of years.
7. These experiments using cold conditions to make biomolecules defy conventional wisdom. Heat is the usual path to drive chemical reactions, but freezing has the advantage of solving the problem of concentration. In this most counterintuitive way, prebiotic cycles of freezing and thawing may have driven the emergence of biomolecules.

III. The Miller-Urey experiments are undeniably important, but many investigators soon recognized two or three seemingly intractable problems.

A. Within a decade of Miller's breakthrough, serious doubts were raised about the true composition of Earth's earliest atmosphere.

1. Miller and Urey took advantage of a highly reactive atmosphere of methane, ammonia, and hydrogen. That composition had come from Urey's atmospheric models of the early 1950s.
2. By the 1960s, new geochemical calculations and data from ancient rocks pointed to a much different and less reactive

early atmosphere of nitrogen and carbon dioxide. Those two gases do almost nothing of interest in a Miller-Urey apparatus.

3. Miller and his many supporters continue to point out that life's biomolecules match those of the original Miller-Urey experiment amazingly well. That fact alone may be sufficient to argue for an early atmosphere rich in reactive methane and ammonia.
4. Geological evidence disagrees, and most geochemists discount the possibility of more than a trace of atmospheric methane or ammonia at the time of life's emergence.

B. Another concern is that all the basic molecular building blocks of life created by Miller and his colleagues represent only the first small steps on the long chemical pathway to life. All those molecules must be carefully selected, then bonded together to form the vastly more complex cell membranes, proteins, DNA, RNA, and many other so-called *macromolecules*. The prebiotic ocean was an extremely dilute solution with a complex mixture of many thousands of different kinds of organic molecules. Most of those molecules played no known role in life's origins. By what emergent processes were just the right molecules selected and organized?

C. Yet another nagging problem is that the macromolecules essential to life have a nasty tendency to break down when exposed to such energetic phenomena as lightning and ultraviolet radiation.

1. These forms of energy ionize molecules. Because they disrupt chemical bonds, they are great for making reactive molecular fragments that recombine into modest-sized molecules, such as amino acids. However, this kind of violent energy destroys the orderly chainlike molecules, such as DNA or proteins.
2. Recall from Lecture Eight that emergent complexity relies on a flow of energy, but not too much energy. This situation makes it doubtful that any life could have emerged in the harsh glare of daylight, which was harmful to life before the development of the complex chemical process of photosynthesis.
3. Could a more benign origins environment have existed?

Essential Reading:

Hazen, *Genesis: The Scientific Quest for Life's Origin*, ch. 6.

Wills and Bada, *The Spark of Life: Darwin and the Primeval Soup*, ch. 2.

Supplemental Reading:

Davies, *The Fifth Miracle: The Search for the Origin and Meaning of Life*, ch. 3.

Dick and Strick, *The Living Universe: NASA and the Development of Astrobiology*, ch. 2.

Fry, *The Emergence of Life on Earth: A Historical and Scientific Overview*, ch. 7.

Lahav, *Biogenesis: Theories of Life's Origin*, part III.

Questions to Consider:

1. Who should receive the most credit for the Miller-Urey experiment: Urey, who suggested the experiment? Miller, who carried it out successfully? Löb, who did essentially the same experiment almost 50 years earlier? Does it matter?
2. Is the fact that the Miller-Urey experiment worked so well good evidence that that's the way biomolecules formed on the early Earth?

Lecture Nine—Transcript
The Miller-Urey Experiment

We have now completed one-third of this exploration of life's origins. What have we learned?

After the first two introductory lectures, in Lecture Three I considered the question, "What is life?" That would seem to be a pretty basic step in discovering the origins of life. But once again, we do not get to have a universal definition of life, nor will we have one until we know a lot more about the range of systems that might be said to be alive. My view is that we need a much richer taxonomy of states than simply "alive" or "not alive." A self-replicating chemical system, for example, may not be alive in the sense of modern cellular life, but it is much more complex in structure and behavior than the typical assemblage of non-living chemicals.

Then in Lectures Four through Seven, I looked at various "top-down" paleontological approaches to understanding primitive life. Yet again the conclusions are uncertain and controversies rage. Scientists have uncovered all kinds of scrappy evidence for ancient life; but every tantalizing trace, whether purported fossil cells, carbon isotope ratios, or biomolecules, is subject to varied interpretations.

In the last lecture, I presented the concept of emergence as a compelling framework for studying the origin of complex systems like life. Life is the quintessential complex emergence system. Unfortunately, we don't have a clear definition of "complexity," much less a law of emergence, but we're learning more all the time.

Well, I hope you're ready for more. What you are seeing here is a science in its vibrant infancy. We're in the most exciting phase, that of rapid discovery and intense debate. All too often, science is presented as a collection of absolute truths—facts set in stone and passed down from professors to students to be memorized and spit back. Well, that's not the science I know. The process of scientific discovery is dynamic; it is frustrating, exhilarating and creative. Every day we go into the lab and have no idea what we're going to find out. We publish our results, promote our points of view, debate, argue, form alliance and sometimes even create enemies. The give and take of this part of science is a whole lot of fun.

I understand it may seem a little frustrating to have so many scientific cliffhangers, but by the time you're through with this lecture series, you'll be poised to share in all the incredible discoveries that are about to come. I absolutely guarantee there will be exciting discoveries in the quest for life's origins.

I think it is safe to say that the "top-down" study of fossils is fascinating, but rather limited, at least in our efforts to understand life's origins. Fossils don't necessarily tell us much about primitive biochemistry. After all, even the most ancient fossil cells, molecules and isotopes are indistinguishable from those of contemporary life forms.

Even so, common sense tells us that the essential first step in life's emergence is pretty obvious. The first step had to be the synthesis and accumulation of lots of carbon-based molecules, key molecules such as amino acids, sugars and hydrocarbons. These are the vital building blocks from which life would eventually be made.

In the beginning, life's raw materials consisted of rocks, oceans and a few simple volcanic gases such as carbon dioxide, nitrogen, maybe a little methane and ammonia. All it took was to add energy to the mix, and a host of organic molecules emerged. Indeed, we'll see that the emergence of simple biomolecules is arguably the best understood aspect of the origins of life.

The experimental pursuit of life's origins began just over a half century ago at the University of Chicago. It was there, in the early 1950s, that an unknown graduate student, Stanley Miller, and his world-famous mentor, Harold Urey, designed what is now known as the Miller-Urey experiment. Actually, a lot of people call it the "Urey-Miller Experiment" in deference to the senior collaborator. In any event, together Stanley Miller and Harold Urey demonstrated for the first time that organic synthesis could have occurred quite readily, as Earth's primitive atmosphere and ocean were subjected to bolts of lightning and the Sun's intense radiation.

In the next few lectures, we'll see that such molecular synthesis is actually quite simple. Chemical reactions in deep space make such molecules, as do processes in the solar nebula. Biomolecules are generated during asteroid and comet impacts. They even form at the high temperatures and pressures of Earth's deep interior. Decades of research have thus led many scientists to accept the theoretical predictions of Alexander Oparin and J.B.S.

Haldane. By four billion years ago, Earth's oceans must have become a dilute soup of life's building blocks. This chemical system was not alive, but it was poised to undergo a sequence of increasingly complex emergence steps of molecular organization and evolution.

That's what we realize today. About a half century ago, life's chemical origins were a complete mystery. The one thing scientists did understand was that there are three apparently universal requirements for life: water, energy and chemical elements. These ingredients hint at a simple experimental approach for studying the origins of life.

The inspiration for such landmark experiments was articulated in the early 1950s by chemist Harold Urey, who lived a long life from 1893 to 1981. Urey won the 1934 Nobel Prize in Chemistry for his landmark discovery of the heavy isotope of hydrogen known as deuterium. His expertise in separating isotopes was actually critical to the Manhattan project, during which he helped to isolate uranium-235, which was a key step in the development of the atomic bomb. Following World War II, Harold Urey devoted much of his effort to modeling Earth's early oceans and atmosphere. His conclusion was that the atmosphere of four billion years ago was very different from today. Rather than the modern mix of nitrogen and oxygen, he suggested that three gasses—hydrogen, methane, and ammonia—dominated the atmosphere. In 1952, Harold Urey was revered as a giant of American science and his ideas about the early Earth carried a lot of weight.

One of the many chemists who were struck by Urey's atmospheric models was a second-year University of Chicago graduate student, the 23-year-old Stanley Miller. Miller's story has been told in some detail by Jeffrey Bada, who was a Miller student in the 1960s, and continued to be a firm proponent of his teacher's ideas in subsequent ideas. Bada and his coauthor, Christopher Wills, reveal some of the behind-the-scenes history of the historic Miller-Urey experiment in a book published in 2000 called *The Spark of Life*.

Urey set the stage for the classic experiment in the fall of 1951 when he presented a seminar that was attended by Miller. As a beginning graduate student, Miller was searching for a promising Ph.D. thesis project. In Urey's lectures, and in his related publications, he proposed that biomolecules might have been synthesized in abundance in a plausible, primitive atmosphere of hydrogen, methane, and other relatively reactive gases.

In fact, Miller's first thesis project was an unsuccessful foray into nuclear theory. It was only after a year of fruitless work that Miller asked Urey if he could attempt an experiment with the objective of simulating Urey's atmosphere and synthesizing amino acids. Wills and Bada revealed that, at first, Urey was reluctant to agree. Instead, he wanted Miller to try an easier and safer research project. One can spend a great deal of time, and fail synthesis experiments, and graduate students don't have the luxury of time. But Miller continued to press his case, and Urey ultimately relented. He gave his student a limit of one year to make headway.

More than anything else, scientists love simple and elegant experiments. That's exactly what Miller and Urey did. They simulated a primitive Earth in a piece of glassware sitting on a tabletop. The Miller-Urey apparatus was basically a flask of heated water from which samples could be extracted. That flask was connected through tubes and condensers to a second larger flask containing a mixture of gases. For those of you listening on audio, you might want to refer to the illustration in your Teaching Company booklet at this point. Miller used a 5-inch-diameter, spherical flask, which he filled about two-thirds full of water, which represented the primitive ocean. He then linked the water flask through glass tubing to a 10-inch diameter, gas-filled flask, situated off to the aside and above. Miller's primitive atmosphere consisted exclusively of three relatively reactive gases—methane, or CH_4, which is sold commercially as natural gas; ammonia, or NH_3, that's the strong smelling component of ammonia-based cleaners; and hydrogen, or H_2, which is the lighter-than-air gas on the Hindenburg that caught fire and exploded.

In addition to a plausible chemical environment, Miller and Urey had to introduce some kind of energy. They used two simultaneous energy inputs. First, two pointed tungsten electrodes penetrated the upper flask to simulate lightning by sparking. In addition, a flame gently heated the water-filled flask to mimic the ocean's continuous evaporation. Sending sparks through a potentially explosive mix of hydrogen and methane is not for the timid. I'm quite sure that Miller's unshielded glass apparatus would have failed today's rigid government safety standards, but there's no question that they devised an elegant experimental design.

Following academic tradition, the student, Miller, performed the experiments. This was in late 1952. Urey, his mentor, would have kept busy with innumerable academic responsibilities. Miller began by pumping out the system with a vacuum pump to remove atmospheric oxygen. Then

Miller filled his glassware with water in a 2:2:1 gas mixture of methane, ammonia, and hydrogen. He used a burner to heat the water, and then he set off a continuous sequence of tiny electric sparks in the gases to simulate lightning. And he waited.

If asked how long it might take to produce something interesting, I imagine Miller would have guessed weeks, months, if at all. You can imagine the scene. When the experiment began, the water was pure and clear. But within a couple of days, the solution had begun to turn yellowish and a black residue had begun to accumulate near the electrodes. Can you imagine Stanley Miller's excitement? In a matter of days, the system had begun to undergo dramatic chemical changes. The atmospheric gases must have reacted to produce organic molecules. With intense anticipation, Miller cut short his first experiment to analyze the tantalizing products. In particular, he wanted out find out if he had produced amino acids.

Organic chemical analysis was a lot more difficult in the early 1950s than it is today. Miller had to resort to a technique called paper chromatography. That's an old and relatively fast procedure that can separate different molecules into distinctive colored spots on a square piece of absorbent paper that is typically a few inches on a side. A modern version of the same thing is this 10-inch-square glass plate that has an absorbent chemical coating. That white coating holds and concentrates liquids onto the surface, just like the old paper version.

Here's what Miller did. First he opened a valve in his glass apparatus and extracted some of the yellow solution. He was concerned that microbes might begin to grow in the nutrient-rich solution. That would confuse his results, so he added some toxic mercury chloride to kill all the bacteria. The next step was the tedious routine of drying down his sample and concentrating it into one small droplet. Then Miller carefully placed the small yellow drop near one corner of the chromatography paper. That solution dried as a small yellow dot.

Then comes the tricky part: Miller suspended the paper above a narrow trough that he had filled with a solution of alcohol and water. He gently lowered the paper so that one edge close to the yellow spot just barely touched the solution. That solution then slowly soaked up into the paper and rose up the sheet by capillary action.

Remember, what Miller was trying to do was separate and identify whatever molecules he made that turned his solution yellow. Some of those molecules were soluble in water or alcohol. So, as the alcohol solution moved up the paper, it carried along some of the unknown molecules. It only took a few minutes for the yellow chemical spot to become a narrow, three-inch-long, faint yellow streak of unknown, mostly invisible chemicals. Different types of molecules were concentrated at different points along this streak.

So the standard procedure Miller followed was to let his paper dry, then rotate it 90 degrees and repeat the process with a second solvent. This time Miller used a phenol solution. You can imagine that molecules move differently through the paper when you use different solvents. So this second step spread out that narrow three-inch streak into a diagnostic, two-dimensional pattern of spots just a fraction of an inch across. Miller dried his paper for a second time.

At this point, Stanley Miller wet a white piece of paper that looked more or less as it had before the experiment began. So one final step was required; he treated the paper with a chemical called ninhydrin. Ninhydrin is useful because it stains otherwise invisible amino acids into distinctive colors.

Almost immediately, Miller saw a distinctive purple spot that appeared exactly at the expected position for the amino acid glycine. Glycine is the smallest and simplest of life's amino acids, and the one you're most likely to see in most prebiotic synthesized experiments.

Can you imagine his excitement? In a couple of days, his apparatus had transformed simple gases into one of the most basic building blocks of life. The graduate student and his advisor were ecstatic. The tabletop experiment had worked much faster than anyone would have expected. This was the type of experiment that could make one's career.

In such a dramatic and newsworthy experiment, once is not enough. So Miller repeated the entire experiment, this time for a full week's duration. He also cranked up the burner to bring the water to a slow boil. The results were just amazing. As before, the water quickly turned yellow, but it kept changing. It gradually turned intriguing shades of pink, and ultimately to a deep red color. At the same time, a gooey black residue formed near the electrodes and oozed down the sides of the larger flask. Clearly, the added time and temperature led to an even richer accumulation of organic molecules. Miller analyzed the contents of the flask as before, and this time he found a complex mixture of potential biomolecules, including at least half a dozen essential amino acids. Once again the message was clear. Reactions of water and air—a plausible model of primitive earth—had produced organic molecules in abundance.

Harold Urey told Stanley Miller to write up and publish his results immediately. Miller obliged, and submitted a concise manuscript to the prestigious journal *Science* in mid-February of 1953. That was just five months since the project's inception. In a move that I still find amazingly generous, Harold Urey also withdrew his name from the paper. Otherwise, everyone would have focused on the Nobel Prize winner, and the graduate student would have received little of the credit.

That first publication by Stanley Miller was an absolute triumph. The two-page article appeared in the May 15, 1953 issue of *Science*. The title summarized the contents, "A Production of Amino Acids Under Possible Primitive Earth Conditions." He began his paper with a now-famous passage:

The idea that the organic compounds that serve as the basis of life were formed when the earth had an atmosphere of methane, ammonia, water and hydrogen instead of carbon dioxide, oxygen and water was suggested by Oparin and has been given emphasis recently by Urey and Bernal. In order to test this hypothesis, an apparatus was built....

I should put this work into context. Many other chemists had synthesized amino acids, not to mention most of the other essential molecules of life. What's more, they've done so with simple molecules subjected to energy. What was different about the Miller-Urey experiment was that, for the very first time, an experiment was designed specifically to mimic the early Earth. This experiment wasn't just another foray into synthetic chemistry; it was the first convincing experiment in the origins of life.

As you can imagine the press had a field day. This was big news. On May 17, *The New York Times* printed a feature story, "Life and a Glass Earth." Many other respected newspapers followed suit. Meanwhile, the tabloids pushed the boundaries by speculating about synthetic life crawling out of test tubes. It would be an understatement to say that few graduate students get that much publicity.

Thanks to that one experiment, Stanley Miller's future was assured. He would soon become the leader in origins-of-life research. Miller and an impressive list of his students soon dominated the teaching and, perhaps more importantly, the funding of origins research. This situation worked pretty well for a while, until perhaps the mid 1970s.

Miller may have had too much invested in his own origins experiment to accept, or even consider, any rival hypothesis. As a result, he and his students rarely encouraged other perspectives to enter the field. He himself opposed funding for lots of interesting new ideas. In fact, it wasn't until NASA got into supporting origins-of-life research in the mid-1970s that the field really opened up. But I'm getting ahead of myself. We'll come to later developments in the next few lectures.

Anyway, as you can see, in 1953, the landscape of research on the origins of life had changed forever. The Miller-Urey experiment absolutely transformed the science of life's chemical origins. For the first time, an experimental protocol mimicked plausible life-forming processes.

Given such an exciting result, lots of other groups jumped at the chance to duplicate the Miller-Urey experiment. At first it was simple replication to the amino-acid synthesis. Within a couple of years, chemists at the University of Bristol in England and at my own Carnegie Institution in Washington, DC, confirmed Miller's findings.

People also conducted a careful search of the older scientific literature, and they turned out a similar series of experiments that had been done almost a half-century earlier by a German chemist named Walter Löb. It turns out that Löb also synthesized amino acids by using sparks in a gas mixture. As an aside, it's interesting to consider why Löb's papers aren't now recognized as the landmark contribution in the field. After all, he did exactly what Miller did. I think there are at least three reasons. First and foremost, Löb's research was not really designed to mimic prebiotic environments or the chemistry of life's origins. It was just an experiment in chemical synthesis, which isn't very newsworthy. In addition, Löb died at a relatively early age, and he wasn't able to promote his findings in the same way that Miller did. A third reason was that Löb was German, and he published in the German language. There was still a lot of anti-German sentiment in the U.S. in the 1950s after World War II, and it's perhaps not surprising that his work was ignored.

In any case, the Miller-Urey experiment pointed the way for lots more experiments. Indeed, thousands of more experiments have been done during the past half-century. I think it's safe to say that this work has shown both the promise and the limitations of the idea that life arose as a chemical process at the surface of the earth. On the one hand, the Miller-Urey kind of experiment works. Time and time again, scientists have tried variations of the process. They used ultraviolet radiation, or different gas mixtures, or powdered minerals, and they almost always got the synthesis of life's most basic building blocks. These experiments are pretty easy to run and these days they're easy to analyze. It's even done in high school labs. So these experiments continue to be a sort of cottage industry in origins-of-life research circles.

This kind of experiment produces amazing results. Almost every kind of biomolecule has been made, including most of life's amino acids, membrane-forming hydrocarbons, energy-rich sugars, and other carbohydrates and metabolic acids as well. The early inventories for Miller-Urey experiments even included some of the building blocks for the genetic molecules DNA and RNA. In fact, the only important molecules not made were the five-carbon sugar ribose—that is the "R" of RNA—and two polycyclic molecules, adenine and guanine; those molecules are essential to both DNA and RNA.

So how do all these molecules form? A lot of the molecular variety occurs because electric sparks and ultraviolet radiation fragment the gases to form very reactive chemical species—molecules such as hydrogen cyanide (HCN) and formaldehyde (CH_2O). These reactive molecules easily bond to other molecules. For example, Miller suspected that the amino acids he made arose when hydrogen cyanide reacts with formaldehyde and ammonia in a chemical process known as the Strecker synthesis.

You can imagine the growing enthusiasms as other scientists discovered new promising ways to make biomolecules. In 1960, chemist John Oró of the University of Houston discovered that when you heat a concentrated hydrogen-cyanide solution, you get lots of adenine. Now, recall that adenine is one of those essential components of DNA and RNA that hadn't been made in the Miller-Urey-type experiments. But Miller-Urey experiments did produce hydrogen cyanide. So that's just one additional step.

Other chemists conducted similar experiments, starting with relatively concentrated solutions of formaldehyde. That's also a molecule produced in

Miller-Urey experiments, and it's thought to be pretty common in some prebiotic environments. Well, these experiments produced a variety of sugars, including a modest yield of that exclusive sugar, ribose. So gradually, through this kind of specialized experiments, they were able to fill in the gaps in the prebiotic inventory of life's molecules.

This is not to say that everyone was convinced. The biggest problem with the experiments of Oró and others was that they used relatively concentrated solutions of reactive organic molecules, such as hydrogen cyanide and formaldehyde. The Miller-Urey experiments never yielded these molecules in anywhere near those concentrations—certainly never enough to make adenine or ribose. But there is a possible way around this problem.

In the mid-1960s, a discovery in the lab of Salk Institute pioneer Leslie Orgel pointed to an intriguing solution. Geologists suspect that the early Earth was not uniformly hot, just like today's Earth is colder at the poles than it is at the Equator. So the early Earth may have had ice caps. It may even have periodically experienced more extensive "ice ages." Orgel realized that such icy conditions might promote intriguing organic reactions. This is because when you freeze a solution of water and organic molecules, only the pure water crystallizes into ice. As the ice crystals grow, the dwindling reservoir of liquid becomes increasingly concentrated organic brine. As we saw in Lecture Eight on emergence, a higher concentration of "agents" often promotes emergence. That seems to be what's happening here.

Orgel and co-workers exploited this idea by gradually freezing containers of a dilute hydrogen cyanide solution to about -20°C. As the water froze, they produced tiny volumes of extremely concentrated hydrogen cyanide. Over a period of weeks to months, this concentrated HCN reacted to produce larger compounds that linked up to four HCN molecules together. That result, in and of itself, wasn't too interesting from a prebiotic chemistry point of view, but the phenomenon became the inspiration for another, much longer experiment. In fact, it was certainly one of the longest experiments in the history of origins research.

Sometime in the mid-1970s—no one seems to remember the exact date—Stanley Miller, who was then a well-established professor at the University of California, San Diego, repeated the Orgel experiment. He put a frozen flask of hydrogen-cyanide solution in the back of a freezer and just left it there. I gather they sort of forgot about it. Then, in the late-1990s, more than two decades later, he and his colleagues found the old container and removed it. It was obvious something interesting had happened. The frozen solution had formed intriguing dark regions that were rich in organic molecules. When they analyzed the products, they discovered an abundant production of adenine, which forms when you link five HCN molecules together. The take-home lesson was clear; it takes a lot of time for reactions to work in extremely cold temperatures, but the primitive earth had millions of years.

These experiments, using cold conditions to make biomolecules, defy conventional wisdom. Heat is the usual path to drive chemical reactions. After all, you can't make cookies by freezing the dough. But freezing has the advantage of solving the very real problem of concentration. The open ocean was just too dilute to promote many critical reactions. So researchers have tried lots of other freezing experiments, and they've succeeded in making amino acids and lots of other interesting biomolecules. In this most counterintuitive way, prebiotic cycles of freezing and thawing may have driven the emergence of biomolecules.

The Miller-Urey experiments are undeniably important and exciting, but many investigators soon recognized two or three seemingly intractable problems. For one thing, within a decade of Miller's breakthrough, serious doubts were raised about the composition of Earth's earliest atmosphere. Miller and Urey took advantage of a highly reactive atmosphere of methane, ammonia and hydrogen. That composition came from Urey's atmospheric models of the early 1950s. But by the 1960s, new geochemical calculations, and pretty unambiguous data from ancient rocks as well, pointed to a much different and less reactive early atmosphere of nitrogen and carbon dioxide. Those two gases do almost nothing of interest in the Miller-Urey apparatus.

So there's a big disagreement in the field of origins research. Miller and his many supporters continue to counter with an argument that's difficult to discount. They point out that life's biomolecules match those of the original Miller-Urey experiment amazingly well. That fact alone may be sufficient to argue for an early atmosphere rich in reactive methane and ammonia. Harold Urey is said to have often quipped, "If God did not do it this way, He missed a good bet." Even so, geological evidence disagrees, and the majority of geochemists discount the possibility of more than a trace of atmospheric methane or ammonia at the time of life's emergence.

There's also another concern about the Miller-Urey scenario. All of those basic molecular building blocks of life that were created by Miller and his colleagues represent only the very first, small steps on the long chemical pathway to life. All of those small molecules have to be carefully selected and then bonded together to form the vastly more complex cell membranes and proteins and DNA and RNA, and lots of other so-called "macromolecules." As you can imagine, the prebiotic ocean was an extremely dilute solution, with a complex mixture of many thousands of different kinds of organic molecules. Most of those molecules played no known role in life's origins. So we have to ask, by what emergent processes were just the right molecules selected and organized?

As if that weren't enough, the Miller-Urey scenario suffers from yet another nagging problem. The macromolecules essential to life have a nasty tendency to break down when exposed to such energetic phenomena as lightning and ultraviolet radiation. These forms of energy ionize molecules. They disrupt chemical bonds so they are great for making reactive molecular fragments that recombine into modest-sized molecules, such as amino acids. But this kind of violent energy destroys the kind of orderly chainlike molecules, such as DNA or proteins.

Recall from Lecture Eight that emergent complexity relies on a flow of energy, but not too much energy. This situation makes it doubtful that any life could have emerged in the harsh glare of daylight, which was really harmful to life before the development of the complex chemical process of photosynthesis. Could there have been a different, more benign origins environment?

And so, faced with such an impasse, a few maverick scientists began to look at other plausible venues for the cradle of life. As we'll see in the next lecture, there are, after all, other environments of Earth that have carbon and water and energy. Indeed, virtually any place on Earth a mile or two below your feet has all the necessary ingredients for life.

Lecture Ten

Life from the Bottom of the Sea

Scope: By the late 1970s, enough problems and questions had been raised about the Miller-Urey experiment that alternative hypotheses were proposed. One of the first and most influential of these competing origin models was the idea that life might have arisen in the deep ocean at a hot hydrothermal vent. Miller and his colleagues immediately raised objections, but by the early 1990s, the hypothesis that life began in a deep, hot environment had become widely accepted as a plausible, if unproven, alternative to the Miller-Urey surface scenario.

Outline

I. The dynamic field called *prebiotic chemistry* involves the effort to synthesize life's essential molecules in environments that mimic what might have occurred on the primitive Earth.

- **A.** The Miller-Urey experiment, which attempted to reproduce the conditions of the early atmosphere and ocean, was wildly successful in this regard, but questions remain.
  1. Was the initial atmosphere of methane and ammonia realistic?
  2. How could the molecules become sufficiently concentrated?
  3. And how was the prebiotic soup protected from destructive lightning and ultraviolet radiation?
- **B.** Given these and other questions, a number of researchers began to propose alternative origin-of-life scenarios, including the idea that life arose in the depths of the ocean, far from the influence of the Sun.

II. Before we consider this idea, however, let's review some basic chemistry.

- **A.** We start with a single carbon atom.
  1. Carbon is element 6, so it has six protons in its nucleus, contributing a charge of +6.
  2. To balance that +6 charge, six negative electrons surround the nucleus. Those electrons are arranged with two in the first

inner shell and four in the second outer shell. The second shell can actually hold up to eight electrons.

3. Carbon, like all elements, is "happier" with a full shell. Every carbon atom wants to acquire four more electrons. This is the key to understanding carbon chemistry.

METHANE

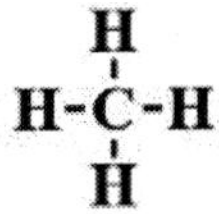

4. Four hydrogen atoms added to a carbon atom make the molecule methane, or CH_4. The carbon atom now has a full second electron shell, and each hydrogen atom is "happy" with two electrons.

ETHANE

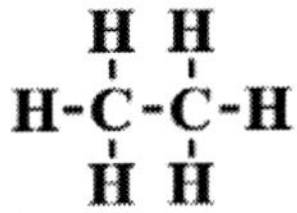

5. The element carbon can also bond to itself. Remove one of the hydrogen atoms in methane and add another carbon atom, and the result is the molecule ethane, which has two carbon atoms, each one bonded to three hydrogen atoms and the other carbon atom.
6. We can make the chain longer just by adding more carbon atoms. We can also make the chain branch by having one carbon atom bond to two others, so some carbon atoms go off to the side. A chain of carbon atoms can even loop back on itself to make a ring structure. Further, rings and chains and branches can be combined in almost any imaginable form.
7. Carbon can bond to a second atom by sharing not one but two electrons, resulting in a *double bond*. In a molecule of carbon dioxide, for example, carbon is double-bonded to each of two oxygen atoms in a linear oxygen-carbon-oxygen arrangement: **O=C=O**. We can also mix the bond types, as in formaldehyde. Formaldehyde has one double bond between carbon and oxygen, plus two single bonds to two hydrogen atoms.

FORMALDEHYDE

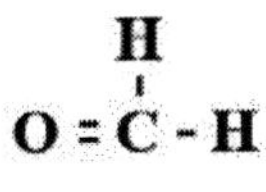

B. Life requires other common chemical elements.

WATER

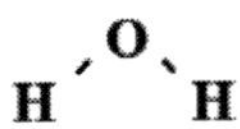

1. Hydrogen is element 1. Hydrogen atoms need one extra electron to fill their first shell. One simple way to do that is to

have two hydrogen atoms bond to each other in hydrogen gas, or H_2. And, as we've seen, hydrogen can share electrons with carbon, as in methane, which is CH_4.

2. Oxygen is element 8, so it needs two more electrons to fill the second shell. Again, hydrogen can fill the bill to make water, H_2O. Or we can make a double bond to carbon, as in carbon dioxide or formaldehyde.
3. Nitrogen is element 7, so it needs three more electrons. One of the common nitrogen molecules is ammonia, NH_3.

AMMONIA

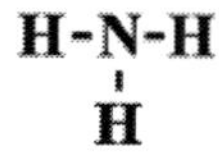

4. This kind of bonding strategy, in which atoms share electrons to fill their outer shells, is called *covalent bonding.*

C. Consider some of life's essential building blocks.
   1. Amino acids start with a central carbon atom to which is added one hydrogen atom; a carboxyl group, or COOH; and an amino group, NH_2. Every amino acid has three parts—hydrogen, carboxyl group, and amino group. The fourth bond to that central carbon atom can be anything: another hydrogen atom, a carbon atom, an NH_2 amino group, or a much larger molecule with a dozen or more atoms. Hundreds of amino acids are possible.

BASIC AMINO ACID STRUCTURE

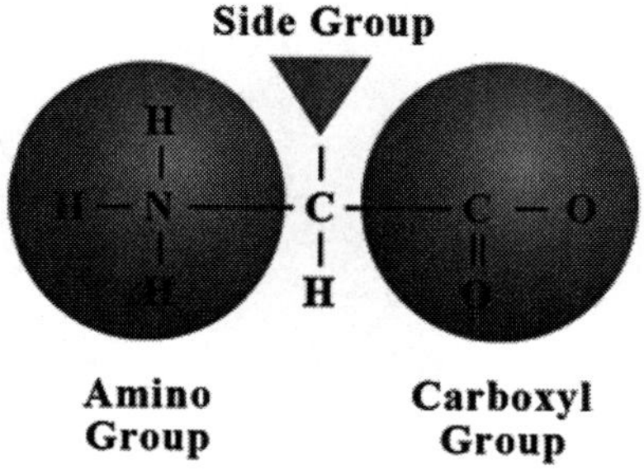

   2. Sugars and other carbohydrates are made from carbon, hydrogen, and oxygen, usually in about a 1:2:1 ratio. For example, ribose has 5 carbon atoms, 10 hydrogen atoms, and 5 oxygen atoms.

3. Once again, bonding comes into play. Each carbon atom bonds with four extra electrons, each oxygen atom with two extra electrons, and each hydrogen atom with one extra electron. Hundreds of sugars exist; every one features these kinds of bonds.

D. Consider one more molecule—the ring-shaped molecule adenine—and start with hydrogen cyanide, one of the very reactive molecules produced in the Miller-Urey experiment.

ADENINE

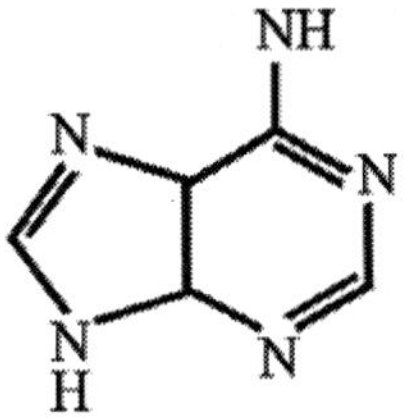

1. In hydrogen cyanide, we see a new bond, the triple bond, because nitrogen and carbon share three pairs of electrons: **H–C≡ N.**
2. If six of these molecules link together, end-to-end, the result is the critical biomolecule adenine.

E. One more bit of organic chemistry is an essential kind of chemical reaction called *polymerization.*

1. In polymerization, smaller molecules link up, usually end-to-end, to form a larger chain-like molecule.
2. Sugar molecules can form polymers, most notably cellulose; some biologists estimate that most of the biomass on Earth is in the form of cellulose.

CELLULOSE

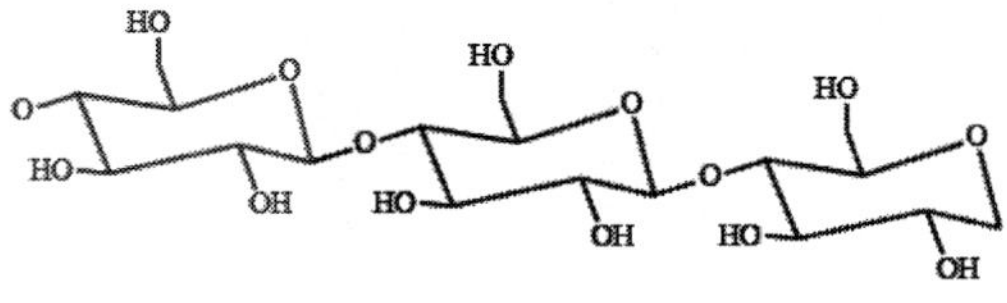

3. In cellulose and other natural polymers, the individual molecular building blocks bond end-to-end in a process called *addition polymerization.* That is what the six H–C≡ N (hydrogen cyanide) molecules do when they form adenine—they polymerize by linking end-to-end.

4. Another essential kind of polymerization is important to every cell. When many amino acids bond together to make a protein, they undergo what is known as a *condensation polymerization.* In this case, two amino acids line up so that the amino group NH_2 of one amino acid is next to the carboxyl group COOH of a second. One hydrogen atom of the amino group combines with the OH of the carboxyl group to form a molecule of water, H_2O. That water molecule dissolves into the surroundings. Then, the nitrogen bonds to the carbon, and the two amino acids are linked by a carbon-to-nitrogen bond. Chemists call this a *peptide bond.*

CONDENSATION POLYMERIZATION

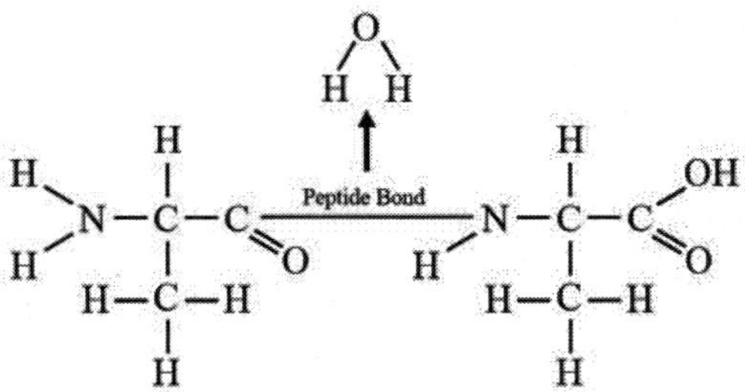

5. Every time a peptide bond forms, a molecule of water is released. In the presence of a quantity of water, peptide bonds tend to break down by a process known as *hydrolysis*. It is easier to form peptide bonds in dry environments, away from water.
6. Of course, that provides a challenge to cells, because they must constantly synthesize proteins, but they are filled with water. It is also a challenge to origins researchers, because the conditions that support synthesis of amino acids are not necessarily the conditions that support the formation of peptide bonds and vice versa.

F. The biggest mysteries in origins-of-life research are not how to make simple organic molecules but how to select and concentrate just the right ones and to make polymers.

III. The history of science has frequently seen the overthrow of what had been thought to be obvious. We must consider whether our intuitive view of life's essential energy source might also be in error.

A. Until recently, one apparent certainty was the role of the Sun as a precept in biology—all life depends on its radiant energy. In the past two decades, however, remarkable discoveries have been made of life at the darkest ocean depths, and microbes have been discovered buried miles beneath Earth's surface in solid rock.

B. Our view of life on Earth was changed forever in February 1977 thanks to marine geologist Jack Corliss and two of his crewmates. They were investigating a volcanic terrain about 8,000 feet under the sea called the East Pacific Rise, located near the Galápagos Islands, where new ocean crust is being formed.

1. Corliss and his crew were looking for a submarine hydrothermal vent, but they found an unexpected ecosystem. They observed new species of albino crabs, giant clams, and 6-foot-long red tubeworms. The area was also cloudy with abundant microbes that coated the rock surfaces. This vibrant community of organisms is at crushing pressures, more than a mile and a half beneath the sea. None of these organisms ever sees the light of the Sun.
2. How does this ecosystem sustain itself? Microbes serve as the primary energy producers. They play essentially the same role as plants do on Earth's surface. These one-celled vent organisms exploit the chemical instability among the cold ocean water, the hot volcanic water, and the various sulfur-rich mineral surfaces over which the two kinds of water mix. Slowly, these microbes obtain energy by altering unstable minerals. These reactions are sped along by the oxygen in ocean water—the indirect result of photosynthesis by plants.
3. Corliss and his colleagues soon elaborated on their findings and proposed that deep-sea hydrothermal vents were the actual location of life's origin.
4. Their thesis is elegantly simple and has become influential. Modern organisms thrive in deep hydrothermal ecosystems, and fossil microbes have been found in 3.5-billion-year-old hydrothermal deposits.
5. Even without the energy of sunlight, nutrients and potential sources of chemical energy abound in hydrothermal systems.

As a consequence, the scientists claimed that hydrothermal systems are the "ideal reactors for abiotic synthesis." Based on this claim, they outlined a sequence of chemical steps for the potentially rapid emergence of life.

C. The controversial paper outlining this thesis was not eagerly received. The official objections included the following:

1. Hydrothermal temperatures are so high that amino acids and other essential molecules decompose rapidly.
2. The dynamic fluid flow through vent systems was not realistically modeled.
3. Corliss and his coworkers also were said to have the ancient ocean chemistry incorrect. Modern hydrothermal ecosystems depend on oxygen-rich ocean water—an indirect consequence of plants and photosynthesis. By contrast, the prebiotic ocean would have had little if any oxygen. The proposed life-sustaining chemical reactions would have proceeded slowly, if at all.

D. Eventually, the hydrothermal origin hypothesis began to catch on among origins experts.

1. New support for the idea kept building, because thriving hydrothermal ecosystems were found in many places along ocean ridges, in both the Atlantic and Pacific oceans.
2. In the early period of Earth history, the deepest ocean ecosystems would have been somewhat shielded from the impacts of comets and meteorites. Hydrothermal vents would have been a much more protected location than the surface for life's origin.
3. New discoveries were made of diverse primitive microbial life deep in the continental shelf.

E. By the early 1990s, the hypothesis that life began in a deep, hot environment had become widely accepted as a plausible, if unproven, alternative to the Miller-Urey surface scenario.

Essential Reading:

Hazen, *Genesis: The Scientific Quest for Life's Origin*, ch. 7.

Wills and Bada, *The Spark of Life: Darwin and the Primeval Soup*, ch. 4.

Supplemental Reading:

Davies, *The Fifth Miracle: The Search for the Origin and Meaning of Life*, ch. 7.

Dick and Strick, *The Living Universe: NASA and the Development of Astrobiology*, ch. 5.

Fry, *The Emergence of Life on Earth: A Historical and Scientific Overview*, ch. 10.

Lahav, *Biogenesis: Theories of Life's Origin*, part III.

Questions to Consider:

1. Do you find the idea that life arose from the Sun's energy more appealing than an origin based on energy from volcanic vents? If so, why?
2. Were Stanley Miller and his colleagues justified in objecting to the hydrothermal origins hypothesis? What steps are reasonable in blocking publication of a scientific idea to which you disagree?

Lecture Ten—Transcript
Life from the Bottom of the Sea

In the last lecture, we got a taste of the dynamic field called prebiotic chemistry, which is the effort to synthesize life's essential molecules in environments that mimic what might have occurred on the primitive Earth. The Miller-Urey experiment, which attempted to reproduce the conditions of the early atmosphere and ocean, was wildly successful in this regard. But there are still lots of questions: Was their initial atmosphere of methane and ammonia realistic? How could their molecules become sufficiently concentrated? How was the prebiotic soup protected from destructive lightning and ultraviolet radiation?

Naturally, with these and other questions, a number of researchers began to propose alternative origin-of-life scenarios. In this lecture, I want to explore one of the most intriguing of those ideas that is the hypothesis that life arose in the depths of the ocean, far from the influence of the Sun. First, I need to step back and consider some basic chemistry.

I think you may have gotten the sense from the last lecture that a lot of origin-of-life research is firmly rooted in the field of chemistry. After all, the first steps to life's emergence are the synthesis and assembly of carbon-based chemicals. So, with your indulgence, I'd like to take a few minutes to review some basic organic chemistry—that's the chemistry of the element carbon. By the way, you can get more information on this topic in Lecture Twenty-Two of my Teaching Company course, *The Joy of Science*.

Let us start with a single carbon atom. Carbon is element six, so it has six protons in its nucleus. Protons are always positively charged, so that's a plus six charge; and to balance that plus six charge, we need six negative electrons surrounding the nucleus. Those electrons are arranged with two in the first inner shell and four in the second outer shell. That second shell can actually hold up to eight electrons. Carbon, like all elements, is happier with a full shell. Every carbon atom wants to acquire four more electrons. This is the key to understanding carbon chemistry. I should make a note about notation here. You can show every electron with a dot, but some atoms have a lot of electrons so it is easiest just to show the outer shell. Carbon, for example, has four electrons in the outer shell. Carbon wants four more electrons to fill its outer shell. The simplest way to do this is to bring in four more atoms, each one to contribute one electron. So we can add four hydrogen atoms to carbon to make a molecule of methane, such as CH_4;

that's what you pay for as natural gas. By adding the four hydrogen atoms, the carbon atom now has a full second electron shell. What is more, each hydrogen atom is also happy with two electrons. As we add hydrogen atoms the number of dots adds up, so pairs of dots are usually just shown as a short little line segment like this. Every line represents a chemical bond with two shared electrons.

The neat thing about the element carbon is that it can also bond to itself, so we can remove one of the hydrogen atoms in methane, and stick on another carbon atom; so you get the molecule called ethane, which has two carbon atoms, each one bonded to three hydrogen atoms and the other carbon atom. We can make the chain longer and longer just by adding more carbon atoms.

What's more, we can make the chain branch by having one carbon atom bond to others, so some carbon atoms go off to the side. Or a chain of carbon atoms can even loop back on itself to make a ring structure. You can combine all sorts of rings and chains and branches in almost any imaginable form.

There is another way that carbon can bond to a second atom, by sharing not one, but two electrons. This situation is called a "double bond" where two atoms share four electrons. In a molecule of carbon dioxide, for example, carbon is double bonded to each of two oxygen atoms in a linear oxygen-carbon-oxygen arrangement. We can also mix the bond types, as in formaldehyde, which we met in the last lecture. Formaldehyde has one double bond between carbon and oxygen, plus two single bonds to two hydrogen atoms; that gives CH_2O.

Life requires a couple of other common chemical elements as well. Let us see how they bond. Hydrogen is element one, so hydrogen atoms need one extra electron to fill their first shell. One simple way to do that is to have two hydrogen atoms bond to each other in hydrogen gas, or H_2. As we've seen, hydrogen can share electrons with carbon, as in methane, which is CH_4.

Oxygen is element eight so it needs two more electrons to fill the second shell. Again, hydrogen can fill the bill to make water, H_2O. Or you can make a double bond to carbon, as in carbon dioxide or in formaldehyde.

And nitrogen, another important element in life, is element seven, so it needs three more electrons. So one of the common nitrogen molecules that we met in the last lecture is ammonia, NH_3. This kind of bonding strategy, in which atoms share their electrons to fill their outer shells, is called covalent bonding. Do you see how this kind of bonding arrangement satisfies each and every atom's desire to have a filled outer shell of electrons? That means a total of two electrons in hydrogen; or a total of 10 electrons, as in the cases of carbon, oxygen or nitrogen.

So, let us consider some of life's essential building blocks. For those of you listening on audio, there are illustrations in the companion volume. First we have the amino acids. You start with a central carbon atom and you add one hydrogen atom to that carbon atom. Next add an amino group, $NH_{2;}$ that is what gives amino acids their name. Then you attach what's called a carboxyl group, or COOH, to that central carbon atom.

Every amino acid has at least three parts—hydrogen, carboxyl group and amino group. Then, the fourth bond to that central carbon atom can be anything. It can be another hydrogen atom—that makes glycine, which is the amino acid that Stanley Miller made in his first experiment. But it can be another carbon atom, or another NH_2 amino group, or a much larger molecule with a dozen or more atoms. There are hundreds of different possible amino acids.

What about sugars and other carbohydrates? They're made from carbon, hydrogen and oxygen, usually in about a one-to-two-to-one ratio. The sugar ribose, for example, has five carbon atoms, 10 hydrogen atoms, and five oxygen atoms. Again, it follows all the rules. Each carbon atom sees four extra electrons, each oxygen atom sees two extra electrons, and each hydrogen atom sees one extra electron. Every atom is happy. There are hundreds of different kinds of sugars, but every one features these kinds of bonds.

Let us look at one more molecule, the ring-shaped molecule adenine. But this time I want to start with hydrogen cyanide, HCN. Recall that hydrogen cyanide is one of the very reactive molecules that were produced in the Miller-Urey experiment. In hydrogen cyanide we meet a new bond, the triple bond, because nitrogen and carbon share three pairs of electrons. So, if we link five of these molecules together end to end—lo and behold—we make that critical biomolecule adenine.

There is one more bit of organic chemistry that I want to review with you. That is an essential kind of chemical reaction called polymerization. In polymerization, smaller molecules link up, usually end to end, to form a larger, chain-like molecule. There are numerous synthetic polymers, such as nylon and rayon and polyester. But life also uses many different polymers. For example, sugar molecules can form polymers, most notably cellulose. Cellulose makes the stringy fibers of celery, as well as stems, leaves, and other plant tissues. In fact, some biologists estimate that most of the biomass on Earth is in the form of cellulose.

In cellulose, and lots of other natural polymers, the individual molecular building blocks just bond to each other end-to-end. That's a process called "addition polymerization." In fact, that's just what the six HCN molecules do when they form adenine—they just polymerize by linking end to end.

But there's another essential kind of polymerization that's extremely important to every cell. When lots of amino acids bond together to make a protein, they undergo what is known as a condensation polymerization. In this case, two amino acids will line up so that the amino group, NH_2, of one amino acid is next to the carboxyl group, COOH, of a second amino acid. What happens is this. One hydrogen atom of the amino group combines with the OH of the carboxyl group to form a molecule of water. That water molecule goes away into the surroundings. Then the nitrogen bonds to the carbon; and the two amino acids are now linked together in a carbon-to-nitrogen bond. Chemists call that a peptide bond. Peptide bonds join amino acids in all sorts of proteins. Your hair is made of protein. Your skin is made of protein, as are your muscles, and literally thousands of other proteins as well.

There's one important point about peptide bonds that I'll return to in later lectures. Every time a peptide bond is formed, a molecule of water is released. By the same token, if there's lots of water around, peptide bonds tend to break down by a process known as hydrolysis. So, it turns out that it's much easier to form peptide bonds in dry environments, away from water. Of course, that provides a real challenge to cells, because they constantly have to synthesize proteins, but they're filled with water. It's also a challenge to origin-of-life researchers, because the conditions that support synthesis of amino acids are not necessarily the conditions that support the formation of peptide bonds, and vice versa.

Well, that's a pretty quick tour of organic chemistry, but I think you can get an idea of how rich this field can be. You can keep making larger and larger molecules by adding a few atoms here and there. Because carbon can link to four neighboring atoms, the possibilities are endless. In fact, as I mentioned in the last lecture, the biggest mysteries in origins of life research are not how to make simple organic molecules—that's pretty easy—it's how to select and concentrate just the right ones, and to make the polymers. Before we consider complex molecules in later lectures, we need to look again at where simple organic molecules come from.

You've probably gotten the sense that there's a lot of debate and uncertainty in the origin-of-life field. But surely a few ideas are well established. Until recently, one of those certainties would have been the role of the Sun. For centuries the ultimate source of life's energy has been as well established as any precept in biology. It's intuitively obvious that all life depends on the radiant energy of the Sun.

Well, it turns out that it's not so simple. In the past two decades, there have been remarkable discoveries of life at the darkest ocean depths. Microbes have been discovered buried miles beneath Earth's surface in solid rock. Life abounds in places that are forever beyond the Sun's influence. The Sun is not life's only energy source.

If the last 500 years of science has taught us anything, it's that some of our most cherished ideas about the natural world and our place in it are just plain wrong. The Sun's a perfect example. You see the Sun rise in the morning and set at night. The obvious conclusion that was reached by almost everyone until relatively recently is that the Sun is going around the Earth—that's what we see. Yet we now know that the sunrise and sunset are simply consequences of the fact that the Earth rotates on its axis every 24 hours. In fact, it's Earth that goes around the Sun once every year. We aren't the center of the universe, our galaxy or even our solar system.

Think about Earth's mountains and oceans. We see these features as majestic and unchanging attributes of our planet. It's true that on the scale of a human lifetime, the mountains and oceans are essentially permanent. Yet geologists have discovered that every topographic feature on Earth is ultimately destined to disappear, while the inexorable processes of plate tectonics constantly rework the surface, forming new mountains and opening new oceans.

The incredible power of science is that it allows us to discover characteristics of the physical universe that are not self-evident. Over and over again, the history of science has seen the overthrow of what had been thought to be obvious. So, we have to consider whether our intuitive view of life's essential energy source might also be in error.

Scientists define energy as the ability to do work, the potential to change. It's easy to see why every living cell requires a continuous source of energy. Organisms need energy to seek out and consume food, to manufacture their cellular structures, to send nerve impulses from one place to another, and much more. Without energy, a cell cannot grow, move or reproduce. Cells even need energy to maintain their integrity, since they are constantly subjected to damage by nuclear radiation, toxic chemicals, and other environmental hazards.

The means by which an organism obtains and uses energy is called metabolism. Metabolism is an ancient chemical process that has to take place in every living cell, and that includes all of the tens of trillions of cells in your body. Until very recently, biologists were certain that the metabolic pathways of virtually every life form on Earth depends one way or another on photosynthesis. This fact was underscored by the pyramid-like structure of food webs. At the base of the food chain, we always find green plants and a host of one-celled organisms, all of which use the Sun's light energy directly to convert to water and carbon dioxide into the chemical energy of sugar molecules plus oxygen. That's photosynthesis. Plants are machines that manufacture carbohydrates—such as the starch of potatoes and the cellulose of celery—to build their leaves, stems, roots, and other physical structures. Plants also are able to process the sugar molecules they make to provide the chemical energy necessary to power all of the plant cells molecular machinery.

Animals and other non-photosynthetic life higher up the food chain have adopted a different strategy. They can't use the Sun to make sugar, so they have to find another source. That's why animals eat plants, or else they eat animals that eat plants. There's a neat symmetry to this arrangement. On the one hand, plants synthesize sugar molecules and oxygen from the Sun's energy plus water and carbon dioxide. On the other hand, animals convert sugar molecules along with oxygen to generate energy plus water and carbon dioxide. And, I have to confess, this view of the biological world, with the Sun as life's only important energy source, is elegant and simple. But it's also wrong.

That brings us to the story of three scientists at Oregon State University in the late 1970s. Our view of life on Earth was changed forever in February

1977. That's when marine geologist Jack Corliss and two crewmates went on a dive in the research submersible *Alvin.* They were investigating an undersea volcanic terrain about 8,000 feet down called the East Pacific Rise. This undersea ridge is located near the Galápagos Islands, and it's been known to be an active volcanic zone. It's a place where new ocean crust is being formed. In fact, the East Pacific Rise is a lot like the more famous Mid-Atlantic Ridge that runs right down the middle of the Atlantic Ocean.

In a sense, this was supposed to be a pretty routine dive. *Alvin* had logged hundreds of dives to examine interesting spots on the sea floor. In this case, they were looking for a submarine hydrothermal vent. That's the kind of undersea geyser where hot water jets up into the surrounding cold ocean water. But on this dive, Corliss and his crewmates found an amazing and totally unexpected ecosystem. They observed new species of spindly albino crabs, and there were football-sized giant clams, and bizarre-looking 6-foot-long, bright red tubeworms. The area was also cloudy with abundant microbes that coated the rock surfaces. This vibrant community of organisms is at crushing pressures, more than a mile and a half beneath the sea. None of these organisms ever sees the light of the Sun.

So how does this ecosystem sustain itself? In these deep undersea zones, it's the microbes that serve as the primary energy producers. They play essentially the same role as plants do on Earth. The way these one-celled vent organisms get energy is really interesting. They exploit the fact that there's a chemical instability among the cold ocean water, the hot volcanic water, and the various sulfur-rich mineral surfaces over which the two kinds of water mix. This situation is rather similar to the instability between a piece of coal and the oxygen-rich air. These deep microbes obtain energy by slowly altering unstable minerals, just as you can heat your house or run a factory by burning coal. As an aside, there's an interesting wrinkle here. These reactions are sped along by the fact that ocean water contains a lot of oxygen, which is indirectly resulted by plant photosynthesis. So, to be precise, these systems do get some, though not all, of their energy from the Sun.

The discovery of this exotic ecosystem was completely unexpected, and it understandably made the news. But Corliss and his colleagues at Oregon State soon elaborated on the finding. Three scientists, Jack Corliss and his fellow professor, John Baross, along with a graduate student, Sarah Hoffman, proposed that deep-sea hydrothermal vents were the actual location of life's origin. Now, I've spent a lot of time talking to the main figures in this story, and the details have certainly become clouded by a quarter century of what has become revisionist history. Here are the basic facts to which everyone seems to agree.

Together, the three scientists collaborated to develop a theory of life's origin at the vents. This thesis is elegantly simple, and so it has become correspondingly influential. There's some pretty compelling evidence. Modern organisms thrive in deep hydrothermal ecosystems. Fossil microbes have been found in 3.5 billion-year-old hydrothermal deposits. Even without the energy of sunlight, all sorts of nutrients and potential sources of chemical energy abound in hydrothermal systems. As a consequence, the Oregon State scientists claimed that hydrothermal systems are the "ideal reactors for abiotic synthesis." Based on this claim, they outlined a sequence of chemical steps for the potentially rapid emergence of life.

Many of us now read the paper that first presented these ideas, and see a highly original, plausible scenario for life's origins. There were many unresolved questions to be sure, but it was certainly a worthy effort, deserving of publication in a highly visible venue. Nevertheless, the controversial manuscript was not eagerly received. It bounced around the review circuit for the better part of a year. First it was rejected by *Nature*, then by *Science*. You must remember that at the time, Stanley Miller and his many students and admirers completely dominated the field of origin-of-life research. Admittedly, the origin-of-life field had seen more than enough crackpot theories.

The official objections were properly couched in the formal jargon of the field. Hydrothermal temperatures are so high that amino acids and other essential molecules decompose rapidly. The dynamic fluid flow through vent systems was not realistically modeled. Corliss and coworkers also had the ancient ocean chemistry incorrect, they said. Modern hydrothermal ecosystems depend on oxygen-rich water, which is an indirect consequence of plants and photosynthesis. By contrast, the prebiotic ocean would have had little, if any, oxygen. The proposed life-sustaining chemical reactions would have proceeded slowly, if at all.

Their ultimate conclusion? Many years of Miller-Urey experiments provide clear support for what is intuitively obvious: life began at the surface. Miller's supporters saw no reason to confuse the issue by bringing in a deep origin. In public discourse, Stanley Miller was a lot blunter. To one reporter, he said, "The vent hypothesis is a real loser. I don't understand why we even have to discuss it."

Well, eventually the Corliss, Baross, and Hoffman manuscript was published, but in about as obscure a place as you could imagine. It was printed as a supplement to the relatively unknown periodical *Oceanologica Acta.* That's a journal that not one in a hundred origin-of-life researchers would have seen. Still, you can't suppress a good idea. Copies of the paper, which was called "An Hypothesis Concerning the Relationship Between Submarine Hot Springs and the Origin of Life on Earth," began circulating as copies. I own a copy of a copy of a copy. My copy is also pretty ratty by now, but everyone I know has a copy that's similar to mine.

As I hinted at before, there's a dark side to this paper, and the complete story is still not clear. But there's a real question about who had the idea first, and who should get credit. Now, Jack Corliss claims the idea for himself. In the 1990s, he told a reporter, "I began to wonder what all this might mean, and this sort of naïve idea came to me. Could the hydrothermal vents be the site of the origin of life?"

But a very different history comes from other people who were close to the story. According to John Baross, the second of the three coauthors on the paper, the hydrothermal vent theory of life's origins was proposed and developed originally by a perceptive Oregon State graduate student named Sarah Hoffman. Hoffman evidently wrote the outline of the hypothesis in 1979 as a term paper for a seminar in biological oceanography. Prof. Charles Miller, who was another OSU oceanographer, taught that seminar.

Charlie Miller, who is not related to Stanley Miller, got very excited about the paper and showed it around. Jack Corliss was one of the people to see it. So Corliss, who was the senior person in the field, asked Sarah Hoffman to polish the work for publication. Hoffman agreed, and in frequent consultation with Baross, expanded on the hypothesis as it would appear in print. Today, both Baross and Hoffman claim that Corliss, who had the most contacts in the field of deep-ocean vents, appropriated the paper as his own. He allowed Baross and Hoffman to be his coauthors, and asked them to expand and polish the prose to conform to the conventions of scientific publishing. But, according to this story, he took charge of the publicity, and he submitted the manuscript with his name first. That's significant, because with three coauthors in that order—Corliss, Baross, and Hoffman—the paper would come to be known as "Corliss, et al." I've heard people expressing the opinion that Corliss got the fame, while Hoffman and Baross were relegated almost to footnote status for a time.

Whatever the truth, this incident affected all three coauthors' lives. Only John Baross remained active and influential in the field of origins research. In 1985, he accepted a prestigious professorship at the University of Washington, and he's developed a leading research effort on hydrothermal life. In fact, he's at the forefront of the highly publicized research field of "extramophile" microbes—those are microbes that live in all kinds of extreme environments, and he's done a lot of provocative work on the possible nature of life on other worlds. This research continues to be supported in part by NASA's Astrobiology Institute.

Sarah Hoffman's graduate work in geochemistry was interrupted by illness, and, after her recovery, she pursued a career in singing. Her experience with Jack Corliss may have been so disillusioning that she dropped out of science altogether.

As for Jack Corliss, he left Oregon State in 1983 for Hungary and the Central European University in Budapest. He worked only briefly on the deep-origins hypothesis, but then changed directions and studied the more abstract field of complex systems. Corliss also served a three-year stint as Director of Research at the controversial Biosphere 2 environmental station in Arizona. He has long since abandoned his studies of the deep ocean.

Whoever deserves the credit, the hydrothermal origin hypothesis began to catch on. New support for the idea kept building, because thriving hydrothermal ecosystems were found in many places along ocean ridges, in both the Atlantic and Pacific Oceans. Several origins experts pointed out another reason to support the idea. At the early period of Earth history, when the surface was being blasted by a continuous rain of comets and meteorites, the deepest ocean ecosystems were somewhat shielded from big impacts. Hydrothermal vents would have been a much more protected location than the surface for life's origins and early evolution. Added to this idea were new discoveries of diverse primitive microbial life deep in the continental shelf. By the early 1990s the hypothesis that life began in a deep, hot environment had become widely accepted as a plausible, if unproven, alternative to the Miller-Urey surface scenario.

These lectures are about the quest to understand life's chemical origins, which is certainly one of the great unanswered questions in science. But I'm also fascinated by the sociology of science. Why do people do science? Who's allowed in the "club"?

If you've been listening carefully, you may have been struck, as I have been, by the paucity of women scientists in the history of this field. Sarah Hoffman is, to the best of my knowledge, the very first woman in this field, and she was, in effect, driven away from science. Moreover, until the 1990s, hardly a single female appeared as coauthor on any origins paper. As a scientist and as an educator, I find this history distressing and somewhat embarrassing. I suspect that at least two factors played a role in this unfortunate bias.

First of all, the origin-of-life research field has always been small, at least until recently, and it was just bad luck that several prominent leaders during the 1950s through the 1970s were male professors who seem to have been unsupportive of women students. Every young scientist needs the support and encouragement of a mentor and role model. Women, lacking this essential support system, would have felt excluded from the origins club. It's really only within the past decade that the research environment has changed.

A second reason for the lack of women in the field may well be that women scientists also chose to stay away from the origins field during the decades when it did not receive large-scale funding and it still held a rather dubious status in some scientific circles. Everyone seems to be interested in how life began, but we've seen that unambiguous experiments and firm conclusions are difficult to come by. It was hard enough for women to be taken seriously as scientists in the 1950s, 1960s and even into the 1970s, and entering a suspect area of research would have made a career even more difficult.

I'm happy to report that times are changing. At a recent NASA astrobiology conference, more than 30 percent of the attendees were women scientists. Female enrollment in some astrobiology Ph.D. programs has exceeded 50 percent. So I have to say, if you or your children or your grandchildren are interested in this dynamic field, it's wide open to everyone. We need good people with good ideas!

To summarize this lecture, by the late 1970s, enough problems and questions had been raised about the Miller-Urey experiment that alternate hypotheses were proposed. One of the first and most influential of these competing origin models was the idea that life might have arisen in the deep ocean at a hot hydrothermal vent. Miller and his colleagues immediately raised objections, but the hydrothermal origins hypothesis would not go away. In the next lecture, we'll see that many other research groups began to explore deep origins, with theory, with experiments, and with observations on deep microbial life.

Lecture Eleven

The Deep, Hot Biosphere

Scope: Is it really possible that life emerged from the hot volcanic zones of the deep ocean? Might the Miller-Urey scenario of a prebiotic soup, as successful as it seems, be wrong? In this lecture and the next, we'll review some of the extraordinary discoveries that have expanded our view of life on Earth and that lend at least some credence to the deep-origins hypothesis.

Outline

I. Recent research has expanded our view of life on Earth and of the deep-origins hypothesis.

A. Recall from Lecture Ten the hypothesis put forward by Jack Corliss and his colleagues: The protected and energy-rich hydrothermal zones of the mid-ocean ridges were the most likely location for life's origins.

B. The hydrothermal-origins hypothesis prompted numerous scientists to look for life in deep, warm, wet environments. And everywhere they looked—in deeply buried sediments, in oil wells, even in volcanic rocks more than a mile down—they found abundant microbes.

1. Microbes live under a mile of Antarctic ice and deep in dry desert sand. They thrive on mineral surfaces anywhere there is water and chemically unstable rocks, such as basalt, which can provide the chemical energy for life.
2. Systematic study of subsurface microbes began in 1987 when the Department of Energy drilled several 500-meter-deep boreholes in South Carolina near the Savannah River nuclear processing facility. Researchers cut away the outer rinds of the drill cores to reveal pristine rock samples and shipped them to analytical facilities and learned that the deep sediments were loaded with microbes.
3. Numerous subsequent drilling studies revealed that microbes live in every possible deep environment—in granite, in basalt, in all kinds of sediments, and in metamorphic rocks.

4. Collecting microbes from mining operations is another way to study life deep underground. This effort is the focus of a new breed of scientists called *geobiologists*. Mine tunnels enable researchers to investigate microbial populations in their native habitat.
5. No better place to collect underground microbes exists than the fabled gold mines of South Africa's Witwatersrand District. From the groundwater there, Princeton geologist Tullis Onstott and his team collect microbes that have never seen light.
6. It is hard to imagine how these cells survive on the meager supply of underground chemical energy of a few kinds of minerals, especially iron-rich minerals that are unstable.
7. Evidence suggests that this deep life lives at an incredibly slow pace. A single cell may persist for thousands of years without doing anything. Cell division almost never happens.

II. If life can only exist at the surface, energized by the Sun, then the Miller-Urey scenario is really the only game in town. But if life can survive at extremes of temperature and pressure, we must rethink our approach to the origin and distribution of life in the universe. Deep origins become plausible, and the study of organic chemistry at those extremes becomes a vital part of origins research.

A. Discoveries of abundant subterranean cells, which somehow thrive far from the light of the Sun, have inspired novel models for life's origins. Of all the scientists who have thought about the subject, probably none was more influential or displayed more imagination than the astrophysicist and general iconoclast Thomas Gold.

1. Gold's foray into deep life began in 1977; his ideas focused on the origin of petroleum, about as well-established a process as any in geology. For more than a century, geologists had known that petroleum forms when vast amounts of decaying cells accumulate over millions of years. When this carbon-rich matter is buried and processed by Earth's heat and pressure, it becomes black gold.
2. Petroleum is always found in ancient sedimentary rocks that once held abundant life; petroleum is always rich in a variety of distinctive biological molecules; and the carbon isotope ratios of petroleum also conform to a biological origin. These

and a dozen other persuasive lines of evidence made the case for petroleum as a fossil fuel open and shut.

3. Gold raised nagging unresolved issues, including an argument that centers on the lighter-than-air gas helium. Petroleum holds many distinctive biomolecules, but oil fields also contain a good deal of helium gas. Given that helium is a light gas and quickly escapes into space from the surface, it must have come from sources deep within the Earth. There is no simple way to reconcile the deep source of helium with the surface origin of the oil.
4. Gold concluded that most of the organic molecules that eventually become petroleum are from deep underground, where they are synthesized by chemical, not biological, processes. Only later, as the organic molecules accumulate in reservoirs near the surface, are they modified by the action of subsurface microbes. That microbial activity produces a characteristic biological overprint on the primarily nonbiological petroleum.
5. In Gold's scenario, the Earth's mantle holds vast sources of primordial hydrocarbons, which form the major molecular components of oil. These hydrocarbons inexorably rise toward the surface, constantly refilling petroleum reservoirs in the process, because they're lighter than the surrounding rocks. In this unproven view, oil is more like a renewable resource than a finite commodity.
6. Gold's hypothesis predicts that petroleum should occur in many different types of rock, though oil fields have been found only in layers of sedimentary formations. Gold countered, quite logically, that conservative petroleum geologists had never looked for oil anywhere but in those sedimentary formations.
7. In 1983, Gold asked the Swedish State Power Board to drill an oil well in a unique granitic mass called the Siljan Ring impact site in central Sweden. Geologists had long recognized tantalizing hints of petroleum in the fracture zones there. Seven years and $40 million later, drilling there led to only a few barrels of crude.
8. However, the modest production of methane gas and oily sludge from the Siljan Ring inspired Gold to embellish his

theory. In a series of popular articles and a 1998 book, he reviewed the growing and accepted body of evidence for deep life in many types of rocks, from sandstone to basalt. He concluded that deep microbial life, much of it nourished by hydrocarbons rising from the mantle, accounts for fully half of Earth's total biomass.

9. He acknowledged that microbes represent only a tiny fraction of the mass of Earth's crustal rocks, but the volume of the crust is so vast—a few billion cubic kilometers—that it nurtures unimaginable numbers of microbes. Our view of life has been skewed because we live on the sunlit surface.
10. Reactions to the deep, hot biosphere were, to say the least, mixed. Biologist Lynn Margulis was sympathetic in her review of Gold's book, but there was universal contempt from petroleum geologists.

B. At about the same time as Gold's controversial drilling experiment, a new proposal emerged for life's deep origin. In 1988 a brilliant, combative German chemist named Günter Wächtershäuser contributed a seminal paper, entitled "Before enzymes and templates: Theory of surface metabolism."

1. The sweeping theory of organic evolution that Wächtershäuser has erected relies on the chemical properties of minerals—primarily compounds of iron, nickel, and sulfur, which are abundant at many deep-sea volcanic vents. According to Wächtershäuser, these minerals provide energy-rich surfaces that catalyze the synthesis and assembly of life.
2. Wächtershäuser's scheme for life's origin is rooted in the philosophy of science, and he was a friend and admirer of the influential philosopher of science Sir Karl Popper. Popper's principal concept was that scientific theories, to be acceptable, must make specific, testable predictions. By extension, the more precise and, therefore, the more testable a theory is in its predictions, then the more believable that theory becomes.
3. Most previous origins researchers had fallen into a trial-and-error approach to devising hypotheses. Origin theories developed around particular sets of observations, rather than any unified falsifiable theory. Wächtershäuser approached the problem in a different way: In his origin-of-life model,

virtually every one of the proposed chemical steps can be tested by experiments.

4. In 1988, Wächtershäuser formulated his theory, starting with a rejection of the "primordial soup" concept based on a simple argument: The primordial soup was much too dilute, and it contained an overwhelming proportion of molecular species that played no conceivable role in the emergence of life. According to this assumption, the emergence of biomolecules by the Miller-Urey process was irrelevant to life.
5. Wächtershäuser's second assumption was equally controversial. Because the prebiotic soup was dilute and unreliable, the very first living organism had to make all of its own molecular building blocks from scratch.
6. His third assumption is that the first life-form used chemical energy from minerals, *not* from the Sun. Earth's early sunlight seemed problematic. Wächtershäuser argued instead that surfaces of unstable minerals, which release energy when they react, provide a simple synthesis strategy that is similar to cellular processes today.
7. The fourth assumption is that life began with metabolism—that is, with a chemical process that converts available energy and simple molecules into the essential chemicals of life. This idea was a real slap in the face to many biologists, convinced that life began with some kind of self-replicating genetic molecule.
8. These four central assumptions are connected by one additional precept that's common to every origin theory—the principle of continuity, or *congruity*. This principle states that the biochemistry we see in cells today, no matter how chemically intricate it may be, evolved in a continuous path from primordial geochemistry. A good example of this principle is Wächtershäuser's reliance on iron and nickel sulfide minerals. Many of the enzymes that today's living cells use to process energy have, at their very core, a mineral-like cluster of iron or nickel atoms and sulfur atoms.

C. In several papers published from 1988–1992, Wächtershäuser detailed more than 100 pages of specific chemical reactions. Each of these reactions is a testable step. Each step leads from simpler geochemical raw materials to the molecules of biology.

1. According to his central chemical idea, minerals at hydrothermal zones on or beneath the ocean floor play multiple roles in life's origins: The minerals catalyze the formation of biomolecules, they act as templates to organize those molecules, and they provide energy to power the whole process.
2. Wächtershäuser sees life emerging spontaneously from the simplest of volcanic molecules, such as carbon monoxide and hydrogen. Molecules bound to mineral surfaces begin feeding off each other, and eventually, they form a chemical cycle that makes copies of itself. That system of self-replicating chemicals fits the most common definitions of life. Wächtershäuser sees this process as inevitable and fast.

III. The idea that life may have arisen in a deep, hot zone of Earth's crust is now a mainstream rival hypothesis to models of surface origins. Thomas Gold's idea of a deep, hot biosphere and Günter Wächtershäuser's iron-sulfur world model are now subjects of intense research.

A. The key to acceptance for any theory of life's origin must rest on chemical experiments. The Carnegie Institution's Geophysical Laboratory is well equipped with apparatus that mimics the high-temperature and high-pressure conditions deep inside the Earth, enabling us to test various hypotheses on the deep origin of life.

B. The problem was that none of us knew very much about the origins of life, nor had any of us worked with organic chemicals at these extreme conditions. In the next lecture, I'll tell you how a group of geologists and geophysicists entered the field of origins research.

Essential Reading:

Hazen, *Genesis: The Scientific Quest for Life's Origin*, ch. 7.

Supplemental Reading:

Dick and Strick, *The Living Universe: NASA and the Development of Astrobiology*.

Knoll, *Life on a Young Planet: The First Three Billion Years of Evolution on Earth*.

Questions to Consider:

1. Why is it difficult to confirm that microbes survive deep underground?
2. What role does the philosophy of science play in Günter Wächtershäuser's theory? In particular, how is it influenced by the work of Karl Popper?

Lecture Eleven—Transcript

The Deep, Hot Biosphere

Is it really possible that life emerged from the hot, volcanic zones of the deep ocean? Might the Miller-Urey scenario of a prebiotic soup, as successful as it seems, be wrong?

In the last lecture, I introduced the heretical hypothesis of Jack Corliss, John Baross, and Sarah Hoffman involving the undersea volcanoes. In 1981, they proposed that the protected and energy-rich hydrothermal zones of the mid-ocean ridges were the most likely location for life's origins. While that idea was met by much ridicule by Stanley Miller and his associates, many other scientists warmed to the concept.

In this lecture and the next, I'll review some of the extraordinary discoveries that have expanded our view of life on Earth, and that lend at least some credence to the deep origins hypothesis.

So, back to the 1980s; the hydrothermal-origins proposal prompted numerous scientists to look for life in deep, warm, wet environments. And everywhere they looked—in deeply buried sediments, and in oil wells, and even in volcanic rocks more than a mile down—they found abundant microbes. Microbes live under a mile of Antarctic ice and deep in dry desert sand. They thrive on mineral surfaces anywhere there're water and chemically unstable rocks, such as basalt, which can provide the chemical energy for life.

It's not easy drilling for life in solid rock. To be sure, the oil industry has perfected the practice of deep drilling, thanks to decades of experience and vast infusions of cash. But the problem for geoscientists looking for microbes a mile down or more isn't how to drill. It's how to drill without contaminating the depths with surface bugs. Bacteria are in the air, in the water, and in the mud used to lubricate diamond drill bits. It's not difficult to bring up rock cores from a mile down, but that rock will be exposed to surface life by the drilling process. So what can we do?

One common trick is to put a colorful dye in the lubricating fluid. When drillers extract a deep core, the dye shows what parts of the drill core have been contaminated. Porous sediments or highly fractured formations will soak up the dye, which is a good indication that the rocks are unsuitable for analysis. But many rocks are quite impermeable to the dye, and thus are ideal for recovering deep life.

The systematic study of subsurface microbes began in 1987. That's when the Department of Energy drilled several 500-meter-deep boreholes in South Carolina near the Savannah River Nuclear Processing Facility. They wanted to make sure the ground wasn't contaminated, and they needed to know how future contamination might affect any life forms. As cores were brought up, the drillers isolated them in sterile plastic enclosures. Researchers cut away the outer rinds of the drill cores to reveal pristine rock samples, and these samples were shipped to various analytical facilities. The results were amazing. The deep sediments were just loaded with microbes.

Numerous subsequent drilling studies reveal that microbes live in every possible deep environment—in granite, in basalt, in all kinds of sediments and in metamorphic rocks, too. Anywhere you live, if you drill a hole down a mile and pull up the drill core, the chances are you'll find microscopic life living on the rocks.

It turns out that drilling isn't the only way to study life deep underground. It's also possible, under just the right circumstances, to collect microbes from deep mining operations. In fact, this effort is the focus of a new breed of scientists, called geobiologists. Mine tunnels have a big advantage, because researchers can go down and investigate microbial populations in their native habitat.

There's no better place to collect underground microbes than the fabled gold mines of South Africa's Witwatersrand District, which are Earth's deepest mines. I want to give you a sense of this heroic and potentially dangerous work. Princeton geologist Tullis Onstott leads the principal effort. Onstott and his students have a base camp at the East Driefontein Mine, which is about 60 miles southwest of Johannesburg. Like most South African gold mines, this facility is a vast underground network of workings that extend more than two miles deep. An army of miners labor around the clock for the gold.

The conditions are brutal. The mines use one of the largest air-conditioning systems in the world; but still, the Earth's heat keeps the deepest tunnels at about 140°F. The temperature inside the Earth typically increases about 20 degrees Fahrenheit for every kilometer of descent. Caves near the surface may be cool, but that's not the case as you continue going deeper underground in the Earth's crust. The mines are so deep, that the air pressure is about two times that of the surface. Tullis Onstott tells of his first, scary trip into the mine's depths, "You really have to be careful. It was 'Don't step there, don't touch that,'" he said. "All I knew is that it was deep and dark and hot."

So why would anyone do research there? The payoff is that, every so often, the miners will break away rock and find a small flow of water. That's groundwater that has spent countless thousands of years filtering down from the surface and accumulating in cracks and fissures of the rock, and as we've seen, if you have water and rocks, the chances are you have microbes as well. Those deep, protected rocks in South Africa's mines are actually home to a tenuous ecosystem of microbial life. If you're quick and efficient, it's possible to collect microbes from the groundwater.

Here's what Onstott's team does. They typically have a half dozen students and post-docs on alert at a surface camp. They store all sorts of sterile sample-collection hardware and protective gear. When news of a fresh water flow gets relayed to the surface, they get to the site as fast as possible. They typically have to navigate though miles of elevators and tunnels, so the trip can take almost an hour. Then they have to work pretty fast, in part, to avoid any extended disruption of the mining operation, but also because you just can't survive prolonged exposure to such hellish conditions.

They have the routine down pat. First, photograph the site of the water flow, make a record of its location and the immediate geological setting, and then collect as many gallons of uncontaminated water as possible from the flow. Contamination is always a major concern, because the mine tunnels are exposed to surface microbes in the air, in lubricants, in sweat and more. But the Princeton team benefits from the positive water pressure of the seep. By collecting the water right from the rock, it's possible to avoid much back contamination. Once the site is recorded and the samples collected, they lug the heavy water-filled bottles to the surface for further investigation.

These results still amaze me. Every sample from the South African mine contains microbes that have never seen light. It's hard to imagine how these cells survive on the meager supply of underground chemical energy. That's the energy of a few kinds of minerals, especially iron-rich minerals that are unstable, and the evidence suggests that this deep life lives at an incredibly slow pace. Measurements of carbon and nitrogen isotopes reveal the rate of metabolism. It turns out that a single cell may persist for thousands of years without doing anything. Cell division almost never happens, so these colonies of organisms just sit there, some of them isolated from the surface for millions of years. It seems a very odd lifestyle, but deep rocks do provide a safe and reliable environment. There are no predators and few surprises two miles underground.

This research has an important take-home message for studies of life's origin. If life can only exist at the surface, energized by the Sun, then the Miller-Urey scenario is really the only game in town. But if life can survive at extremes of temperature and pressure, undreamed of even a couple of decades ago, then we have to rethink our entire approach to the origin and distribution of life in the universe. Deep origins become plausible, and the study of organic chemistry at those extremes becomes a vital part of origins research.

Well, as you can imagine, all these discoveries of abundant subterranean cells, which somehow thrive far from the light of the Sun, have inspired novel models for life's origins. But, of all the scientists who have thought about the subject, probably none was more influential or displayed more imagination than the astrophysicist and general iconoclast, Thomas Gold.

Tommy Gold was born in Vienna, Austria, in 1920, and educated at the best schools in Switzerland and England. He was already recognized as a brilliant scientist and one of the world's leading experts in radio astronomy, when he became the head of Cornell University's Center for Radiophysics and Space Research. That was in 1959. It was there that he made his biggest scientific splash. He proposed the truly inspired theory that pulsars, which are steady pulsating radio sources from deep space, are actually rapidly rotating neutron stars. You can learn more about that extraordinary discovery in the Teaching Company's course, *Understanding the Universe*, by Prof. Alex Filippenko. In any case, Tommy Gold soon received a slew of honors, including election to the Royal Society of London and the National Academy of Sciences in Washington, DC.

Most scientists are perfectly happy to contribute to just one favorite field, but Gold was never content to stay put. Over and over in his career, he ventured into new academic territory, and often offered bold ideas. In the 1940s, when he was still in his 20s, Gold did controversial experiments on hearing and on the structure of the inner ear. Only much later were his ideas accepted as being correct. He published speculative papers on cosmological models of the universe, and on instabilities of Earth's rotation, and on the potentially grave dangers to lunar astronauts from deep powdery soils.

Gold's highly publicized foray into deep life began in 1977, when he rattled the wealthy and conservative field of petroleum geology. He was, by then, a safely tenured professor at Cornell, not to mention a National Academy of Sciences member, and he could rattle people with some impunity. Gold's ideas focused on the origin of petroleum, which was about as well established a process as anything in geology. For more than a century, geologists had known that petroleum is a fossil fuel. It forms when vast amounts of decaying cells accumulate over millions of years. When this carbon-rich matter is buried and processed by Earth's heat and pressure, it becomes black gold. And the evidence for this point of view really is overwhelming: petroleum is always found in ancient sedimentary rocks that once held abundant life; petroleum is always rich in a variety of distinctive biological molecules; the carbon isotope ratios of petroleum also conform to a biological origin. These, and at least a dozen other persuasive lines of evidence, made the case for petroleum as a fossil fuel—open and shut.

Well, as you've probably gathered, Gold disagreed. He raised a laundry list of nagging unresolved issues. Perhaps the most compelling of these arguments centers on the lighter-than-air gas, helium. Petroleum does, indeed, hold lots of distinctive biomolecules, but oil fields also contain lots of helium gas. That light gas quickly escapes into space from the surface, so the helium must have come from sources deep within the Earth. There's no simple way to reconcile the deep source of helium with the surface origin of the oil. Gold concluded that most of the organic molecules that eventually become petroleum are from deep underground, where they are synthesized by chemical, not biological, processes. Only later, as the organic molecules accumulate in reservoirs near the surface, are they modified by the action of subsurface microbes. That microbial activity produces a characteristic biological overprint on the primarily non-biological petroleum.

In Gold's intriguing scenario, the Earth's mantle holds vast sources of primordial hydrocarbons, which form the major molecular components of oil. These hydrocarbons inexorably rise toward the surface, constantly refilling petroleum reservoirs in the process, because they're lighter than the surrounding rocks. In this unproven, heretical view, oil is more like a renewable resource than a finite commodity.

In a flamboyant and unconventional move, Gold introduced this novel idea, not as a scientific paper, but as an op-ed piece in the June 8, 1977, issue of the *Wall Street Journal.* That was smack in the middle of the oil crisis. You can imagine how an oil-hungry nation sat up and took notice of the radical hypothesis.

So who's right about the origin of oil? Is this just a "he-said/she-said" debate with no resolution? Well, a scientific hypothesis is useful only if it's testable, and Gold was a good scientist. He soon proposed a test of truly dramatic proportions. On the one hand, Gold's oil-from-below hypothesis predicts that petroleum should occur in many different types of rock. On the other hand, all known oil fields have been found in layers of sedimentary formations; just those rocks that would have accumulated lots of remnants of past life. Gold countered, quite logically, that conservative petroleum geologists had never looked for oil anywhere but in those sedimentary formations. After all, it costs a lot of money to drill a well, and no one wants to look stupid by drilling a dry well in igneous granite or metamorphic gneiss. Perhaps, Gold said, there are immense new oil fields just waiting to be found in igneous and metamorphic rock.

Gold came to this debate armed not with just a provocative theory, but also a persuasive, dynamic oratorical style. So in 1983 he made a simple (and very expensive) proposal to the Swedish State Power Board. He asked them to drill an oil well in solid granite, which is, of course, the very last place on Earth a well-trained petroleum geologist would look. Gold's proposal was quite specific. He targeted a unique granitic mass, called the Siljan Ring impact site, which is located in Central Sweden. The Siljan Ring is actually a highly-fractured mass of granite that was formed 368 million years ago when an asteroid smashed into the crust. And there are good reasons to drill there. Geologists had long recognized tantalizing hints of petroleum in the fracture zones; you find carbon-filled cracks and flammable methane seeps. Well, you can imagine the scene: you had Gold's enticing rhetoric; increasingly optimistic brochures from Dala Deep Gas, which was the company formed to do the drilling; and, of course, the energy-poor Swedish population. Well, the slick promotion lured thousands of people into spending millions of dollars on exploratory holes.

To make a long story short, nobody got rich. By 1990, after seven years and $40 million later, the company had drilled a 6.8-kilometer-deep hole and they had produced modest amounts of oil-like hydrocarbons and methane gas—nothing commercial, but a few barrels of crude. So that was a small enough yield for all the oil experts to say "I told you so," but it was also a large enough yield to convince Tommy Gold that his theory was correct.

Everyone saw exactly what they expected to see. Gold kept asking the experts, "Where else could those tantalizing traces of organic molecules have come from if not the mantle?" But most scientists saw the Siljan experiment as a complete failure that vindicated the oil companies' skepticism. Whatever the truth, no one is likely to drill again for oil in granite, at least not anytime soon.

Tommy Gold was unflappable, and the modest production of methane gas and oily sludge from the Siljan Ring inspired him to embellish his theory. He extrapolated far beyond any of his peers, to introduce a brave new biological world, *The Deep Hot Biosphere*. In a series of popular articles, and a popular 1998 book of that title, he described an unseen domain of life. The oil fields, which lie at the heart of Gold's 1977 deep petroleum hypothesis, are a wonderful food source for deep microbes. In thoughtful arguments and entertaining prose, he reviewed the growing and accepted body of evidence for deep life in many types of rocks, from sandstone to basalt. Gold came to the dramatic conclusion that deep microbial life, much of it nourished by hydrocarbons rising from the mantle, account for fully half of Earth's total biomass. He acknowledged that microbes represent only a tiny fraction of the mass of Earth's crustal rocks, but the volume of the crust is so vast—a few billion cubic kilometers—that it nurtures unimaginable numbers of microbes. Our view of life has been skewed because we live on the sunlit surface, while these microscopic life-forms lie completely hidden from everyday view. In one of his most memorable statements, Tommy Gold said, "It is we who live in the extreme environments."

Reactions to *The Deep Hot Biosphere* were, to say the least, mixed. Biologist Lynn Margulis was quite sympathetic in her book review for *Physics Today*. She said, "Gold's claims must be debated, evaluated, tested and understood rather than rejected out of hand. If correct, his grand scheme will fundamentally alter the depth of our relationship to prodigious Mother Earth." But there was pretty universal contempt from the petroleum geologists. Alton Brown, an oil geologist from Texas, wrote a scathing review in *American Scientist*. He was blunt, "I do not recommend this book," he said, and he entitled his review "Upwelling of Hot Gas."

In spite of these criticisms, Tommy Gold continued his scientific crusade against the oil companies until his death in 2004. He loved to lecture on the subject, and in April of 1998, he visited the Geophysical Lab, where he gave a regular Monday morning seminar. He was terrific—impassioned, eloquent, persuasive, funny. Throughout the lecture, he would use wonderful rhetorical flourishes, like a lawyer arguing his case. He said his theory provided "the only possible explanation." The evidence "persuades one completely," he would say. Well, while he may not have persuaded anyone completely, everyone at the Geophysical Lab admired his creativity and unswerving conviction, and, as much as anything, Tommy Gold helped to remind all of us how very much we don't know about the insides of our planet just a few miles below our feet.

About the same time that Tom Gold's controversial drilling experiment in Sweden was yielding commercially insignificant, yet scientifically intriguing results, a new proposal emerged for life's deep origin. In 1988, a brilliant, combative, German chemist by the name of Günter Wächtershäuser contributed a seminal paper, entitled "Before Enzymes and Templates: Theory of Surface Metabolism."

Wächtershäuser is not your usual scientist. He works in Munich as a patent attorney by day, and his work on life's origin is a sideline, though one he pursues with great passion. In my lifetime, I've known of only a handful of people who can successfully pursue science as a hobby, but Wächtershäuser fits that description.

The sweeping theory of organic evolution that Wächtershäuser has erected relies on the chemical properties of minerals—primarily compounds of iron, nickel and sulfur, which are abundant at many deep-sea volcanic vents. According to Wächtershäuser, these minerals provide energy-rich surfaces that catalyze the synthesis and assembly of life.

Wächtershäuser's grand scheme for life's origin is unusual in many ways. For one thing, his ideas are rooted in the philosophy of science. I have to confess, most researchers like myself tend to ignore the philosophy of science. That field seems incredibly abstract and removed from the practical everyday details of doing experiments. It's just hard to imagine philosophy dictating which chemicals to mix in the lab.

But Wächtershäuser takes a much broader view. It also helps that he was a friend and admirer of the influential philosopher of science, Sir Karl

Popper, who died in 1994. Popper's principal concept is that scientific theories, to be acceptable, must make specific, testable predictions. It must be possible, at least in principle, to prove the theory false by empirical tests. Indeed, if a theory doesn't make testable predictions, then it's nothing more than pseudoscience, according to Popper.

By extension of this principle, the more precise and, therefore, the more testable a theory is in its predictions, then the more believable that theory becomes. Wächtershäuser and Popper talked about these issues. So, Popper encouraged the development of a new theory on life's origins. As Wächtershäuser told a reporter in 1997, "During breakfast I mentioned my ideas on an alternative, and he urged me to work it out."

It turns out that the philosophy of science really is important in this context. Most previous origins researchers had fallen into a kind of trial-and-error approach to devising hypotheses. They would cook up some likely geochemical recipe and run an experiment. If the experiment did something interesting, then they patched together a theory around those observations. Stanley Miller's experiment is a case in point. The Miller-Urey experiment produced amino acids, so his followers became convinced that that's how it happened long ago in nature. Corliss and co-workers, who found life thriving in hydrothermal zones, fell into the same kind of trap, around a different set of observations, and so origin theories developed around particular sets of observations, rather than any unified falsifiable theory. Wächtershäuser approached the problem in a very different way.

Let me tell you a little about a lecture that Günter Wächtershäuser gave at the Carnegie Institution in 1998. At the time, none of us in my lab knew him, but his ideas had greatly influenced our own experimental program. We had been focusing on the possibility that metabolism can proceed without catalysts in the presence of iron and nickel minerals, and that is a central part of Wächtershäuser's theory.

He really talks like a courtroom lawyer, and his seminar was quite intense. The room was packed, and Wächtershäuser delivered his lecture in a combative, persuasive style. His model is very complex, so an hour wasn't nearly enough to cover everything. But a lot of us had read and reread his papers, so we followed sort of the way you do a favorite bedtime story. Not that we accepted everything he proposed on faith; indeed, many of our experiments were firmly rooted in the competing hypothesis of a primordial soup. It was amusing the way he handled the more controversial assumptions of his model. He'd mention opposing views only briefly and then just dismiss decades of conventional wisdom as obviously irrelevant. But Wächtershäuser's own origin-of-life model is wonderfully crafted and rich in detail. Virtually every one of his proposed chemical steps can be tested by experiments. So, whether he's right or wrong, we all admired the effort and paid close attention.

He formulated his theory from scratch, by starting with several assumptions about the most likely characteristics of the first life-form. Right from the start, in 1988, he had made waves by rejecting the "primordial soup" concept. This first assumption of Wächtershäuser's model is based on a simple argument: random prebiotic synthesis can't possibly have played an essential role in life's origin, because the primordial soup was much too dilute, and it contained an overwhelming proportion of molecular species that played no conceivable role in the emergence of life. According to this assumption, the emergence of biomolecules by the Miller-Urey process was irrelevant to life. So, in one sweeping assumption, Wächtershäuser threw out decades of origins research.

Wächtershäuser's second assumption was equally controversial. He argued that the first life form made its own molecules. Now, almost all other workers in the field have assumed that the first living cell obtained its essential molecules, such as amino acids and sugars, from its surroundings. That's a strategy called heterotrophy, from the Greek for "other nutrition". That seems to make sense because it's a lot easier to rely on chemicals that are already present in the environment. Fungi, many bacteria and all animals are heterotrophs, but Wächtershäuser says, "No, the prebiotic soup was dilute and unreliable, so the first life must have been autotrophic—that means "self-nutrition"—like a plant. The very first living organism had the ability to make all of its own molecular building blocks from scratch."

However, photosynthesis in modern plants is an incredibly complex series of chemical reactions that relies on lots of specialized molecules. So Wächtershäuser wasn't done shaking the establishment. His third assumption is that the first life form used chemical energy from minerals, <u>not</u> from the Sun. As we've seen, sunlight and lightning are so energetic that they can disrupt cells. Furthermore, these violent energy sources are unlike the chemical energy sources that most cells use today. Earth's early sunlight seemed problematic. Wächtershäuser argued instead that surfaces of unstable minerals, which release energy when they react, provide a simple synthesis strategy that's similar to cellular processes today.

The fourth assumption of Günter Wächtershäuser is that life began with metabolism; that is, with a chemical process that converts available energy and simple molecules into the essential chemicals of life. This idea was a real slap in the face to many biologists. Most biologists are convinced that life began with some kind of self-replicating genetic molecule, something like RNA, as we'll see in Lectures Twenty-One and Twenty-Two. The problem with that scenario is that even the simplest genetic molecule is much more complex than anything in the prebiotic soup. Wächtershäuser, instead, thinks life began with a relatively simple cycle of chemical reactions.

These four central assumptions are connected by one additional precept that's common to every origin theory—the principle of continuity, or "congruity." This principle states that the biochemistry we see in cells today, no matter how chemically intricate it may be, evolved in a continuous path from primordial geochemistry. A good example of this principle is Wächtershäuser's reliance on iron and nickel sulfide minerals. You see, many of the enzymes that today's living cells use to process energy have, at their very core, a mineral-like cluster of iron or nickel atoms and sulfur atoms.

Well, that's a lot of assumptions, and each one of the first four is controversial. But Wächtershäuser's theory is a lot more than a list of assumptions. In several papers published between 1988 and 1992, he detailed more than 100 pages of specific chemical reactions. Each of these reactions is a testable step. Each step leads from simpler geochemical raw materials to the molecules of biology. "You don't mind if I brag a little," he told a reporter a few years ago, "but something like this has never been done in the entire field."

We're going to take a much closer look at Günter Wächtershäuser's theory in Lecture Twenty after we've covered more of the territory that his theory aims to bring together. For now, we've gotten a glimpse of his central chemical idea. He says that minerals at hydrothermal zones, on or beneath the ocean floor, play multiple roles in life's origins. The minerals catalyze the formation of biomolecules. They act as templates to organize those molecules, and they provide energy to power the whole process. Wächtershäuser sees life emerging spontaneously from the simplest of volcanic molecules, such as carbon monoxide and hydrogen. Molecules bound to mineral surfaces begin feeding off each other, and eventually they form a chemical cycle that makes copies of itself, and that system of self-replicating chemicals fits the most common definitions of life. Wächtershäuser sees this process as inevitable and fast. "It takes maybe two weeks," he once told me.

Wächtershäuser's theory is controversial, so there's been lots of discussion and analysis, both pro and con. As you might suspect, Stanley Miller and his students are uniformly dismissive. Jeff Bada stated for the record, "There's probably nothing there because, otherwise, people would have found it already." Bada and another Miller student, Antonio Lazcano, also dismiss Wächtershäuser's originality. They wrote in *Science* that, "It's not a new idea." Well, that's probably an unsustainable conclusion. Whatever you might think of Wächtershäuser's elaborate model, it seems very hard to see how it could be dismissed as a restatement of some earlier theory.

The idea that life may have arisen in a deep, hot zone of Earth's crust is now a mainstream rival hypothesis to models of surface origins. Thomas Gold's idea of a deep, hot biosphere and Günter Wächtershäuser's iron-sulfur world model are now subjects of intense research, as we'll see in the next two lectures, and later in Lecture Twenty.

The key to acceptance for any theory of life's origin must rest on chemical experiments. My laboratory at the Carnegie Institution's Geophysical Laboratory is particularly well equipped with apparatus that mimics the high-temperature and high-pressure conditions deep inside the Earth. So we were poised to test various hypotheses on the deep origin of life. The problem was that none of us knew very much about the origins of life, nor had any of us worked with organic chemicals at these extreme conditions.

So, in the next lecture, I want to tell you how a group of geologists and geophysicists entered the field of origins research.

Lecture Twelve

Experiments at High Pressure

Scope: To explore the deep-origin hypothesis, we need a new breed of experiments. Some of the first high-pressure experiments to investigate the origins of life took place in my own laboratory at the Carnegie Institution of Washington.

Outline

I. In February 1996, Professor Harold Morowitz asked me to collaborate with him on some new experiments in the field of life's origins.

A. As a biological theorist who has focused on the origins of life and metabolism, Morowitz is one of the foremost proponents of the field of emergence. He argues persuasively that all modern cells carry clues about life's earliest biochemical processes. That means that molecular "fossils" persist in all of us, and these molecules may help us understand the chemistry of life's origins.

B. Morowitz had found that the combination of high pressure of about a thousand times that of Earth's atmosphere and a high temperature of 350°C could drop the dielectric constant of water from 80 to 20. "That's like an organic solvent!" he told me. What *was* he talking about? Let's take a step back and look at a water molecule.

1. Water is the essential molecule, with one oxygen atom that's linked to two hydrogen atoms, H_2O. A key aspect of water is that the molecule is bent at an angle, and this shape gives water a property called *polarity*: The side of the molecule with hydrogen atoms has a slightly positive charge, whereas the side with oxygen has a slightly negative charge. This polarity leads to many critical properties.

POLARITY OF WATER MOLECULE

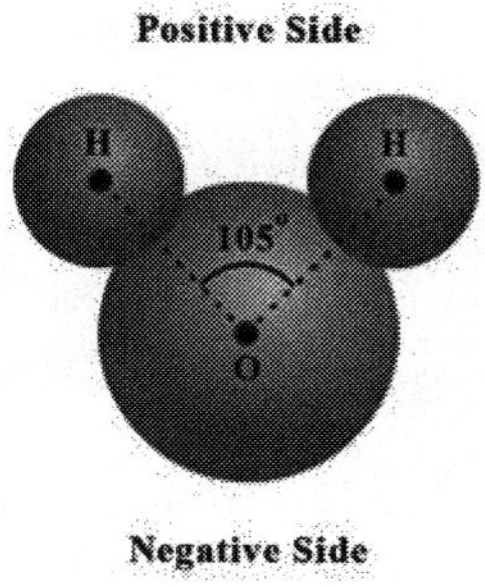

2. Water quickly dissolves salts, sugars, and amino acids. Over time, water dissolves many kinds of rocks and minerals. Water molecules also stick strongly together because the positively charged parts are attracted to the negatively charged parts. Water has strong surface tension and forms droplets, which is why it rains. If water were not polar, then droplets could not form. There wouldn't be clouds or an efficient way to move water from the oceans to the land.
3. The dielectric constant is a measure of polarity. At room temperature, air has a dielectric constant of just over 1, and glass has a dielectric constant of 5 to 10. The dielectric constant of water is among the highest values known: about 80 on the chemist's scale. However, at pressures a thousand times greater than the atmosphere and at temperatures of 200°C or 300°C, water's dielectric constant drops down to values in the 10 to 20 range.

C. For years, Morowitz had thought about the role of water, which poses a persistent problem in most scenarios of life's origin. Water is the liquid of life, whereas carbon provides the framework for all biomolecules, but carbon and water don't mix. Most of the key chemical steps in assembling carbon-based biomolecules don't work in water. That makes it hard to understand how life could have started on a wet planet.
   1. One intriguing possibility is that some of life's critical chemical reactions occur much more easily at high pressure and temperature.

2. Water changes properties at high temperature and pressure. As pressure rises, so does the boiling point: At higher pressures, water will remain liquid at much higher temperatures. Morowitz had assembled a table of the properties of water that held a possible clue. The numbers specified the range of temperatures and pressures that might allow carbon assembly to take place in water.

II. The Geophysical Laboratory in Washington, DC, has an impressive arsenal of high-temperature and high-pressure apparatus. Because I had specialized in studying chemical reactions at extreme conditions, the study of the chemistry of carbon at such conditions seemed as if it would be easy.

A. Morowitz suggested that we start with pyruvate, a molecule with 3 carbon atoms, 3 oxygen atoms, and 3 hydrogen atoms. Pyruvate was a natural choice for our collaborative experiment, because it plays a central role in metabolism, that is, the process by which cells gather energy and atoms to sustain themselves, to grow, and to reproduce. The pyruvate molecule is a key ingredient in the metabolism of every cell, and it plays at least two important roles.

1. Most cells in plants and animals get energy by splitting a 6-carbon sugar, called glucose, into two pyruvate molecules. This process is called *glycolysis*.

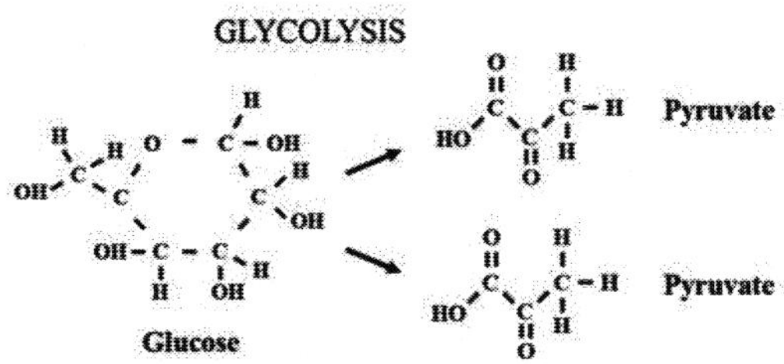

2. Another, much more primitive metabolic step also uses pyruvate. The *reverse citric acid cycle* is a loop of molecules that builds up in a series of steps from small molecules to citric acid, a molecule with 6 carbon atoms, 7 oxygen atoms, and 8 hydrogen atoms. The citric acid molecule splits into two smaller molecules and the process starts over, but this time

one cycle becomes two, two become four, and so on in a self-replicating system of molecules—a metabolic cycle.

THE REVERSE CITRIC ACID CYCLE

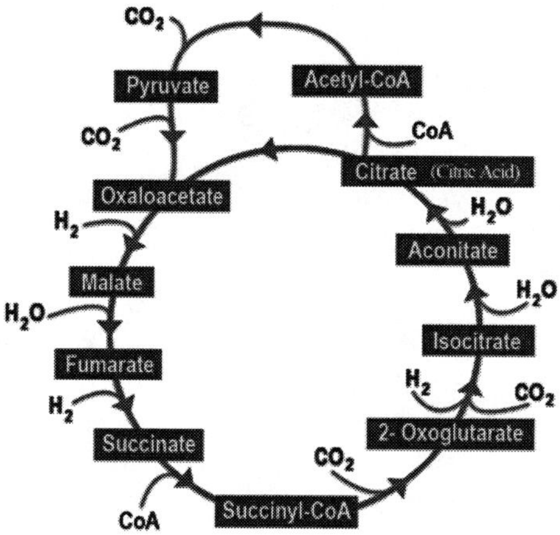

3. Add one carbon dioxide molecule to pyruvate, which has 3 carbon atoms, and the result is another key molecule in the cycle, oxaloacetate, which has 4 carbon atoms.
4. This chemical reaction is absolutely essential to life as we know it, but it never works in water at room pressure without a complex biological catalyst—that's a protein molecule that dramatically increases the rate of the chemical reaction. Without all of a cell's chemical machinery, pyruvate just breaks down into fragments with only 1 or 2 carbon atoms.

III. We tried a variety of experimental conditions, with temperatures ranging from 150°C to 300°C and pressures ranging from 500 to 2,000 atmospheres. Those are conditions similar to what is found in hydrothermal systems both at the sea floor and deeper in the crust.

A. At this point, I consulted with Hatten S. Yoder, Jr., director at the Geophysical Laboratory. Although Yoder's equipment could handle more extreme conditions, 1,000 or 2,000 atmospheres is incredible pressure. Even pressures as low as a few atmospheres can be extremely dangerous and explosive.

B. To do these experiments, we used a classic metal-capsule technique to study chemical reactions at high temperatures and pressures. We sealed the chemical reactants into tiny cylindrical gold tubes. Because the gold is soft and chemically inert, it crushes down on the chemical reactants and provides a clean, isolated, high-pressure environment. Then we heated the assembly to promote reactions.

1. In the first experiments, I used about 50 milligrams of water and 3 milligrams of pyruvate—that's just a tiny droplet.
2. We then added a white-powdered chemical called oxalic acid dihydrate, which breaks down to a mixture of water plus carbon dioxide above 100°C, and we welded the capsules shut.
3. Next, Yoder inserted three identical gold capsules into a platinum sample holder; then he put the sample holder into a foot-long metal cylinder that held an electric heater.
4. After the capsules were inserted, he loaded the metal cylinder with a ceramic filler rod, thermocouple wires to measure temperature, and a ceramic end cap.
5. Then he packed the entire assembly with a white, sand-like powder of aluminum oxide. He used the filler material because his bomb uses expensive argon gas as the pressure medium. He wanted to save time and money, and reduce the danger, because there would be less gas to explode.
6. Yoder prepared the sample holder and sealed it inside the bomb made of hardened steel. A steel plug, called a head, sealed in the pressure and provided insulated channels for wires to carry electric current for the heater and temperature sensors.
7. With the bomb properly sealed, we were ready for an experimental run at high pressure and temperature. The idea was to fill the bomb with pressurized argon gas from a big tank to bring the system to about 100 atmospheres of pressure. Then, a compressor pumped the bomb up to 2,000 atmospheres in a matter of minutes. Once the pressure was stabilized, Yoder programmed the computerized furnace controls to increase the temperature up to 250°C.
8. Inside the steel bomb, the three gold tubes were being crushed and heated to conditions similar to those found several miles

beneath Earth's surface. Our hope was that under such an extreme environment, the pyruvate would react in an interesting way. Perhaps we'd make oxaloacetate, as Morowitz hoped.

9. We came back two hours later to do the final step, what's called "quenching" the run. The heating current shut off automatically at the programmed time, and the temperature dropped rapidly. The whole system cooled to about 100°C in a minute or so. Then Yoder released the gas pressure by opening an escape valve.
10. When I opened up one of the capsules, it was immediately obvious that the pyruvate *had* reacted. We had made some interesting organic molecules, but they did not look remotely like oxaloacetate, which is a colorless, odorless solid.

IV. What did we learn? I think that complexity is the key. We were wrong to expect that one simple chemical reaction would dominate the experiment. There are just too many different reaction pathways.

A. We found that pyruvic acid polymerizes with itself to make clusters of two, three, or more molecules.

B. Obviously, some dynamic organic reactions proceed quite rapidly at hydrothermal conditions, but knowing where to begin experimentation with this chemistry is difficult. We were faced with choosing from among thousands of simple carbon-based molecules as starting materials. We also needed to select a plausible environment from a wide range of pressure, temperature, and other experimental variables.

V. In the next lecture, we'll see some of the many possible directions that subsequent experimental research has taken to understand the possibility of life under hydrothermal conditions of high pressure.

Essential Reading:

Hazen, *Genesis: The Scientific Quest for Life's Origin*, Preface.

Morowitz, *The Emergence of Everything*, chapter 12.

Supplemental Reading:

Lahav, *Biogenesis*, Part IV.

Lehninger et al., *Principles of Biochemistry*, chapters 14 and 15.

Questions to Consider:

1. What was the rationale for the experiment that Morowitz proposed on pyruvic acid?
2. Given the results we obtained, was the Morowitz experiment a success or a failure?

Lecture Twelve—Transcript
Experiments at High Pressure

The origin of life is an amazingly rich field of scientific study. A central guiding principal of science as a way of knowing, and what sets it aside from other scholarly pursuits is that it is based on reproducible observations and experiments. Let us say you carefully measure and record some phenomenon, and you present those measurements in the form of a scientific paper. Then I can reproduce those same measurements, using the same methods you have described. That is the basis of scientific progress.

So far in this course, I have described quite a few observations of the natural world— meteorites and fossils, deep-sea hydrothermal vents and microbes in gold mines—and all of these observations inform our thinking on the origins of life. But life began as a chemical process, and if you are going to understand chemical processes, you had better do experiments.

We are almost at the halfway point in this course, and I have really only talked about the Miller-Urey experiment and a few follow-up investigations. What is more, all of this work was done at low temperature and room pressure. That makes sense if you think the origin of life occurred at or near the Earth's surface, but what about the deep-origin hypothesis? For that, you need a new breed of experiments. That is why I would like to tell you about some of the first high-pressure experiments to investigate the origin of life. Those experiments took place in my own laboratory at the Carnegie Institution of Washington.

I first became involved in origins-of-life research in February of 1996. That was when my good friend and colleague Harold Morowitz came to my office to make a seemingly simple request. Harold is also a professor at George Mason University. He moved there in 1988 after almost half a century as a student and then professor at Yale University. Harold's chief claim to fame is as a biological theorist who has focused on the origins of life, particularly the origins of metabolism. He is one of the foremost proponents of the field of emergence, which I talked about in Lecture Eight. Harold argues persuasively that all modern cells carry clues about life's earliest biochemical processes. That means that molecular "fossils" persist in all of us, and these molecules may help us understand the chemistry of life's origins. I have been captivated by his writings, and I have to confess, almost everything I tell you about life's origins in this series of lectures has been influenced by him.

So, in February of 1996, Harold walked into my office and asked me to look at a Xeroxed piece of paper with a table of lots of numbers. At the top of the table were the words "Dielectric Constant," and there were columns of numbers for temperature and pressure. I really wasn't sure what he was getting at, but it turned out to be a table of the properties of water.

Harold was really excited. He had found that at high pressure of about one thousand times that of Earth's atmosphere and a high temperature of three hundred and fifty degrees Celsius could drop the dielectric constant of water from 80 to 20. "That's like an organic solvent," he told me. So, what was he talking about? Let us take a step back and look at a water molecule.

Water is the essential molecule with one oxygen atom that is linked to two hydrogen atoms; that is, H_2O. A key aspect of water is that the molecule is bent at an angle. This shape gives water an important property called polarity; the side of the molecule with hydrogen atoms has a slightly positive charge, while the side with oxygen has a slightly negative charge, and this polarity leads to all kinds of crucial properties.

For one thing, water is an amazing solvent. Water quickly dissolves salts and sugars and amino acids. Over longer time periods, water can dissolve many kinds of rocks and minerals. That is why the ocean is salty. Water molecules also stick strongly together because the positively charged parts are attracted to the negatively charged parts. So water has strong surface tension and forms droplets. That may not seem important, but that is why it rains. If water was not polar, then droplets could not form. There would not be clouds and there would not be an efficient way to move water from the oceans to the land. So, for these and other reasons, the polarity of water is really important.

The dielectric constant is of interest because it is a measure of polarity. At room temperature, air has a dielectric constant of just over one, and glass has a dielectric constant of five to ten. The dielectric constant of water, which again is a measure of water's polarity, is among the highest values known, about 80 on the chemist's scale. But at pressures a thousand times greater than the atmosphere, and temperatures of two or three hundred degrees Celsius, water's dielectric constant drops way down, to values in the 10 to 20 range. That is a huge difference.

For at least twenty years, Harold had thought about the role of water, which poses a persistent problem in most scenarios of life's origin. We all know that water is the liquid of life, while carbon provides the framework for all biomolecules. However, carbon and water don't mix. Most of the key chemical steps in assembling carbon-based biomolecules do not work in water. For example, you just cannot link amino acids together to make peptide bonds and form a protein in any obvious water-rich environment. That makes it hard to understand how life could have started on a wet planet.

As you have probably guessed, one intriguing possibility is that some of life's critical chemical reactions occur much more easily at high pressure and temperature. The evidence I've described in the last couple of lectures, including the discovery of hydrothermal ecosystems and the deep, hot biosphere, raises the stakes in this idea.

So, the table of numbers Harold Morowitz showed me held a possible clue. Water changes properties at high temperature and pressure. The most obvious change is that, as pressure rises, so does the boiling point; at higher pressures you can have liquid water at much higher temperatures. The numbers in that table specified the range of temperatures and pressures that might allow carbon assembly to take place in water. Harold was especially optimistic. I can still remember his big smile, "So maybe Jack Corliss is right; maybe it *is* the vents," he said, referring to the deep, hot zone of a hydrothermal ocean vent.

Then Harold Morowitz asked me that simple question, "So, can you do the experiments?" Harold knew that the Geophysical Laboratory in nearby Washington, DC, had an impressive arsenal of high-temperature and high-pressure apparatus. We had specialized in studying chemical reactions at extreme conditions, so it was logical to think that we might be able to study the chemistry of carbon at such conditions. It actually seemed like it would be an easy experiment.

Let me set the stage a bit. At the time, my research was almost entirely in the field of mineralogy and crystallography. I had helped to develop the specialized discipline of high-pressure and high-temperature crystallography. That's where we measure the atomic structures of minerals at the conditions similar to those that occur in Earth's deep interior. At the time Harold came to my office, I had been doing that work for a quarter century, and frankly, I was more than ready to try something new.

So I knew basically nothing about origins research, and I was thoroughly delighted to have this excuse to collaborate with Harold Morowitz. He's one of the real leaders in the origins field, and his name provided the kind of instant credibility that it might take a decade or more to build. Remember, the origins field has had a lot of controversy, and perhaps more than its share of polarizing points of view.

Now, a successful experiment has to be planned with great care, and you can imagine that Harold Morowitz had already spent a lot of time thinking about a research strategy. "Let us start with pyruvate," was his suggestion. So, what is pyruvate? Pyruvate is a molecule with three carbon atoms, three oxygen atoms, and three hydrogen atoms. Pyruvate contains a lot of energy, and it was a natural choice, because it plays a central role in metabolism. Remember, metabolism is the process by which cells gather energy and atoms to sustain themselves, to grow and reproduce. The pyruvate molecule is a key ingredient in the metabolism of every cell, and it plays at least two important roles.

On the one hand, most cells in plants and animals get some of their energy by splitting a 6-carbon sugar, called glucose, into two pyruvate molecules. That's the process of "glycolysis." Glycolysis itself is interesting, because oxygen isn't needed; and getting energy in this simple way is common across so many kinds of organisms. After glycolysis breaks glucose into pyruvate, the pyruvate molecules are broken into even smaller molecules to release even more energy.

But there is another, much more ancient and primitive metabolic step that uses pyruvate. In fact, it is the most primitive known metabolic cycle. It is called the reverse citric acid cycle, because it reverses the cycle by which modern organisms release energy. We will be coming back to the reverse citric acid cycle several more times, but let's look at the basic idea.

In this cycle, we want to take the simplest molecules on the early Earth—water from the oceans and carbon dioxide from volcanoes—and build up all the other molecules of life. That's the way Harold Morowitz thinks that life began. This reverse citric acid cycle is a loop of molecules that builds up in a series of steps from small molecules to citric acid. Citric acid is a molecule with 6 carbon atoms, 7 oxygen atoms, and 8 hydrogen atoms. Then the citric acid molecule splits into two smaller molecules and you start over, but this time one cycle becomes two, two become four, and so on. What you have is a self-replicating system of molecules—a metabolic cycle.

Now, let's go back to our friend, the molecule pyruvate. If you add one carbon dioxide molecule to pyruvate, which has 3 carbon atoms, you get another key molecule in the cycle, called oxaloacetate. Oxaloacetate has four carbon atoms. It is just basic chemical math: 3 + 1 = 4. But here is the catch; this chemical reaction is absolutely essential to life as we know it, but it never works in water at room pressure unless you have a complex biological catalyst—that is a protein molecule that dramatically increases the rate of the chemical reaction. Without all of a cell's chemical machinery, pyruvate just breaks down into fragments with only one or two carbon atoms. How can you jumpstart life if pyruvate tends to breaks down?

So Harold had this great idea. He wanted us to try the reaction at higher pressures and temperatures, where water behaves differently. Morowitz thought that pressure and temperature might reverse the trend and cause 3-carbon pyruvate plus carbon dioxide to react and form 4-carbon oxaloacetate. Now, if we could show in a simple experiment that hydrothermal conditions promote this critical metabolic reaction, then a lot of scientists would be much more persuaded by the hydrothermal origins hypothesis. I have to tell you that experimentalists dream of opportunities like this. It was a conceptually simple experiment with a really big potential payoff.

We decided to try a variety of experimental conditions, with temperatures ranging from 150°C to 300°C, and pressures ranging from 500 to 2,000 atmospheres. Those are conditions similar to what's found in hydrothermal systems, both at the sea floor and also deeper in the crust.

Reaching such conditions is easier said than done, so I decided to talk with Hat Yoder, who is one of my scientific heroes. Every scientific career, as you can imagine, is influenced by teachers and collaborators and employers. Well, for me, Hatten S. Yoder, Jr. filled all three roles. He was Director at the Geophysical Laboratory, and hired me in 1976; and in subsequent years, he taught me how to do some really neat high-temperature, high-pressure experiments.

As I said, Hat is my scientific hero. He was a big, handsome man, with strong hands and powerful arms. He came to the Carnegie Institution of Washington after serving as an officer with the Pacific Fleet in World War

II. Then he built a high-pressure lab from scratch and, for more than half a century, he kept the facility running, as he studied the origin of basalt and other volcanic rocks. In the 1950s, he became the youngest Earth scientist ever to get into the National Academy of Sciences. That's a huge honor for any scientist.

When I came to him in 1996, he was 75 years-old and still going strong. Not only was he still doing lots of experiments, but he also was an officer in the prestigious American Philosophical Society, and he served on numerous visiting committees and advisory boards. When I described Harold Morowitz's experiment on pyruvate, he jumped at the chance to try something new.

Hat Yoder was first and foremost an experimentalist, so he raised technical issues right away. Naturally, he wanted to know the range of pressure and temperature. Hat's high-pressure lab was designed unlike any other. He had engineered his apparatus to achieve really extreme conditions—pressures of 10,000 atmospheres at more than 1,000°C—and here I was suggesting experimental runs at a measly 2,000 atmospheres and 250°C. But Hat knew a good experiment when he saw one, so he just shrugged his shoulders and let out a soft, "Humph!," and he was ready to go.

Although Hat Yoder's equipment could have handled more extreme conditions, these were still incredible pressures we were talking about here—one or two thousand atmospheres. Even pressures as low as a few atmospheres can be extremely dangerous and explosive. You might be surprised to learn that your pressure cooker sustains no more than about two atmospheres of pressure, and your car's tires hold less than three atmospheres—you know how they can blow out violently. Well, Hat's hand-built device, which is quite aptly called a "pressure bomb" in the business, works at many thousands of atmospheres. It has an internal volume about the same size—and with about the same explosive power—as a stick of TNT. If he ever had a catastrophic, explosive failure at those pressures, it could easily kill anyone in the room, and knock out a corner of the lab building. But his machine was designed to withstand many times the pressures we were going to need, so we really weren't too worried.

Let me tell you about how we did these experiments. We used a classic metal-capsule technique to study chemical reactions at high temperatures and pressures. What you do is seal the chemical reactants into tiny cylindrical gold tubes. The gold is soft and chemically inert, so it crushes down on the chemical reactants and provides a clean, isolated, high-pressure environment. Then you can heat the whole assembly to promote the reactions.

I made the sample capsules from a cylinder of gold. It is just a hollow tube, much like a soda straw. I cut the cylinder into one-inch lengths and then welded each short piece shut at one end. When finished, I had a bunch of cylindrical sample holders made of gold.

Next, I weighed out exact amounts of water and pyruvate, which are both liquids, and it was easy to load them into the capsules with a syringe. In those first experiments, I used about 50 milligrams of water and 3 milligrams of pyruvate—that is just a tiny droplet. Now, we also needed to put carbon dioxide into the capsule, but it is really difficult to weld shut a tube containing gas. So what we did was adapt an old experimental trick and used a white powdered chemical called oxalic acid dihydrate. The trick is that oxalic acid dihydrate breaks down to a mixture of water plus carbon dioxide above 100°C. So all the reactants—water, carbon dioxide and pyruvate—were loaded.

Then comes the tricky part of welding the capsules shut. What you do is crimp the open end shut with a pair of pliers and then clamp that end in a vise, so a little bit of gold sticks up above the vise. The idea is to weld that end shut with a super-hot, carbon-arc welder and just a flick of the wrist. If you do it right, the gold just melts and flows, zipping shut the capsule.

My own first attempts at welding shut a capsule were just a complete mess. My motion was clumsy, and the capsule heated for a fraction of a second too long. Pyruvate is quite volatile, and some of it just heated up and boiled away. So instead of a smooth weld, I got this burp of smelly gas, and there was an ugly, gaping hole in the weld. Of course, all the carefully weighed ratios of reactants were messed up, too. So that was a bust, and I had to start over.

It took a lot of practice, but I eventually learned to weld one end of a capsule while the other end was kept immersed in an ultra-cold liquid nitrogen bath. So, if I was fast and careful, all the volatile reactants stayed frozen during the welding. In time, I got so that more than half of the capsules worked.

So, after a lot of trial and error, I had a set of capsules ready for high pressure and high temperature. I took them to Hat Yoder, and he began to teach me the many steps in doing an experimental run. First, Hat inserted three identical gold capsules into a platinum sample holder. Then he put the sample holder into a foot-long metal cylinder that held an electric heater. I had to watch carefully, because Hat had decades of experience, and he had done this operation a thousand times. After the capsules were inserted, he loaded the metal cylinder with three more things—a ceramic filler rod, thermocouple wires to measure temperature and a ceramic end cap. Then he packed the entire assembly with a white, sand-like powder of aluminum oxide. The reason he used all that filler material is because his bomb uses expensive argon gas as the pressure medium. You want to fill up as much of the volume as possible with solids to minimize the amount of argon gas that you have to pump in. That saves time, gas, and it is a lot safer, too, because there is less gas to explode.

So Hat had prepared the sample holder, which was about a foot long and three-quarters of an inch in diameter. That assembly had to be sealed inside the bomb made of hardened steel. A fat steel plug, called a head, sealed in the pressure, and still provided insulated channels for wires to carry electric current for the heater and temperature sensors. So Hat inserted the cylinder into his massive metal bomb. Then he sealed the bomb with a giant nut; it is six inches long and has a six-sided head. He tightened that giant nut with an impressive three-foot-long, 20-pound wrench. He even used a three-foot-long metal pipe as an extension for added torque on that nut.

You should have heard the noise as Hat banged away at the unwieldy wrench. The racket was horrendous. By way of apology, he repeated a line I had heard him use before. He said, "It is always good to make a lot of noise in the lab. That way the director will know you are working."

With the bomb properly sealed, we were ready for an experimental run at high pressure and temperature. We retreated behind a wall of battleship gray naval armor which Hat Yoder had salvaged that after the War. I watched as he opened and closed a complex sequence of valves. The idea was to fill the bomb with pressurized argon gas from a big tank. That brought the system up to about 100 atmospheres of pressure. Then a big compressor pumped the bomb up to 2,000 atmospheres in a matter of minutes. Once the pressure was stabilized, Hat programmed the computerized furnace controls to increase the temperature up to 250°C. And that was it. Inside the steel bomb my three gold tubes were being crushed and heated to conditions similar to those found several miles beneath Earth's surface. Our hope was that under such an extreme environment, the pyruvate would react in some interesting way. Perhaps we would make oxaloacetate, as Harold Morowitz hoped.

We came back two hours later to do the final step, what is called "quenching the run". The heating current shut off automatically at the programmed time, and then the temperature dropped rapidly. The whole system cooled to about 100° C in a minute or so. Then Hat Yoder released the gas pressure by opening an escape valve. There was a big *whoosh.* He unscrewed the big nut, pulled out the sample assembly, and dumped the cylinder's contents into a shallow metal tray. There in the tray were the end cap, the thermocouple, a filler rod, the platinum holder and lots of fine white sand. Finally, there were the three fat, shiny gold capsules. It was a success. The welds had held and the capsules were intact, and you can imagine how we both were itching to know what was inside. I had done three capsules so we could try different analytical procedures if necessary. So two of the capsules went straight into the freezer, and I took the third upstairs for analysis.

There is usually nothing fancy about opening a gold capsule. It is just a question of snipping one end open and pouring out the contents. You do not want to be sloppy. So I carefully cleaned the outside of the capsule in organic solvents. That was to avoid any possible contamination by machine oil or fingerprints, which would confuse the analysis. Then I held the capsule in liquid nitrogen for about 30 seconds to freeze the contents so they would not leak out during the snip. I also carefully positioned the capsule over a glass vial to catch the gold and its contents.

So it was just a little snip with the clippers. But this time there was a bang and the gold clipping blasted off into some remote part of the lab. There must have been a release of several atmospheres of internal gas pressure; that is why the gold capsules were so puffed up. I was actually quite surprised and a little bit shaken. I had not expected anything like that, and then I dropped the opened capsule into the bottom of the vial. It just lay there for a few seconds, but then it began to hiss and foam; and this yellow-brown, oily substance started frothing out of the capsule. It formed a coating on the gold and on the bottom of the vial. There was also a strong odor that smelled a lot like Jack Daniels.

Well, one thing was immediately obvious: the pyruvate had reacted. We had made some really interesting organic molecules, but they did not look remotely like oxaloacetate, which is a colorless, odorless solid. To find out what in the world we had made, we had to analyze our products with the modern hardware of organic geochemistry.

I needed an expert in organic chemical analysis, so I did what anyone else would have done at the Carnegie Institution; I consulted with George Cody. Cody was the newest scientific staff member at the Geophysical lab. He's an organic geochemist, which means he is trained to analyze the kind of messy, oily stuff we had made. I had already alerted him to the experiment, and when I showed him our sample of smelly goo, he knew exactly what to do next.

What Cody proposed was that we should analyze our unknown run products using a combination of two analytical techniques: gas chromatography and mass spectrometry, or GCMS for short. That is a procedure that has been in use since 1956. In GC, as gas chromatography is generally known, you send the unknown chemical products, along with a chemically inert gas, through a long, thin column that has the shape of a tube. The thin column is filled with specially prepared organic molecules. What happens is that different kinds of molecules travel through this column at different speeds depending on how they interact with the chemicals inside the column. So, on average, smaller, less-reactive molecules move through the column faster than the larger, "sticky" molecules. So you wait maybe 30 or 40 minutes, and the gas chromatograph sorts a collection of many different molecules into lots of separate little spikes. Each spike represents a different kind of molecule. That's how you separate a complex mix of molecules into different chemical species.

Next comes the mass spectrometer part; that is the MS of GCMS. We have already met mass spectrometry in Lecture Six, where I described the measurement of carbon isotopes. Remember, you could separate carbon-12 from carbon-13, because they have slightly different weights. Well, in George Cody's GCMS, the machine first breaks each kind of molecule into characteristic fragments. His machine measures the relative masses of the molecules and all their little fragments. So each little peak from the GC can then be analyzed separately as a suite of characteristic mass fragments. Those mass fragments provide a kind of fingerprint of the product molecule. Virtually every kind of organic molecule has its own, unique fingerprint.

It took about three hours to prepare the sample for analysis. That is because most of our molecules were oily liquids, and they had to be converted into a gaseous form to travel through the gas chromatograph. We prepared our oily liquids by a chemical process called derivatization, which converted our oily molecules into ones that easily evaporated at high temperature. Ultimately, we ended up with a vial of volatile yellow liquid that would be easy to evaporate.

George filled a syringe with a small amount of this liquid, and he injected a tiny drop into his GCMS with a practiced, swift motion. The machine heated up the drop and boiled it, so that the gas began its journey through the long column. Then we just sat back and watched as a spectrum of peaks gradually appeared on the computer monitor.

The horizontal axis of the spectrum is time in minutes, so that different peaks appear at different times. The vertical axis indicates how much of that compound is present. So, the first small peak showed up at 10.79 minutes. We knew the first peak had to represent a small molecule with probably only two or three carbon atoms. Small molecules travel faster. Then there was another peak at 11.71 minutes, and another at 11.96. Then the peaks began coming faster and faster. Numerous peaks appeared, piling in on top of each other. Every one of these spikes represents a different molecular product. By the 20-minute mark, we had literally hundreds of peaks, and a broad hump of products was emerging. George just muttered, "Humpane," in disgust. Analytical chemists call something "humpane" when they end up with such a morass of different molecules that there was absolutely no hope of sorting it all out. There was almost nothing we could do with that molecular mess.

Well, pyruvate had certainly reacted in our capsules. However, instead of what we had hoped for—that a reaction of simple 3 carbons + 1 carbon = 4 carbons—we had produced an explosion of molecules. There must have been tens of thousands of different kinds of molecules. We could not find a trace of oxaloacetate, but instead we had made a bewildering array of other molecular species. It might take a lifetime to figure out what molecules were in this messy molecular suite.

On the one hand, Morowitz's hypothesis had failed; pyruvate did not react with carbon dioxide to form oxaloacetate under these extreme conditions. But we had caught a glimpse of a surprisingly robust and complex emergent carbon chemistry in a hydrothermal environment.

So, what did we learn? I think that complexity is the key. We were wrong to expect that one simple chemical reaction would dominate the experiment. There are just too many different reaction pathways, so the result is lots of different products. For one thing, we found that pyruvic acid polymerizes with itself to make clusters of two, three, or more molecules. Then these clumps can twist around to form ring structures. In addition, some pyruvate breaks down to release carbon dioxide and other fragments, which can react in other ways. So the reactions balance between making larger molecules on the one hand, and fragmenting them on the other. Obviously, some very dynamic organic reactions proceed quite rapidly at hydrothermal conditions. This is certainly chemistry worth exploring.

It was hard to decide where to begin. We were faced with choosing from among thousands of simple carbon-based molecules as starting materials. What's more, we needed to select a plausible environment from a wide range of pressure, temperature, and other experimental variables. We also knew that rocks and minerals were an integral part of any deep geochemical environment; but there are literally hundreds of different rocks and minerals to choose from. We couldn't just blindly strike out, trying combinations of conditions at random. This work could easily devour a hundred scientific lifetimes. So, at that point, in the summer of 1996, we began to realize how much more we'd need to learn.

In the next lecture, we'll see some of the many possible directions that subsequent experimental research, both at the Carnegie Institution and elsewhere, has taken to understand the possibility of life under hydrothermal conditions of high pressure.

Lecture Thirteen

More Experiments Under Pressure

Scope: As work on the possibility of hydrothermal origins is more widely publicized, scientific objections from those committed to an ocean-surface model become more intense. Research on rocks and minerals seems to play an important role in answering some of those objections.

Outline

I. Rocks near the surface teem with life, and biology shapes the Earth's surface in many ways. What's more, life's origin is now seen as a geochemical process. It has become essential to merge *geo* and *bio*.

A. The Geophysical Lab's arsenal of furnaces and pressure vessels, which for decades was used to study rocks, now stands ready to heat and squeeze carbon-based biomolecules; these are not intrinsically different from any other chemical we might have chosen to study.

B. From 1996–2005, the Geophysical Lab ran more than 2,000 experiments in hydrothermal organic-synthesis. Our strategy has been simple: We subject a plausible prebiotic chemical mix of water, carbon dioxide, and minerals to a controlled set of conditions that mimics an environment on the early Earth. We try to reproduce a high-pressure, high-temperature environment that's typical of deep-ocean hydrothermal systems.

C. In 1997, we were optimistic that the origins-of-life community would welcome our demonstration of straightforward hydrothermal organic synthesis as a logical extension of the Miller-Urey approach. After all, we had subjected a closed vessel containing water and simple carbon-containing molecules to heat. We had attempted to simulate a plausible prebiotic environment, and we produced synthetic organic molecules.

D. Miller and his scientific colleagues were not persuaded by experiments at extreme conditions. Their reasons for disliking the hydrothermal origin hypothesis were straightforward.

1. It is certainly true that organic synthesis may occur near the 100°C limit of liquid water, they argued, but at the much higher temperatures of our experiments, most biomolecules of interest break down at a higher rate than they can form.
2. Of special importance, amino acids decompose rapidly in hot vents. As a consequence, hydrothermal zones on the early Earth, which were much more vigorous than they are today, would have destroyed many essential biomolecules in the prebiotic ocean.
3. Indeed, calculations suggest that relatively rapid circulation at the vents billions of years ago might have led to a recycling of the entire ocean through the vents every few thousand years. Such hostile zones were, thus, antithetical to life.
4. Attacks and criticism continued. Our starting solutions were much too concentrated—completely unrealistic conditions for the early oceans—and the resulting products were much too dilute to accomplish any interesting subsequent chemistry. Our geological assumptions about vent-water circulation were at fault as well.

E. In many respects, the Miller camp was right. Our first hydrothermal experiments were flawed in several ways.

1. After several hundred experiments under a wide range of pressures and temperatures, we realized that water, carbon dioxide, and nitrogen, by themselves, aren't sufficient to make many useful biomolecules.
2. Many of the key biomolecules were missing from the inventory of our products, and many of the organic molecules we made had no possible role in biology. We needed some other chemical ingredient, or ingredients, to make progress. Mineralogy turned out to be surprisingly helpful.

II. Minerals can promote key chemical reactions that wouldn't happen otherwise. They have a remarkable potential to protect the products of organic synthesis from the ravages of high temperature. It now appears that rocks and minerals—as containers, as templates, as catalysts, and as chemical reactants—are critical to understanding life's emergence.

A. The simplest imaginable role for rocks was as protection. Imagine a shaded cove where the concentration of organic molecules continues to increase and the Sun's rays are shielded by rocky outcrops and overhangs. In this way, rocks might have been Earth's early sunblock.

B. Rocks and minerals might have provided protection on a smaller scale as well. Many volcanic rocks of the early Earth were laced with air pockets left by expanding volcanic gases. Evaporating seawater might have deposited a rich mix of organic molecules in such tiny hollows. Each cavity may have acted like a small test tube, in which further reactions could proceed.

C. Mineralogist Joseph V. Smith and his colleagues have taken high-magnification electron microscope pictures of weathered mineral surfaces and often find microscopic cracks and pores.

  1. Feldspar, also known as alkali aluminum silicate, is by far the commonest of all rock-forming minerals, and it sometimes has millions of these tiny weathered pockets. Each one of the tiny pockets is about the same size and shape as some living cells.

  2. In some microscope images, you can even see microbes living in such pockets. These sites might have provided a place for molecules to congregate.

  3. If so, then each of these pores and cracks becomes a separate experiment in molecular self-organization.

III. The idea that minerals enhance biomolecular stability received a boost from experiments done in 1999–2000 by Jay Brandes, a chemical oceanographer.

A. These experiments showed that nitrogen plus water converts to ammonia in the presence of iron minerals of the type found at vents.

  1. The work identified a promising source of ammonia, necessary for prebiotic synthesis.

  2. Then Brandes showed that ammonia and other compounds can react to produce amino acids at vent conditions, with temperatures between 200 and 250°C.

B. These results did not jibe with studies by Stanley Miller's group on the rapid breakdown of amino acids.

  1. How could amino acids be made at vent conditions if the vents decompose amino acids?

  2. Brandes did another round of experiments that focused on the stability of amino acids under a variety of geochemically relevant conditions, including the presence of minerals.

C. Brandes's new experiments involved the amino acid leucine, which typically breaks down in a few minutes in pressurized water at 200°C.

1. When an iron-sulfur mineral called pyrrhotite was added, the leucine lasted much longer.
2. Pyrrhotite is a mineral commonly found at submarine volcanic vents; Brandes's result suggests that minerals somehow protect the amino acid.
3. The exact protection mechanism is still under study, but Brandes's experiments seemed to demonstrate that minerals have the ability to increase the stability of essential biomolecules.

IV. Two other lines of evidence show how amino acids might have survived.

A. One recent set of experiments relies on fossil bones, which often preserve ancient proteins.

1. Bones are composite materials that combine a strong brittle mineral matrix interweaved with flexible fibers of proteins. Exposed and unprotected, these proteins break down in a matter of centuries.
2. However, some fossil bones are known to preserve their proteins for many millions of years. The secret to such long-term preservation is the strong bonding between the minerals and the amino-acids in the proteins.
3. By locking onto the proteins, minerals can protect and preserve amino acids. And the same mechanism may have helped to preserve amino acids in the ancient hydrothermal vents.

B. Another piece of evidence for the survivability of amino acids comes from a recent set of experiments conducted by Stanford University graduate student Kono Lemke and geochemist David Ross of the U.S. Geological Survey in Menlo Park.

1. Like Brandes, these experimenters put amino acids in a superhot pressure cooker and watched for them to decompose.
2. Lemke and Ross used a flexible "gold bag" apparatus, which they filled with a solution of the amino acid glycine and water. They sealed the assembly, ramped up the pressure and

temperature, ran their samples for weeks, and monitored the solution regularly.

3. Over time, the concentration of glycine gradually declined. Though they did not use minerals in these experiments, they still observed a much slower rate of breakdown than had been reported in other studies at lower pressure.
4. They also found a surprisingly fast rate of amino acids linking together to form molecular chains. This result was odd, because hot water is known to accelerate the breaking of peptide bonds between amino acids.
5. Lemke and Ross found hints of yet another potentially significant behavior. Water readily dissolves single amino acid molecules, as well as small clusters with just two or three amino acids linked together. Longer peptide chains, by contrast, are much less soluble.
6. It is not clear that amino acids are always unstable under hydrothermal conditions. Amino acids may gain stability by sticking to rocks or by sticking to each other. In any event, a lot more work needs to be done on amino-acid stability in hydrothermal systems.

V. Perhaps the most fundamental origins-of-life reactions are known as *carbon fixation* reactions. Carbon fixation is the addition of more carbon atoms into organic molecules, and it continues to be one of the most sought-after goals in prebiotic-synthesis experiments. We find two rather different chemical pathways, depending on what minerals we use.

A. A number of common minerals, including many with the elements iron, copper, and zinc, promote carbon addition by a process known as the *Fischer-Tropsch*, or *F-T, synthesis*.

1. First identified in 1923, F-T synthesis produces long, chainlike organic molecules by exposing carbon dioxide and hydrogen to a hot, dry metal surface.
2. A number of labs have studied F-T synthesis with minerals under hydrothermal conditions. The results show that organic molecules can be built from scratch on a hot metal surface, though much less efficiently than in the industrial process.

B. Molecules get bigger by the insertion of carbon monoxide; that's a molecule with one carbon atom and one oxygen atom.

1. We're still adding one carbon at a time, but the molecule winds up looking different and having different properties.
2. Imagine what happens in building molecules using both these processes. By repeating these simple reactions over and over—adding a carbon atom here and an oxygen or hydrogen atom there—eventually, new, more complex organic molecules emerge.

VI. Studies of natural hydrothermal systems complement these experiments on carbon fixation.

A. Strong evidence shows small quantities of organic molecules still being produced at hydrothermal vents by an F-T-like process. Many of these molecules are hydrocarbons of the type that form petroleum, perhaps some fraction of which forms deep in Earth's crust or mantle.

B. One conclusion seems certain: Mineral-rich hydrothermal systems helped to make a varied inventory of biological building blocks. This must have been chemistry on a grand scale.

C. Several researchers have proposed origin-of-life theories that rely on minerals as solid catalysts to promote the synthesis of life's building blocks. Some of our recent experiments reveal a much more complex behavior for minerals.

1. We find that in high-temperature, high-pressure water, minerals often dissolve. The dissolved atoms from the minerals can become chemical reactants that can then jumpstart new synthesis pathways.
2. The behavior of iron is even more dramatic. Some iron minerals dissolve in water to form brilliant orange and red solutions. Chemists have long known that this kind of iron complex promotes chemical reactions, which may be evidence for a primitive catalyst that promoted the assembly of more complex molecular structures.
3. As we'll see in the next lecture, the molecules of life were synthesized in many different environments.

Essential Reading:

Hazen, *Genesis: The Scientific Quest for Life's Origin*, chap. 8.

Wills and Bada, *The Spark of Life: Darwin and the Primeval Soup*, chap. 3.

Supplemental Reading:

Dick and Strick, *The Living Universe: NASA and the Development of Astrobiology*, chap. 9.

Lahav, *Biogenesis: Theories of Life's Origin*, part III.

Questions to Consider:

1. Do you think that the approximately $20 million of tax money NASA uses to support the Astrobiology Institute each year is well spent?
2. Why did Stanley Miller and his colleagues criticize our experiments at high temperature and pressure? Were they justified?

Lecture Thirteen—Transcript
More Experiments Under Pressure

In the last couple of lectures, we've been exploring what must have been the first emergent steps in life's chemical origin, the synthesis of essential organic molecules. Stanley Miller's experiments certainly demonstrated one likely scenario, with a plausible ancient energy source—lightning near the ocean-atmosphere interface. Even if the Miller-Urey assumptions about the exact atmospheric composition were wrong, surface processes undoubtedly drove some organic synthesis reactions.

Subsequent workers, notably Jack Corliss, Tommy Gold, and Günter Wächtershäuser, proposed models that relied on deeper origins and chemically derived energy sources. It seems only logical that any plausible source of carbon-based molecules should be fair game for origins researchers. Every plausible source deserves rigorous experimental study.

In 1996, when Hat Yoder, George Cody and I did our first experiments on hydrothermal chemistry and the origins of life, we were absolute novices in the origins game. We certainly didn't think we were wasting our time. In fact, we thought that our experimental studies of organic synthesis in hot, high-pressure water might provide some compelling evidence that deep hydrothermal zones could have contributed to the prebiotic production of life's molecules. Of course, Stanley Miller and his students—especially Jeffrey Bada at the Scripps Institution of Oceanography—were still focused on the surface of the oceans and might have been more than pleased to correct our thinking; but in 1996 they didn't even know we were engaged in this research.

In this lecture, I want to describe some more of the experimental efforts that are informing the hydrothermal approach to research on life's origins. As work on the possibility of hydrothermal origins became more widely publicized, scientific objections from those committed to an ocean-surface model became more and more intense. In this lecture, we'll see that research on rocks and minerals seems to play an important role in answering some of those objections.

First I want to talk about money, and I'll use an example from my own experience at the Geophysical Lab of the Carnegie Institution. Science is expensive. Gold tubing, platinum furnaces, special chemicals and high-pressure gear all cost money. That is not to mention the salaries of staff scientists and post-docs. We were very fortunate that the Carnegie Institution was willing to support this work in the beginning. The institution has a modest endowment that allows us to undertake risky science that might not get funded otherwise; but we couldn't go it alone for too long.

Our efforts, like those of several other organizations, received a terrific boost in 1997, when NASA announced the formation of a new Astrobiology Institute. In September of 1997, NASA issued a call for proposals. They requested that large, coordinated groups of researchers submit topical proposals to become founding teams of this institute. NASA's prime objective was to promote the study of life's origins and distribution in the universe. They wanted to achieve this objective with novel, interdisciplinary efforts, including a virtual institute that networked researchers from multiple organizations. They had about $20 million per year to support this new program. That was quite an increase in funding, which previously had supported mostly small grants to one or two researchers.

We jumped at the chance. At that point, people at the Geophysical Laboratory were doing a variety of research on hydrothermal systems in addition to our hydrothermal experiments. There were ongoing studies of the chemistry of deep-sea vents, inventories of microbial populations and studies of ancient hydrothermal processes in meteorites. What's more, scientists at our sister laboratory, the Department of Terrestrial Magnetism, which shares the same campus in Northwest Washington, had major projects in the search for extrasolar planets, the exploration of Venus and Mercury and models of the formation of the solar system. They, too, had a lot to say about the origin and distribution of hydrothermal systems. So we all got together and drafted a proposal to study the physics, chemistry, geology and biology of hydrothermal systems.

It was, to say the least, an interdisciplinary proposal. In addition to the Carnegie members, we included microbiologist John Baross at the University of Washington—he was one of the original three, with Jack Corliss and Sarah Hoffmann, whose 1983 path-breaking paper on geothermal origins came up in Lecture Ten—geochemist Everett Shock at Washington University in Saint Louis and, of course, our original inspiration, Harold Morowitz at George Mason University. Even if the proposal had not been funded, it gave us all a much broader view of the subject and suggested new ways that we could collaborate and couple our efforts.

Out of about 50 proposals we were selected to be one of the first 11 teams in NASA's Astrobiology Institute, along with Harvard University, UCLA, Arizona State, Penn State and others. Let me tell you, being selected was huge. For one thing, it meant about $5 million in funding over the next five years through mid-2003. But a million dollars a year spread out over 14 co-investigators and, yet again, as many students and post-docs, is not an immense sum. It helped, but Carnegie was still footing most of the bill. By the way, we got a five-year extension in 2003, and the effort is still going strong.

What was much more important to us back in 1997, was that membership gave us a lot of recognition and a kind of instant credibility. The Astrobiology Institute opened doors. It attracted new students and distinguished visiting investigators. The NAI, as the Institute has come to be known, also created an interdisciplinary network of scientists in all fields. These are people who really like to talk to each other. So, instead of lots of separate disciplines, each with its own jargon, NAI has fostered an amazing community of people who speak a common language and share common scientific goals. We all want to discover the origins of life, and we want to search the universe for signs of life. I have to say, I'm proud to be part of an organization that supports such lofty ideals.

I suppose that it must have seemed rather odd to some members in the scientific community that a group of geologists and geophysicists were studying the origins of life. The Geophysical Lab has been around for almost a century, and for that entire span it has supported research on the physics and chemistry of Earth materials at extreme temperatures and pressures. Carnegie scientists have melted granite, they've synthesized basalt and they've heated and squeezed just about every possible rock-forming mineral. All of this research was part of an effort to understand how our dynamic planet works. But what does geology or geophysics have to do with life?

You may be surprised to learn how much scientific approach to rocks—and to life—have changed. There was a time, not so long ago, when geology and biology seemed to be completely separate worlds, the dead versus the living. Now it's generally recognized that rocks near the surface teem with life, and that biology shapes the Earth's surface in many ways. What's more, life's origins are now seen as a geochemical process, so it's become essential to merge geo and bio. As a result, the Geophysical Lab's impressive arsenal of furnaces and pressure vessels, which for decades were used to study rocks, now stands ready to heat and squeeze carbon-based biomolecules, which aren't intrinsically different from any other chemical we might have chosen to study.

Between 1996 and 2005, the Geophysical Lab ran more than 2,000 experiments in hydrothermal organic synthesis. Throughout this work, our strategy has been simple and, I think, very much in the spirit of Stanley Miller's original experiments. Just like the Miller-Urey experiments, we subject a plausible, prebiotic chemical mix of water, carbon dioxide and minerals to a controlled set of conditions that mimics an environment on the early Earth. In our lab, we try to reproduce a high-pressure and high-temperature environment that's typical of deep-ocean hydrothermal systems.

Our first experiments with pyruvate and carbon dioxide, proved to be a great way to get our feet wet. In that series of runs, we didn't make oxaloacetate, but we did synthesize sugars, alcohols and a host of larger molecules, many of them incorporating dozens of carbon atoms. We found that under extreme hydrothermal conditions, the pyruvate molecules link together in long chains; that's the process of polymerization that I talked about in Lecture Ten. Indeed, some of the pyruvate molecules combine to form rings and branching structures that are reminiscent of molecules found in modern cells.

In a later set of experiments, we tried sealing water and nitrogen gas, plus a variety of powdered minerals, into the gold tubes. That's a reasonable model for primitive conditions at some hydrothermal vents. In these experiments, the nitrogen reacted to form ammonia, NH_3, which is an essential starting material for synthesizing amino acids and other biomolecules. You may recall that ammonia is one of the gases used in Miller's original experiment. Anyway, our work suggests that hydrothermal vents may have been one of the principal sources of the early Earth's ammonia.

Then, continuing with this line of experiments, we put ammonia into gold capsules with pyruvate and found those two compounds react readily to form the amino acid alanine; that's an amino acid with three carbon atoms. So we had gone from simple molecules to the building blocks of proteins. By 1999, we had established a chemical pathway leading from simple carbon compounds, nitrogen gas and common minerals to key biomolecules.

In 1997, not long after our initial results on pyruvate, George Cody, Hat Yoder, and I submitted our first research paper to *Science* for peer review. At the time, we were optimistic that the origins-of-life community would welcome our demonstration of straightforward hydrothermal organic synthesis as a logical extension of the Miller-Urey approach. After all, we had done something very similar to Miller's work of a half century before. We had subjected a closed vessel, containing water and simple carbon-containing molecules, to heat. We had also followed Miller's example in attempting to simulate a plausible prebiotic environment. We too, produced lots of synthetic organic molecules.

As I have said, Miller and his many scientific colleagues had staked their claim to an origin of life at Earth's surface, and they were not persuaded by experiments at extreme conditions. Their reasons for disliking the hydrothermal origin hypothesis were straightforward and clearly argued. It is certainly true that organic synthesis may occur near the 100°C limit of liquid water, but at the much higher temperatures of our experiments, they would argue that most biomolecules of interest break down at a much higher rate than they can form. Of special importance, amino acids decompose rapidly in hot vents. As a consequence, hydrothermal zones on the early Earth, which were much more vigorous than they are today, would have destroyed many essential biomolecules in the prebiotic ocean. Indeed, calculations suggest that relatively rapid circulation at the vents billions of years ago, might have led to a recycling of the entire ocean through the vents every few thousand years. Such hostile zones were thus antithetical to life.

We were, by no means, the first researchers to suffer these criticisms. Recall that the Oregon State group, who first outlined the hypothesis for hydrothermal origins of life in 1983, following discoveries of deep-sea vents by the submersible *Alvin* in 1977, had found themselves repeatedly thwarted by the argument that life could not survive under such conditions. Their groundbreaking manuscript was rejected more than once as a result.

Stanley Miller had been relentless in his criticisms ever since the hydrothermal origins idea first arose. In a 1988 article in *Nature*, Miller and Jeffrey Bada wrote, "The high temperatures in the vents would not allow synthesis of organic compounds, but would decompose them." Remember that Miller was a lot blunter in a *Discover* feature in 1992, when he called the vent hypothesis "a real loser."

The attacks continued in *The Spark of Life*; a general audience book on origin theories by Jeffrey Bada and Christopher Wills, published in 2000. They call those of us working on the hydrothermal hypothesis "ventists," and they argue, "Vents are more likely to destroy organic compounds than create them... At these high temperatures, amino acids undergo total decomposition on a time scale of only a few minutes." In short, "Vents ... do terrible things to the primordial soup."

There were other criticisms as well. For one thing, our starting solutions were too concentrated, completely unrealistic conditions for the early oceans. Also, the resulting products were much too dilute to accomplish any interesting subsequent chemistry. They criticized our geological assumptions about vent-water circulation as well.

Well, you get the picture, and you can imagine that the so-called "ventists" were defensive and a bit angry. Criticisms from the Miller camp appeared self-serving, just defending old ideas, while discounting any promising new possibilities. But, as much as those of us working on hydrothermal origins hated to admit it, in many respects, the Miller camp was basically right. Our first hydrothermal experiments were flawed in several ways. Our starting mixtures of pyruvate were much more concentrated than any conceivable prebiotic ocean. In fact, pyruvate is pretty unstable and is unlikely to have been an important prebiotic reactant in any environment. Also, there were all sorts of experimental shortcomings. In a sealed gold tube we couldn't control, much less measure, the acidity. That is important, because acidity can drastically alter reaction pathways. Also, we couldn't be absolutely sure that all our products formed at high temperature. It was always possible that some reactions occurred during the rapid cooling, and with sealed tubes we'd never be sure. We never really proved that the gold was an inert container. It was possible, though not likely, that gold might have triggered some of the observed reactions.

In short, it was a very typical start to a complex scientific research program. The art of doing science has never been to avoid mistakes; in fact, just the opposite is true. Progress is often made by making mistakes as fast as you possibly can, though you try to avoid making the same mistake twice. It took us, perhaps, several hundred experiments under a wide range of pressures and temperatures to realize that water, carbon dioxide and nitrogen, by themselves, are not sufficient to make very many useful biomolecules. Sure, under the right circumstances with unrealistically high concentrations of ingredients, we could trigger some interesting organic reactions. However, many of the key biomolecules were missing from the inventory of our products. Furthermore, the molecules we did manage to make were pretty indiscriminate. We made lots of organic molecules with no possible role in biology. Clearly, something was missing. We needed some other chemical ingredient, or ingredients, to make progress. It is in this instance that my original specialty, mineralogy, turns out to be surprisingly helpful.

Minerals have the potential to promote all sorts of key chemical reactions that wouldn't happen otherwise. What's more, minerals have a remarkable potential to protect the products of organic synthesis from the ravages of high temperature. It makes a lot of sense. Before there was biology, the only raw materials on our planet were the ocean, the atmosphere and the rocks. Miller and Urey focused on oceans and atmosphere and, for most of the past half century that was the paradigm. But it now appears that rocks and minerals, in all sorts of roles—as containers, templates, catalysts and as chemical reactants—are critical to understanding life's emergence.

The simplest imaginable role for rocks was as protection. Picture an ancient coastal landscape, with a rugged coastline and tidal flats. Imagine there are rocky outcrops and overhangs, too. So you have these tidal pools where seawater evaporates and, in the process, concentrates organic molecules. That concentration process might have promoted the emergence of more complex groupings, called macromolecules, but the Sun's harsh radiation blasts them as fast as they form. Well, that's where the overhangs come in. Imagine a nicely shaded cove, where the concentration of organic molecules just keeps increasing and the Sun's rays are shielded. In this way rocks might have been Earth's earliest version of sun block.

Rocks and minerals might have provided protection at a smaller scale as well. Have you ever looked at the porous volcanic rock called pumice? Some pumice even floats. Well, many volcanic rocks of the early Earth were also laced with countless air pockets, left by expanding volcanic gases. Evaporating seawater might have deposited a rich mix of organic molecules in such tiny hollows. Then, each cavity acted like a small test tube where further reactions could proceed.

Mineralogist Joseph V. Smith, who is professor emeritus at the University of Chicago, has proposed that we consider even smaller protected environments than that. He and his colleagues have taken very high magnification electron microscope pictures of weathered mineral surfaces, and they often find countless microscopic cracks and pores. Feldspar, which is also known as alkali aluminum silicate, is by far the most common of all rock-forming minerals, and it sometimes has millions of these tiny weathered pockets. Each one of the tiny pockets is about the same size and shape as some living cells. So, each of these sites might have provided a place for molecules to congregate. If that is true, then each of these pores and cracks becomes a separate experiment in molecular self-organization.

Remember, a major objection to the hydrothermal hypothesis, is that biomolecules are destroyed much faster than they can form at those temperatures. So the hypothesis that minerals might enhance the stability of some biomolecules is intriguing. The idea that minerals enhance biomolecular stability, received a big boost from a series of experiments that were done in 1999 and 2000 by a Carnegie post-doc, Jay Brandes. Brandes came to the Geophysical Lab from the University of Washington, where he got his Ph. D. in chemical oceanography. Scientists trained as chemical oceanographers know a lot about the kinds of reactions that might occur in deep ocean environments.

Brandes's first publication at the Geophysical Lab was entitled, "Abiotic nitrogen reduction on the early Earth." It appeared in the September 24, 1998, issue of *Nature,* and it immediately gained a lot of attention. His experiments showed that nitrogen plus water converts to ammonia in the presence of a variety of iron minerals of the type found at vents. His work identified a promising source of ammonia, which is necessary for prebiotic synthesis. Then he showed that ammonia and other compounds could react to produce amino acids at vent conditions, with temperatures between 200 and 250 degrees.

These results just didn't jive with studies by Stanley Miller's group on the rapid breakdown of amino acids. How could you make amino acids at vent conditions, if the vents decompose amino acids? We were making appreciable quantities of amino acids in our runs at 200°C, but these chemicals aren't supposed to be stable at that temperature. So Brandes did another round of experiments that focused on the stability of amino acids under a variety of geochemically relevant conditions, including the presence of minerals.

Brandes's new experimental runs involved the amino acid leucine, which typically breaks down in a few minutes in 200°C pressurized water, but when we added an iron-sulfur mineral called pyrrhotite in the capsules, leucine lasted much longer. Pyrrhotite is a mineral commonly found at submarine volcanic vents, so his result suggested that minerals somehow protect the amino acids. The exact protection mechanism is still under study, but Brandes's experiments seemed to demonstrate that minerals have the ability to increase the stability of essential biomolecules.

At least two other lines of evidence show how amino acids might have survived. One of these recent sets of experiments relies on fossil bones, which often preserve ancient proteins. Bones are composite materials that combine a strong brittle mineral matrix that is interwoven with flexible fibers of proteins. Exposed and unprotected, these proteins are not very stable. They'd break down in a matter of a few centuries. But there are some fossil bones that are known to preserve their proteins for many millions of years. In fact, some recent work in Andrew Steele's laboratory has even revealed hints of fossil protein in dinosaur bones more than 70 million years old, though that work still needs a lot of confirmation. Now, the secret to such long-term preservation is the strong bonding between the minerals and the amino acids in the proteins. By locking onto the proteins, the minerals are able to protect and preserve amino acids. The exact same mechanism may have helped to preserve amino acids in the ancient hydrothermal vents.

Another piece of evidence for the survivability of amino acids comes from a recent set of experiments conducted by Stanford University graduate student Kono Lemke and geochemist David Ross of the United States Geological Survey in Menlo Park. They did experiments a lot like the ones of Jay Brandes. Like Jay, they put amino acids in a super-hot pressure cooker and watched for them to decompose, but they used a different protocol. Rather than using gold tubes to seal in their reactants, Lemke and Ross employed an exotic, flexible "gold-bag" apparatus. In many ways, the gold bag is a great improvement on the sealed-tube experiments.

As that name implies, the centerpiece of the gold-bag apparatus is a flexible bag of gold about the size of a grapefruit. The bag, which opens at one end, must be carefully crafted from pure gold foil. It is attached to a titanium-valve system, with which Lemke and Ross could add their experimental solutions and extract samples at will. Pressure is applied by submerging the entire gold assembly in a water-filled pressure chamber that can be compressed to several hundred atmospheres and heated to several hundred degrees.

Can you see how the gold-bag apparatus is a lot like our smaller gold-tube experiments? After all, both rely on holding a sample in a deformable, chemically inert, gold container. The advantage of the gold-bag setup is that you can extract samples of the reacting fluid during an experiment. Of course, trade offs are a part of science, just as they are in life. The main disadvantage of the gold-bag apparatus is that the thin-walled container is extremely fragile. Just a slight slip and the bag will tear. Not only does that ruin an experiment, but it also requires tedious and exact welding to repair it. Still, once the gold bag is properly filled and loaded in the water bath, it usually works well.

So here's what Lemke and Ross did. They filled the gold container with a solution of the amino acid glycine and water, sealed the assembly, and ramped up the pressure and temperature. They ran their samples for weeks, and monitored the solution regularly. Over time, the concentration of glycine gradually declined. Now remember, they didn't use minerals in these experiments. But even so, they observed a much slower rate of breakdown than had been reported in other studies at lower pressure. They found one other unexpected reaction. They found a surprisingly fast rate of amino acids linking together to form molecular chains. This result was odd, because hot water is known to accelerate the breakdown of peptide bonds between amino acids. In fact, that's a principal reason why some boiled foods are so squishy. Strong bonds between amino acids often give tough foods like raw steak their texture. When you boil these foods, they become tender; the bonds break. The formation of amino acid bonds in the Lemke and Ross experiments thus points to rather more complex behavior than had been expected.

Lemke and Ross found hints of yet another potentially significant behavior in their gold-bag experiments. Water readily dissolves single amino acid molecules, as well as small clusters with just two or three amino acids linked together. But they observed that longer peptide chains are much less soluble. Lemke and Ross imagine a scenario in which clumps of amino acids form rapidly in the hot vents and are then exposed to the cooler seawater. If the concentration of long amino-acid chains is high enough, then these molecules might separate out as a relatively stable separate phase. That kind of separation process is exactly the kind of emergent molecular selection and organization that's required for life's origins. The bottom line is that you can't just say amino acids are unstable under hydrothermal conditions. They may gain stability by sticking to rocks, or by sticking to each other. In any event, a lot more work needs to be done on amino-acid stability in hydrothermal systems.

I want to conclude this lecture by describing experiments on what are perhaps the most fundamental origins-of-life reactions—what are known as "carbon fixation" reactions. Carbon fixation is the addition of more carbon atoms into organic molecules, and it continues to be one of the most sought-after goals in prebiotic-synthesis experiments. After all, the first thing you have to do to make life must be to make bigger molecules out of smaller molecules. Just like Stanley Miller, we want to synthesize amino acids and sugars out of carbon dioxide, ammonia, water and other small molecules. We see that this kind of reaction occurs rapidly in many of our experiments. We also find two rather different chemical pathways, depending on what minerals we use.

On the one hand, we see that lots of common minerals, including many minerals with the elements iron, copper and zinc, promote carbon addition by a process that's vital to industry, known as the Fischer-Tropsch, or F-T, synthesis. First identified in 1923, F-T synthesis produces long, chain-like organic molecules, by exposing carbon dioxide and hydrogen to a hot, dry metal surface. You wind up building a carbon chain, just as we saw back in Lecture Ten; chains can be elongated one carbon atom at a time by this process. So, for example, you have methane, you add a carbon you go to ethane, another carbon goes to propane, butane and so forth. It works. We can build organic molecules from scratch on a hot metal surface, though a lot less efficiently than in the industrial process.

We see a rather different synthetic pathway when we use nickel or cobalt sulfides like this. Then we observe that molecules get bigger by the insertion of carbon monoxide; that's a molecule with one carbon atom and one oxygen atom. We're still adding one carbon at a time, but the molecule winds up looking different and having different properties. So imagine what happens if you build molecules by using both of these processes. Repeat these simple kinds of reactions over and over again—you add a carbon atom here, and an oxygen or hydrogen atom there. Eventually new and more complex organic molecules emerge.

Studies of natural hydrothermal systems complement these experiments. There's strong evidence that small quantities of organic molecules are still being produced at hydrothermal vents by a Fischer-Tropsch-like process. Many of these molecules are hydrocarbons of the type that form petroleum.

Who knows, maybe Tommy Gold was correct and at least some fraction of petroleum forms deep in Earth's crust or mantle.

There's a lot we don't know, but one conclusion seems certain. Mineral-rich hydrothermal systems helped to make a varied inventory of biological building blocks. This must have been chemistry on a grand scale. There are tens of thousands of miles of deep-ocean hydrothermal ridges. There are billions of cubic kilometers of warm, wet crust. Earth had hundreds of millions of years to process all the raw materials. By any estimate, organic molecules must have been produced in prodigious amounts.

I keep saying there's so much we want to know about origins-of-life chemistry, and I'll end this lecture with one more example. Minerals have been regarded primarily as solid and stable platforms on which the chemistry takes place. In this spirit, several researchers have proposed origin-of-life theories that rely on minerals as solid catalysts to promote the synthesis of life's building blocks. But some of our recent experiments reveal a much more complex behavior for minerals.

We find that in high-temperature and high-pressure water minerals often dissolve. In the process, the dissolved atoms from the minerals can themselves become chemical reactants. So, for example, sulfur that was dissolved from sulfide minerals combines with the carbon dioxide and water to form molecules called thiols and thioesters. These are reactive molecules that can jump-start new synthesis pathways.

The behavior of iron is even more dramatic. Some iron minerals dissolve in water to form brilliant orange and red solutions. This color is exciting because it points to the formation of iron complexes—that is, iron atoms surrounded by a starburst of organic molecules, as in Holm's Complex, a classic chemical reagent. Chemists have long known that this kind of iron complex promotes chemical reactions. So we might be seeing evidence for a kind of primitive catalyst that promoted the assembly of more complex molecular structures. That is a real bonus in our efforts to foster chemical complexity in the prebiotic environment.

It has become clear that rocks and minerals play all sorts of roles in life's emergence. Porous rocks acted as convenient containers for organic chemical reactions, while rocky overhangs played the simple role of sun block, by protecting delicate molecules from ultraviolet radiation. Minerals promoted the production of ammonia and other key starting materials for life's chemistry. Some minerals even participated as reactants in the key chemical reactions that led to molecular complexity on the primitive Earth.

Organic synthesis was an inevitable and robust characteristic of the young Earth. Miller-Urey processes in the atmosphere and hydrothermal processes in the deep ocean and crust acted for hundreds of millions of years. As a result, the early Earth abounded in organic molecules. As we'll see in the next lecture, the molecules of life were synthesized in many different environments.

Lecture Fourteen

Deep Space Dust, Molten Rock, and Zeolite

Scope: Origins-of-life researchers now understand that rocks and minerals must have played critical roles. The universe boasts an extraordinary range of productive environments, including three of the most surprising—dust in deep space, igneous rocks, and zeolite crystals.

Outline

I. New research reveals that deep space is loaded with interesting organic molecules, which must have predated Earth by billions of years.

   A. Deep space is home to gigantic molecular clouds, where carbon, hydrogen, oxygen, and nitrogen combine in complex reactions. The molecular clouds are also colder than –100°C. Why don't such high vacuums and low temperatures preclude any sort of chemical reaction?

   B. The key is a combination of mineral dust and energy. In these remote regions, ice-covered dust particles tend to pick up stray atoms and molecules and are subjected to ultraviolet radiation from distant stars. That radiation can make molecules more reactive, just as in the Miller-Urey process. Molecules absorb this radiation and become sufficiently reactive to form larger collections of atoms on the dust grains.

   C. We know about remote molecules through radio astronomy. Each type of organic molecule absorbs characteristic wavelengths of radio waves.

      1. By far, the most abundant molecules are the simplest, such as hydrogen, carbon monoxide, carbon dioxide, and water.

      2. More than 140 different compounds have been identified, including many larger molecules with a dozen atoms or more.

   D. Theorists have explained this diversity of deep-space molecules.

      1. We know how to calculate the efficiency with which small, cold molecules accumulate onto tiny dust particles. That process forms microscopic ice coatings.

2. We can also predict how these icy particles absorb ultraviolet radiation from stars. That radiation shuffles electrons and triggers chemical reactions.
3. We can make plots of reaction cascades. In these plots, dozens of boxes each represents a different kind of molecule, and each box is linked to many other boxes by arrows. Each arrow between two boxes represents a small clump of atoms that gets added or subtracted in a chemical reaction. The net result of all these reactions is that new, larger molecules gradually accumulate in the molecular clouds.

HOW DEEP SPACE REACTION CASCADES FORM LARGER MOLECULES

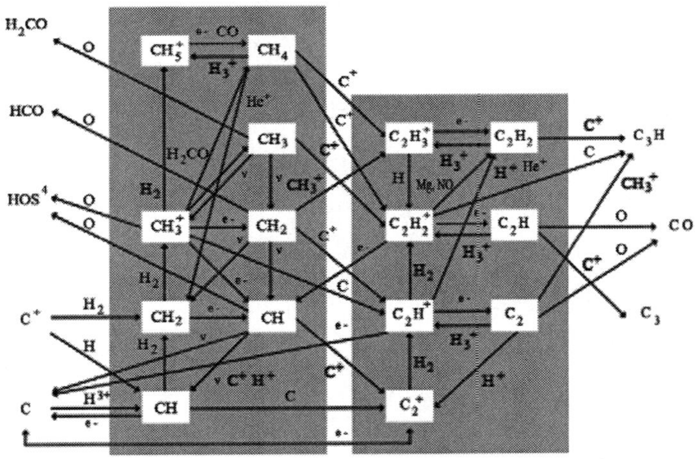

E. Dense molecular clouds are the birthplace of stars and planets.
1. The pervasive inward pull of gravity causes local regions of a molecular cloud to collapse into a new planetary system, with a central massive star and an array of planets and moons.
2. Flat disks of dust surround the new stars. As new planets form in these disks, a steady rain of organic-rich particles contributes to the life-forming inventory.
3. Once one of these *protoplanetary disks* forms, the molecules are much more concentrated. As new radiation forms from the new star, the rate of organic synthesis increases. Theorists

assert that organic molecules constitute a part of any planet-forming mix.

F. A team at the NASA Ames Research Center in California, attempts to simulate deep-space environments. This group, led by astrochemist Louis Allamandola, has produced a variety of interesting molecules in a chilled vacuum chamber.

1. Of special interest are reactive molecules called *nitriles* and *ethers*. *Nitriles* are compounds with nitrogen and carbon. *Ethers* consist of two hydrocarbons linked through an oxygen atom. These molecules can react to form larger molecules of importance to biology.
2. The team has made alcohols and ring-like hydrocarbons, as well as molecules related to amino acids or long chainlike molecules, similar to the building blocks of cell membranes.
3. Samples directly from space—carbon-rich meteorites, in particular—lend support to these experiments. The Murchison meteorite, which fell in 1969, is loaded with organic molecules. Comets are also known to be rich in the molecular precursors of life, as are so-called *interplanetary dust particles*.
4. Theory, observations, and experiments seem to agree: The prebiotic Earth must have been seeded abundantly with all sorts of extraterrestrial organics.
5. One valid question about the organics-from-space hypothesis is: Even if meteorites and comets carry organic molecules, to what extent could these molecules survive their collisions with Earth?
6. We might think that the impact of an asteroid or comet would be totally destructive. Yet carbon-rich meteorites contain amino acids and other potential biomolecules, and experiments suggest that impacts don't destroy all organic molecules.

G. Jennifer Blank and her colleagues at the Lawrence Berkeley National Laboratory used a giant experimental gas gun to study what happens to organic molecules during violent collisions.

1. Blank's experiments began with a flat, cylindrical, stainless-steel capsule, filled with a solution of five different amino acids in water and sealed. She positioned the sample in a metal well at the end of the 40-foot-long gun barrel.

2. A technician powered up the weapon, firing the projectile at more than 4,000 miles per hour. When it hit the end, the sample container was flattened. For a fraction of a second, the amino-acid solution experienced pressures in excess of 200,000 atmospheres at temperatures of almost 1,000°C.
3. Most of the original amino acids were destroyed, but the researchers also found that in every run, some of the amino acid molecules reacted with each other to form pairs of amino acids.
4. The energetic impact induced the formation of some peptide bonds to make larger clusters of molecules—just the kind of bonding that's a crucial step in the assembly of proteins.
5. The conclusion: Big impacts on the early Earth must have reduced the quantity of organic molecules, but they may also have diversified the quality of prebiotic chemical species and helped to select and modify the prebiotic inventory.

II. Chemist Friedemann Freund from the NASA Ames Research Center claims that igneous rocks—formed from an *incandescent melt*—may have been a principal source of Earth's organic molecules. His idea is rooted in the basic chemistry of melts.

A. At temperatures well above 1,000°C, molten rocks contain a variety of dissolved impurities. They incorporate water, carbon dioxide, and nitrogen. Those impurities became the organics in Freund's model.

1. As the melt cools from high temperature, various minerals begin to crystallize, one after another. The first mineral might be olivine, which is rich in magnesium, silicon, and oxygen. That olivine, in turn, will also incorporate a small amount of carbon and nitrogen.
2. Carbon and nitrogen don't easily enter the crystal lattice; these impure elements concentrate at defects in the crystal structure. Such defects are elongated spaces where foreign atoms can bond to each other and, ultimately, form chainlike molecules with a carbon backbone.
3. According to Freund, every mineral in an igneous rock has the ability to concentrate carbon in this way; thus, every mineral has the potential to manufacture such organic molecules.
4. All those organic molecules are locked into the minerals, which are locked into the rocks. Freund says that when rocks

weather away, they release organic molecules into the environment. The amounts of organic molecules produced in this way could be immense.

B. Freund rests his theory on two samples that he has been investigating for more than 20 years.

1. His first sample consists of synthetic crystals of magnesium oxide, which has one of the highest melting temperatures known.
2. Pure MgO crystals are typically clear and colorless, but Freund's crystals have a cloudy appearance that suggests the presence of impurities. What makes these crystals cloudy?
3. One hint comes from spectroscopy. These crystals absorb light at wavelengths characteristic of carbon-hydrogen bonds and oxygen-hydrogen bonds. These absorption features aren't found in pure magnesium oxide, but they are exactly the kinds of bonds we'd expect from organic molecules. When Freund crushed his MgO crystals and analyzed the impurities, he found organic molecules.
4. His second sample was natural gem-quality olivine, one of the most common minerals in igneous rocks. As before, his spectroscopic studies indicated numerous C-H and O-H bonds. And as before, he was able to extract organic molecules from crushed powders.
5. According to Freund, his olivine crystals contain an amazing 100 parts per million of carbon. He claims that much of the carbon occurs in biologically interesting, chainlike organic molecules.
6. By his estimates, the release of organic molecules from weathering rocks could rival the Miller-Urey process as a source of biomolecules.

C. Many scientists remain unconvinced. Even the hardiest organic molecule is fragmented into simple components, such as carbon dioxide and water at 1,000°. Organic contamination is everywhere; any surface quickly becomes coated with adsorbed organic molecules.

III. The formation of organic molecules in high-temperature minerals is a controversial idea, but we know that minerals played a role in other types of synthesis. Fascinating ideas exist about interactions between minerals and molecules.

A. This kind of study requires knowledge of two different kinds of chemical entities: a crystalline solid—that's the mineral, with its regularly repeating atomic structure—and an organic molecule—a tiny cluster of carbon and other atoms.

B. Gustaf Arrhenius, a senior NASA-supported researcher at the Scripps Institution of Oceanography, is an expert in both organic chemistry and minerals, and he has a deep understanding of crystal structures.

1. He and his colleagues have focused on a common group of minerals known as *double-layer hydroxides*. These minerals incorporate magnesium, iron, chromium, nickel, calcium, aluminum, and other elements. Small molecules, such as water or carbon dioxide, can occupy the spaces between the layers.

CRYSTAL STRUCTURE
OF A DOUBLE-LAYER HYDROXIDE

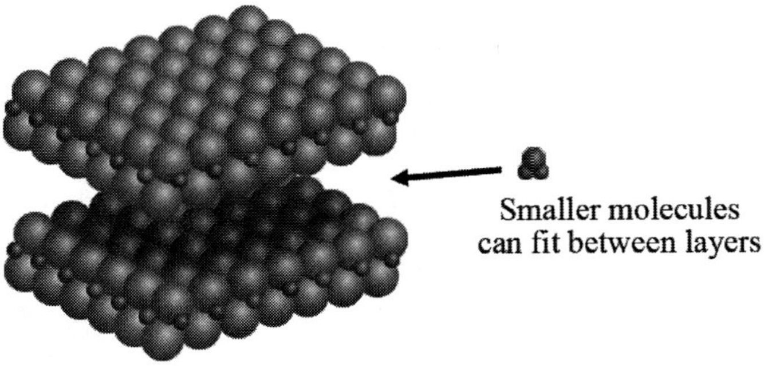

2. These researchers fine-tune the double-layer hydroxides to perform specific chemical tasks by changing their chemical contents.
3. Arrhenius suspects that double-layer hydroxides and other layered minerals may play important roles in the synthesis of biomolecules.
4. This idea is based on the ability of double-layer hydroxides to soak up small organic molecules in the spaces between the layers.

5. Once they're confined and concentrated, these smaller molecules tend to bond to form larger molecular species that might not otherwise emerge from the soup.
6. The most interesting molecules produced so far in Arrhenius's lab are sugar phosphates. A pairing of a sugar molecule with a phosphate group forms the backbone of the genetic molecules DNA and RNA. The sugar-phosphate pair also is integral to a key energy-rich molecule called *ATP*.

IV. Professor Joseph Smith of the University of Chicago has been exploring zeolites. Smith is an authority on this diverse family of natural and synthetic crystals, all of which feature lattice-like frameworks of silicon, aluminum, and oxygen atoms.

STRUCTURE OF A ZEOLITE CRYSTAL

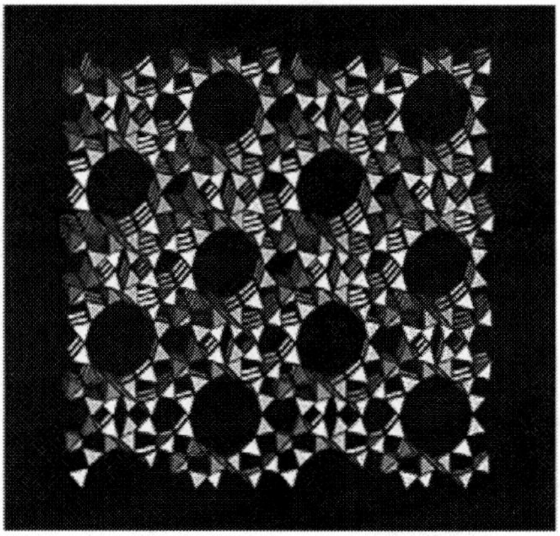

A. Zeolites have a diverse range of open framework structures.
1. The most valuable aspect of these minerals is that their channels are just the right size to accommodate a variety of organic molecules. Hydrocarbon molecules that aren't too large can enter these zeolite channels, whereas larger molecules are excluded.
2. Certain zeolite minerals split larger molecules into uniform-sized fragments. That chemical trick lies at the heart of the multi-trillion-dollar petroleum refining business.

B. Smith imagines that prebiotic molecules concentrate inside zeolite channels, protected and possibly reacting to form more complex chains of molecules. Zeolites do abound in the weathering products of volcanic rocks, and organic molecules were certainly present in the same volcanic environments. Smith suggests that the first cell wall might have been the internal surface of one of these minerals.

V. Minerals and organic molecules could have interacted in many intriguing ways to make organic molecules on the early Earth. Carbon-rich molecules emerged from environments in deep space to the deep crust. Amino acids, sugars, hydrocarbons, bases—all of the essential molecular species—were available. The biggest challenge in understanding the origins of life lies in the selection of just the right biomolecules and their assembly into the vital structures we call *macro*molecules.

Essential Reading:

Hazen, *Genesis: The Scientific Quest for Life's Origin*, chap. 9.

Wills and Bada, *The Spark of Life: Darwin and the Primeval Soup*, chap. 4.

Supplemental Reading:

Lahav, *Biogenesis: Theories of Life's Origin*, part III.

Questions to Consider:

1. What are some of the possible environments for the origin of life's biomolecules? Which ones seem most likely to have played a major role?
2. What are some of the roles that have been proposed for rocks and minerals in the origin of life?

Lecture Fourteen—Transcript

Deep Space Dust, Molten Rock, and Zeolite

In the last five lectures, we have been exploring the first essential step in the origins of life, that is, the emergence of organic molecules necessary for life, such as amino acids, sugars and hydrocarbons. The Miller-Urey experiment was successful in making these molecules, using nothing more than water and a simple atmosphere, and that dramatic success may have deflected many researchers from considering other plausible processes.

I have to remind you that the Miller-Urey process is chemically violent. It works by blasting molecules with lightning or ultraviolet radiation and thereby stripping off electrons. In such a state, small groups of atoms quickly recombine into larger, organic molecules and everyone agrees that interesting molecules emerge from this process. However, lightning and ultraviolet radiation also thwart the formation of larger molecular structures, including the various polymers and membranes that are required by all known life-forms. The Miller-Urey scenario does not seem to have been the entire story.

As a mineralogist, I have been fascinated by the gradual shift in emphasis over the last 20 years. We all know that the first living entity emerged from interactions of air, water, and rock. But rocks and minerals have, until recently, received a lot less attention. That is not surprising because most of the pioneers in origins research were organic chemists, who had little or no training in mineralogy. But a slew of fascinating recent experiments are changing that misperception. Origins-of-life researchers now understand that rocks and minerals must have played a sequence of critical roles. So, in this lecture and the next we will see some of the ways that minerals promote the synthesis of biomolecules and how they help to assemble those molecules into growing and evolving structures.

It is not just about Miller versus the "Ventists" anymore. An explosion of new research has dramatically altered the research community's view of the emergence of biomolecules. You can understand why this change has occurred. In 1953, when Miller announced organic synthesis in his bench top apparatus, the results were almost shocking. A half century ago, nobody realized how easy it is to make amino acids, sugars and other key biomolecules from the atmosphere and oceans, but it now seems that just about anywhere you find energy and simple gas molecules, interesting organic molecules are likely to emerge. The more we study carbon chemistry in a range of likely prebiotic environments, the easier organic synthesis seems to be. The universe boasts an extraordinary range of productive environments and in this lecture we will look at three of the most surprising:—dust in deep space, igneous rocks and zeolite crystals.

On a probability scale of one to 10, you might predict that something of chemical interest would probably happen at the sunlit surface of the ocean. You might give that a seven or an eight. You might even consider deep hydrothermal vents: Perhaps a three or a four on a 10-point scale of probability.

But my guess is the last place you might think to look for life-forming molecules is the black vacuum of interstellar space. That has to be close to zero. But, believe it or not, new research is revealing that deep space is loaded with interesting organic molecules. What's more, these molecules from space must have predated Earth by billions of years.

How is that possible? We now realize that deep space is home to gigantic molecular clouds, where carbon, hydrogen, oxygen and nitrogen combine in complex sequences of reactions. These regions of space are known as "dense molecular clouds," though the words "dense" and "clouds" are terribly misleading. We are talking about immense, and largely empty, volumes of space, much larger than our solar system. The Eagle Nebula is one dense molecular cloud that has been dramatically photographed from the Hubble Space Telescope. It is hard to imagine, but next to an eight-by-11 inch photograph of the Eagle Nebula, our entire solar system, if placed on the scale of that photograph, would be barely visible as a dot.

The density in this cloud is many orders of magnitude less than the best vacuum conditions attainable on Earth. A typical "dense" interstellar cloud contains only about a million atoms per cubic inch. That certainly is denser than deep space, which has a typical density of perhaps one atom per cubic inch, but no one except an astronomer would call these regions of space "dense."

These clouds are also frigid. Temperatures are colder than -100°C. So you have to wonder, why don't such high vacuums and frigid temperatures preclude any sort of chemical reaction? The key turns out to be a combination of mineral dust and energy.

In these remote regions, you find occasional minute ice-covered dust particles. These particles tend to pick up stray atoms and molecules. They also are subjected to ultraviolet radiation from distant stars and that radiation can make molecules more reactive, just as in the Miller-Urey process. So gradually, molecules absorb this radiation and they become sufficiently reactive to form larger collections of atoms on the dust grains.

The reason we know about these incredibly remote molecules is radio astronomy. As you see in this slide, each type of organic molecule absorbs characteristic wavelengths of radio waves. So if you measure the radio waves coming from a distant star on the far side of a dense molecular cloud, then certain diagnostic wavelengths will be weaker than the background. For decades, radio astronomers have recognized the distinctive signatures of numerous organic species in these clouds. By far, the most abundant molecules are the simplest diatomic and triatomic species, such as hydrogen, carbon monoxide, carbon dioxide and water. More than 140 different compounds are now known, including many larger molecules with a dozen atoms or more. That number grows larger every month.

Theorists have little difficulty explaining this diversity of deep space molecules. They know how to calculate the efficiency with which small, cold molecules accumulate onto tiny dust particles. That process forms microscopic ice coatings. They can also predict details of how these icy particles absorb ultraviolet radiation from stars. That radiation shuffles electrons and triggers chemical reactions.

Then they make plots of reaction cascades that look like a complex chain-of-command diagram for some giant governmental agency. In these plots, dozens of boxes each represent a different kind of molecule and each box is linked to many other boxes by arrows. Each arrow between two boxes represents a small clump of atoms that is added or subtracted during a chemical reaction. If you are listening on audio, you can see a picture of one of these diagrams in your course booklet. The net result of all these reactions is that new, larger molecules gradually accumulate in the cloud.

Now, if these clouds were permanent features of our galaxy, if they never did anything else, then the story would not be very important. But dense molecular clouds, like this, are the birthplace of stars and planets. They are the stellar nurseries. The pervasive inward pull of gravity causes local regions of a molecular cloud to collapse into a new planetary system, with a central massive star and an array of planets and moons. The Hubble telescope has taken many pictures of new solar systems in the Eagle Nebula. Flat disks of dust surround the new stars and the surrounding dust is the same kind of material that formed the original cloud. So you can imagine that as new planets form in these disks, there's a steady rain of organic-rich particles that contributes to the life-forming inventory.

It is even more complex than that. Once one of these so-called "protoplanetary disks" forms, the molecules are much more concentrated and there's a lot of new radiation from the new star that's forming, so the rate of organic synthesis goes way up. So, theorists tell us, organic molecules inevitably constitute a significant part of any planet-forming mix.

Regardless of how convincing a theory may sound, experiments are critical in developing a scientific model. That's why a research team at the NASA Ames Research Center at Moffett Field, California, has been trying to simulate these cold, deep space environments. The group is led by veteran astrochemist Louis Allamandola. Allamandola and his colleagues exploit a neat, chilled vacuum chamber. The central part is about eight inches in diameter and it is made of stainless steel and has thick, glass observation ports. They want to produce suites of organic molecules by simulating the conditions of space. First, they pump out the chamber to a high vacuum, though admittedly it's nowhere near the vacuum of space. Then they introduce a fine spray of simple gas molecules, like water, carbon monoxide, methane and ammonia, into the chamber. Those gases quickly freeze onto an aluminum disk. Then they expose the thin ice layer to a beam of ultraviolet radiation. The radiation triggers the formation of larger molecules, many of which match the distinctive molecular emissions from those distant clouds in space.

The NASA team has been able to produce a variety of interesting molecules with their bench-top apparatus. Depending on the starting mix of reactants, they can make all sorts of compounds. Of special interest are reactive molecules called nitriles and ethers. Nitriles are just compounds with nitrogen and carbon, while ethers consist of two hydrocarbons linked through an oxygen atom. These molecules are of special interest because they can react to form other, larger molecules of importance to biology. In addition to nitriles and ethers, they have made lots of alcohols and ring-like hydrocarbons. Other experiments produce molecules related to amino acids, or long chainlike molecules that are similar to the building blocks of cell membranes. Indeed, this is yet another way to make a useful suite of molecules, like the Miller-Urey experiment, or like the hydrothermal experiments I told you about in the last lecture.

The few samples that we have directly from space, the carbon-rich meteorites, in particular, lend support to these elegant experiments. The Murchison meteorite, which fell to earth in Australia just two months after the Apollo 11 moon landing of 1969, is just loaded with organic molecules that are thought to be produced long ago in deep space. Comets are also known to be rich in the molecular precursors of life. So are the so-called interplanetary dust particles. These microscopic bits constantly drift down to Earth's surface from space, adding some dozens of tons of new material every year. So, here is a satisfying case where the theory, the observations and the experiments seem to agree: The prebiotic Earth must have been seeded abundantly with all sorts of extraterrestrial organics.

I have to note that even this well documented and widely acknowledged aspect of life's origin is not universally accepted. Jeffrey Bada comments as follows, "Even if cosmic debris struck the prebiotic Earth at 10,000 times the present levels, the resultant prebiotic soup would still have been much too weak to engender life." Stanley Miller was blunter. In a recent interview, he dismissed the whole idea, saying, "Organics from outer space, that's garbage, it really is."

That criticism may be another example of Miller's group defending their turf but there is one very valid concern about organics from space. Even if we agree that meteorites and comets carry many organic molecules, we have to wonder, to what extent can these molecules survive their catastrophic collisions with Earth? Any large body that collides with Earth is pretty much going to be vaporized by the impact. Wouldn't the intense temperatures and pressures of that process effectively destroy any organic molecules from space, not to mention many organic molecules already on the ground?

Indeed, it is hard to imagine an event more destructive to life and its molecules than the impact of an asteroid or comet. Yet, many carbon-rich meteorites, like the Murchison, contain lots of amino acids and other potential biomolecules. So it seems that impacts don't have to destroy all organic molecules. In fact, recent experiments suggest quite the opposite is true.

I want to tell you about the work of Jennifer Blank and her colleagues at the Lawrence Berkeley National Laboratory in Berkeley, California. They used a giant experimental gas gun that is used to study impacts. The gun can fire disks of metal at extremely high velocity. Those metal impactors then hit a variety of rocks and other samples to see what happens. Jen Blank wanted to know what happens to organic molecules during such violent collisions.

Blank's experiments began with a flat, cylindrical, stainless-steel capsule that was about an inch in diameter. She filled the cylinder with a solution of five different amino acids in water and sealed it tight. She then positioned that sealed sample in a metal well at the end of the 40-foot-long gun barrel. So, that became the target.

When everything was ready, they evacuated the gas gun chamber and a technician powered up the weapon. You can picture the scene. There was a countdown: "Five, four, three, two, one, fire!" Then a loud boom and a big shock wave that shook the entire building. The metal projectile flew down the gun barrel, in this case at more than 4,000 miles per hour. When it hit the end, a microsecond later, it squashed the neatly prepared sample container flat. For a tiny fraction of a second, the amino-acid solution experienced tremendous pressures, in excess of 200,000 atmospheres, at temperatures of almost 1,000°C.

So, what happened to the amino acids? Jen Blank pried out her deformed steel capsule and carefully milled down the metal to extract a few drops of liquid. Right away she could see changes. The originally clear amino acid solution had turned dark brown. To analyze the contents she used the organic chemists' standard analytical techniques: chromatography and mass spectrometry, just like we've seen in previous lectures. As you might expect, in every experiment, most of the original amino acids were destroyed by the heat and pressure. But she also found something I think is quite remarkable. In every single run, some of the delicate amino acid molecules reacted with each other to form pairs of amino acids. In other words, the energetic impact induced the formation of some peptide bonds to make larger clusters of molecules. This is just the kind of bonding that is crucial in the assembly of proteins.

Jennifer Blank's surprising results were highly publicized. The conclusion—big impacts on the early Earth must have reduced the quantity of organic molecules, just as everyone expected but the surprise was that, at the same time, these violent events might have helped to increase the diversity of prebiotic chemical species. In this way, impacts may have helped to select and modify the prebiotic inventory.

I think you can see that there are many intriguing ideas about ways to make biomolecules. Various researchers have explored syntheses at Earth's surface, or deep underground, in deep space or in meteorite impacts. But of all the clever scenarios for the prebiotic production of organic molecules, perhaps the most original and controversial ideas come from chemist Friedemann Freund. Freund is a longtime researcher at the NASA Ames Research Center, where he has been investigating ways that minerals might promote the formation of life's building blocks.

Freund's claim is that igneous rocks—those are rocks that form from an incandescent melt—were a principal source of Earth's organic molecules. His idea is rooted in the basic chemistry of melts. At high temperatures, well above 1000°C, molten rocks inevitably contain a variety of dissolved impurities. They always incorporate a little bit of water, carbon dioxide, and nitrogen. It is those impurities that become the organics in Freund's model. As the melt cools from high temperature, various minerals begin to crystallize, one after another. The first mineral might be olivine, which is rich in magnesium, silicon, and oxygen but that olivine will also incorporate a small amount of carbon and nitrogen.

Carbon and nitrogen are elements that don't easily enter the crystal lattice. These impurity elements concentrate at defects in the crystal structure. These defects are elongated spaces where foreign atoms can bond to each other and ultimately form chainlike molecules with a carbon backbone. According to Freund, every mineral in an igneous rock has the ability to concentrate carbon in this way, so every mineral has the potential to manufacture such organic molecules.

Of course, all those organic molecules are locked into the minerals, which are locked into the rocks, but rocks eventually weather away. Freund says that when they do, they release organic molecules into the environment. Given the vast volumes of igneous rocks that weather away every year, the amounts of organic molecules produced in this way could be immense.

Freund's ideas are not just theoretical. He rests his case on two samples that he has been investigating for more than 20 years. His first sample consists of synthetic crystals of magnesium oxide, or MgO. Magnesium oxide has one of the highest melting temperatures of any known material and cooling a white-hot melt from almost 3,000°C produced these crystals. These incredibly high temperatures were produced in what is called a carbon arc furnace, where high electric current is passed through rods of carbon. As a result, lots of carbon and other elements probably contaminated the melt.

Pure MgO crystals are typically clear and colorless but Freund's crystals have a cloudy appearance that suggests the presence of lots of impurities. So what makes these crystals cloudy? One hint comes from spectroscopy. These crystals absorb light at wavelengths characteristic of carbon-hydrogen bonds and oxygen-hydrogen bonds. These absorption features are not found in pure magnesium oxide but they are exactly the kinds of bonds we would expect from organic molecules. What's more, when Freund crushed his MgO crystals and analyzed the impurities—you guessed it—he found organic molecules.

Now, these MgO samples were, admittedly, synthetic crystals formed at incredibly high temperature. What about natural rocks and minerals, which crystallize at much lower temperatures? For his second sample, Freund focused on a natural sample of gem-quality olivine. That is sold as a beautiful green gemstone known as peridot. This is one of the most common minerals in igneous rocks. As before, his spectroscopic studies indicated lots of C-H and O-H bonds. As before, he was able to extract organic molecules from crushed powders. According to Freund, his olivine crystals contain an amazing 100 parts per million of carbon. He also claims that much of the carbon occurs in biologically interesting, chain-like organic molecules. By his estimates, the release of organic molecules from weathering rocks could rival the Miller-Urey process as a source of biomolecules.

I'm sure that by now you won't be surprised to learn that a lot of scientists are unconvinced by Friedemann Freund's proposal and they think this whole concept of organic molecules from igneous rocks is a wacky idea. They point out that even the hardiest organic molecule is fragmented into simple compounds like carbon dioxide and water at a thousand degrees. In addition, organic contamination is everywhere, so any surface, no matter how well it is cleaned, quickly becomes coated with adsorbed organic molecules. According to the majority view, Freund's observations are simply artifacts of questionable experimental procedures. So for now, the jury is still out.

The formation of organic molecules in high-temperature minerals is, understandably controversial, but there is no doubt that minerals played a role in other synthesis processes. I want to conclude this lecture by describing some fascinating ideas about interactions between minerals and molecules.

One of the challenges in this kind of study is that it requires knowledge of two very different kinds of chemical entities. On the one hand, there is a crystalline solid; that's the mineral, with a regularly repeating atomic structure. Then there is also a molecule, a tiny cluster of carbon and other atoms. One of the problems in origin-of-life research is that most experts have come from the world of organic chemistry. They know a lot about the carbon-based biomolecule part but not so much about the minerals. Therefore, details of what's going on at the atomic scale are often fuzzy.

A notable exception to this pattern is Gustaf Arrhenius, who is a senior, NASA-supported researcher at the Scripps Institution of Oceanography in La Jolla, California. Arrhenius is an expert in both organic chemistry and minerals, and he has a deep understanding of crystal structures. As a result, he and his colleagues have focused on a common group of minerals known as double-layer hydroxides. Double-layer hydroxides are common in nature and come in an almost limitless range of composition, and they always display an elegant layered structure. These minerals incorporate magnesium, iron, chromium, nickel, calcium, aluminum and many other elements in the double-layer structure. There is another complexity because small molecules, like water or carbon dioxide, can occupy the spaces between the layers. Arrhenius and his coworkers have learned to fine tune his double-layer hydroxides to perform specific chemical tasks just by changing their chemical contents.

Arrhenius suspects that double-layer hydroxides and other layered minerals might play important roles in the synthesis of biomolecules. This idea is based on the ability of double-layer hydroxides to soak up small organic molecules in the rather large spaces between the layers. Once they are confined and concentrated, these smaller molecules have a tendency to bond together to form larger molecular species that might not otherwise emerge from the soup.

The most interesting molecules produced so far in Arrhenius's lab are sugar phosphates. Those are molecules in which a sugar molecule is linked to a phosphate group. Arrhenius and his colleagues prepare a solution of the two-carbon molecule glyco-aldehyde phosphate and pour the solution over powdered double-layer hydroxides minerals and just let it sit there. The glyco-aldehyde phosphate molecules link together to form sugar molecules with four or six carbon atoms. Within a couple of days, lots of sugar-phosphate has formed. This is an important result because a pairing of a sugar molecule with a phosphate group forms the backbone of the genetic molecules DNA and RNA. The sugar-phosphate pair also is integral to key, energy-rich molecules like ATP.

There are hundreds of common minerals in nature and there is a good chance that others behave in a similar way. Professor Joseph Smith, of the University of Chicago, has been exploring yet another intriguing group of minerals, called "zeolites." Smith is an authority on this diverse family of natural and synthetic crystals, all of which feature lattice-like frameworks of silicon, aluminum and oxygen atoms. Zeolites have a diverse range of open framework structures. In some zeolites, their open frameworks form long, straight channels that run the length of a crystal. In others, there is a complex three-dimensional pattern of interconnecting cavities, like some futuristic artificial cave. The most interesting and valuable aspect of these minerals is that their channels are just the right size to accommodate a variety of organic molecules.

It turns out that hydrocarbon molecules that are not too large can enter these zeolite channels, while other, larger molecules are excluded. What's more, if you find the right zeolite mineral, it is able to split larger molecules into uniform sized fragments. That chemical trick lies at the heart of the multi-trillion-dollar petroleum refining business.

Well, if zeolites can process oil today, perhaps they played a role in processing prebiotic molecules. Joe Smith imagines that prebiotic molecules concentrate inside zeolite channels, where they are protected and where they might react to form more complex chains of molecules. This is an attractive hypothesis because zeolites abound in the weathering products of volcanic rocks. What's more, organic molecules were certainly present in the same volcanic environments.

It's easy to imagine how one of these minerals could be packed with a long column of molecules. Joe Smith goes farther. He even suggests that the first cell wall might have been the internal surface of one of these minerals. In this case, the experiments to test these ideas have not been done. It is extremely difficult to study what is going on with organic molecules inside a mineral.

However, the message from Friedemann Freund, Gustaf Arrhenius, Joe Smith and other researchers is clear. Minerals and organic molecules could have interacted in a lot of intriguing ways. There were many different ways to make organic molecules on the early Earth.

Let's summarize what we've learned in the last half-dozen lectures. The first step in life's emergence must have been the formation of lots of organic molecules but where did life's crucial molecules form? A large group of researchers agree with Stanley Miller and his associates. According to them, the vast majority of prebiotic molecules formed near Earth's surface through the action of lightning and ultraviolet radiation. These molecules then accumulated in the ancient oceans, where they were poised to undergo further chemical evolution.

Other scientists look to the productive environments of deep-ocean hydrothermal vents, where chemical reactions are driven by the instability of iron and nickel minerals. At first, this idea was met with a lot of skepticism but discoveries of abundant microbial life deep underground, coupled with successful synthesis experiments at high temperatures and high pressures, have convinced a lot of experts that the hydrothermal hypothesis is worth considering.

Then we have the production of organic molecules in deep space, in dense molecular clouds, and their subsequent delivery to the early Earth by carbon-rich meteorites and comets. Impacts may have further increased the diversity of Earth's biomolecules. Then there are all the ways that rocks and minerals might have contributed.

We can all understand how scientists might sometimes fall in love with their own theories. Yet, in spite of the polarizing advocacy of one favored origin environment or another, I think it is safe to say that experiments point to the likelihood that there was no, single dominant source.

So, it is not a matter of Miller versus the Ventists, or deep-space synthesis versus synthesis on Earth. Many different, ancient environments had carbon atoms and energy enough to trigger chemical transformations. Most likely, many different environments contributed to the prebiotic inventory.

The prebiotic Earth must have had an embarrassment of organic riches that accumulated from many different sources. Carbon-rich molecules emerged from all sorts of environments, from deep space to the deep crust. Amino acids, sugars, hydrocarbons and bases—all of the essential molecular species were available.

The biggest challenge in understanding the origins of life is not the emergence of biomolecules. It turns out, the difficult challenge is the selection of just the right biomolecules and the assembly of those biomolecules into the vital structures we call macromolecules.

Lecture Fifteen

Macromolecules and the Tree of Life

Scope: The greatest challenge in understanding life's emergence lies in identifying the processes by which just the right assortment of smaller molecules was selected, concentrated, and organized into key macromolecules, including RNA, and into self-replicating cycles of molecules, especially the citric acid cycle. The chances of having just the right combination of molecules bump into each other and assemble into anything useful were practically zero. In one way or another, a surface must have been involved in life's emergence.

Outline

I. The great mystery of life's origins lies in the gap between simple organic molecules and primitive cells.

A. Living cells depend on the assembly of just the right combination of small organic molecules into much larger macromolecules. Each macromolecule has a specific cellular function: to provide protective barriers, to promote chemical reactions, to store energy, and more.

B. Every living creature is formed from millions of intricate molecular components, which interact through a balanced cascade of chemical reactions. At the heart of every cell is the chemical machinery to grow, develop, reproduce, and respond to changes in the environment.

C. Even the most primitive single-celled organisms are amazingly intricate in their structures and functions. Indeed, chemical complexity seems tied up in the very definition of life. Yet living cells are built from relatively simple molecular parts.

D. One of the transforming biochemical discoveries of the last two centuries is that all known life-forms rely on a few types of chemical reactions, and these chemical reactions result in a handful of molecular building blocks. Life's construction materials are almost exclusively carbon-based organic molecules that bond together to form layered membranes or chainlike polymers.

II. One distinctive trait of the molecules of living organisms is their modular design. Most of life's structures are built from just four key types of molecules—sugars, amino acids, nucleic acids, and hydrocarbons (see Lecture Ten). Here is how each one combines to form macromolecules.

A. Sugars are the basic building blocks of carbohydrates, which include Earth's most abundant macromolecules. Our diets include a number of common sugars that are part of every cell's complement of genetic material. These small molecules incorporate at most a few dozen atoms.

1. All these energy-rich sugars have a ratio of carbon to hydrogen to oxygen of about 1:2:1; for example, glucose is $C_6H_{12}O_6$.

2. Most of life's sugars are not isolated. Typically, they are locked into macromolecules in which millions of sugar molecules link into giant polymers.

B. Amino acids form a second key group of biomolecules. (See Lectures Nine and Ten, including amino acid diagram in Lecture Ten, p. 53.) They link end-to-end with peptide bonds to form proteins—the chemical workhorses of life because of their myriad vital functions.

C. Nucleotides combine end-to-end in long polymers to form DNA and RNA. They are more complicated than sugars or amino acids

1. Each nucleotide is built from three small molecules:

a. a five-carbon sugar—*ribose* in RNA, or its close relative *deoxyribose* in DNA (see diagram of both in Lecture Seven, p. 33; a diagram of ribose is also below in Lecture Fifteen, p. 16);

b. a so-called *base*—one of five closely related, ring-shaped molecules;

c. a *phosphate group*—a tiny cluster made up of a phosphorus atom surrounded by four oxygen atoms.

BASIC NUCLEOTIDE STRUCTURE

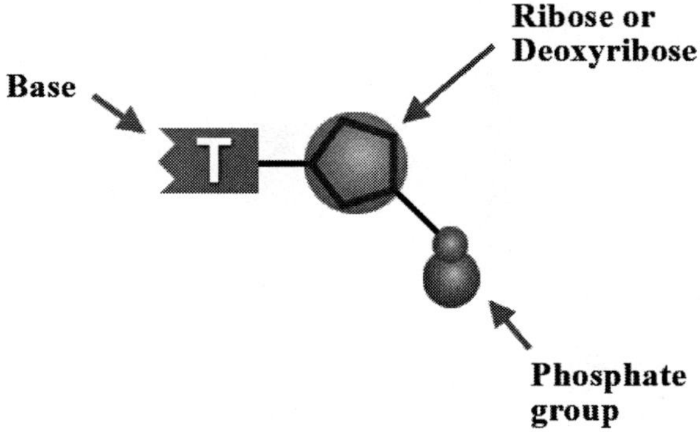

2. RNA is a long chain of single nucleotides, whereas DNA has two chains that link together in a ladder-like structure and twist into the famed *double helix* structure.

D. Lipids are elongated hydrocarbon molecules that include a wide variety of fats and oils. They organize themselves in every cell to form membranes, store energy, and perform other critical functions.

SAMPLE STRUCTURE OF A LIPID

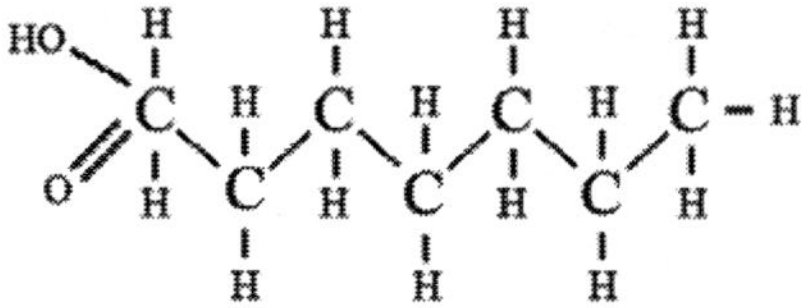

III. The most convincing aspect of Stanley Miller's experiment and the discoveries of subsequent prebiotic chemists was that they synthesized examples or close relatives of all four types of common biomolecules. It is little wonder that so many thought that an understanding of life's emergence was just around the corner. The story isn't quite that simple.

A. It is easy to catalog all the biological molecules synthesized in the early experiments, but this ignores the selectivity that is an equally important part of the prebiotic synthesis story. For every useful molecule that was produced, many other molecular species have no obvious biological use. Sugar molecules are one example.

1. Every living cell relies on ribose and deoxyribose. Chemists have identified plausible prebiotic synthesis pathways that yield small amounts of these sugars. For every molecule produced, many other 5-carbon sugar molecules are also produced.
2. Sugar synthesis typically produces a morass of more than a hundred 3-carbon, 4-carbon, 6-carbon, and 7-carbon sugars, and they occur in chains, in branching structures, and in rings. Life uses only a small handful of these sugars.

B. If that weren't enough, life turns out to be even choosier.

1. Many of the most common biological molecules, including the 5-carbon ribose and deoxyribose sugars, come in pairs related by mirror symmetry. In most respects, these left- and right-handed varieties are chemically identical, but they differ in their shapes.

RIBOSE

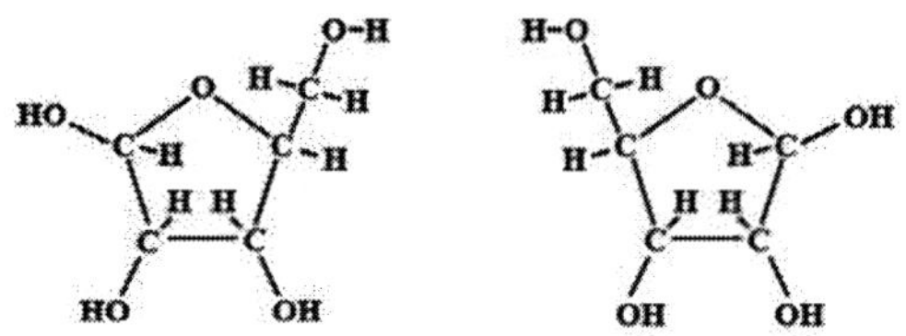

Left-Handed Right-Handed

2. Laboratory synthesis almost always produces equal amounts of left- and right-handed sugars, but life is highly selective. Life only employs the right-handed sugars, never the left.

C. The chemical selectivity of modern life raises a knotty problem. We generally think of the first life-form as being chemically simple compared with the modern scene.

1. Given the rich variety of prebiotic molecules that must have been available, might the earliest life-forms have used a more diverse suite of organic molecules? Over time, might much of this diversity have been stripped away? Is it possible that

ancient life developed with a completely different set of biochemical pathways?

2. To understand the biochemistry of ancient living cells, we turn to the remarkable field of molecular phylogeny.

IV. The central assumption of molecular phylogeny is that genomes in modern cells evolved over billions of years from ancestral cells. The DNA of the last common ancestor gradually mutated and branched out to achieve the diversity of life on Earth today. We can resort to comparative studies of many different genomes and look for shared traits.

DNA

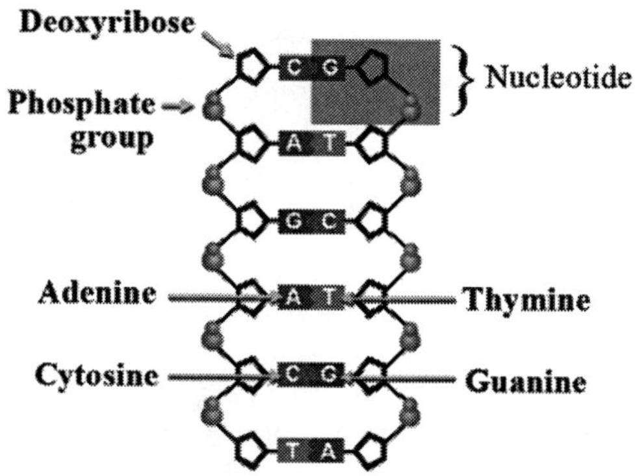

A. Differences among DNA sequences suggest the timing of evolutionary branching. If two DNA sequences are quite similar, they must have branched only recently. If two sequences differ a great deal, they likely split from a common ancestor long ago.

1. Here's the catch: There are some vital chemical functions in every cell that just cannot be changed all that much.
2. DNA sequences that show relatively little variation among many diverse species are likely to represent the most ancient and essential biochemical traits.
3. The same technique has been used to analyze 58 different fifteenth-century manuscript copies of the "Wife of Bath's Prologue" from *The Canterbury Tales* by Geoffrey Chaucer; 44 of the manuscripts fall neatly into five groups, representing five different copying sources.

4. One group of 11 manuscripts that had received much less study from literary scholars proved crucial. These 11 copies, labeled "O," for original, appear to lie much closer to the presumed original manuscript than any of the others.

PHYLOGENY OF "THE WIFE OF BATH'S TALE"

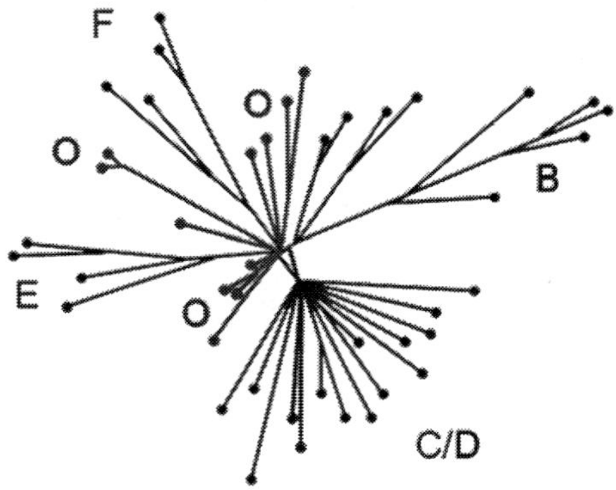

B. New sequencing techniques have allowed us to determine numerous complete genomes, including more than 100 genomes of different kinds of microbes. Molecular phylogenists use powerful computers to compare these genetic texts.

1. The key to success is to find some gene that is present in every cell. Many researchers have zeroed in on *16S-ribosomal RNA*, a sequence of RNA letters that's part of the ribosome, which assembles proteins.
2. This critical structure is one of the most highly conserved regions in genomes. In fact, the 16S-ribosomal RNA sequence in the simple bacterium *E. coli* overlaps yours by about 50 percent.

C. The pioneer and acknowledged master of molecular phylogeny is University of Illinois geneticist Carl Woese. He first started to compare 16S-ribosomal RNA sequences in the 1970s.

1. Before the so-called *Woesian revolution*, life was divided into several kingdoms, including plants, animals, fungi, and various single-celled groups.

2. What Woese found was that all plants, animals, and fungi are genetically similar and, thus, constitute just a single kingdom, or "domain of life," called the *Eukarya.*
3. Most single-celled creatures, on the other hand, display astonishing chemical diversity and fall into two additional domains, called *Bacteria* and *Archaea.*

WOESE'S PHYLOGENY

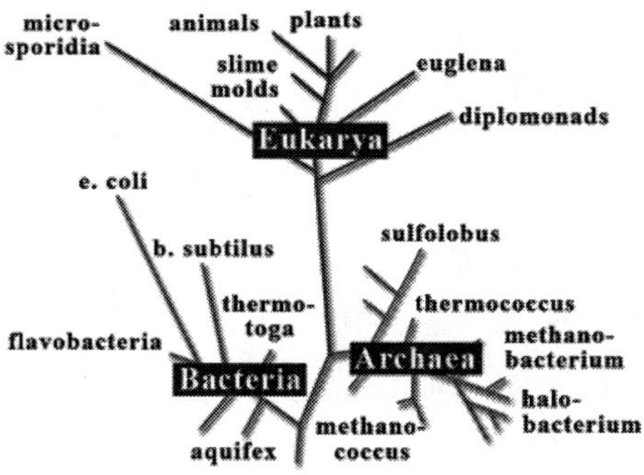

D. The original motivation of Carl Woese was to figure out the evolutionary tree of life in a top-down effort to gain insights into the nature of the last common ancestor.
   1. Woese figured out, for example, that Archaea and Bacteria arose long before Eukarya.
   2. He noticed that many of the most primitive microbes are extremophiles that live at high temperatures. That discovery encouraged advocates of a deep, hot emergence of life.
   3. Molecular phylogeny points to other seemingly primitive traits. For example, the most primitive members of the Archaea domain are usually *autotrophs*—they synthesize their own biomolecules. That result is at odds with the view of many origins specialists that the first cells must have been *heterotrophs* that scavenged molecules from the prebiotic soup.
   4. This controversy cannot be resolved with phylogenetic studies. It is always possible that the deep-dwelling autotrophs evolved from surface heterotrophs and that the autotrophs were the sole survivors of a globe-sterilizing event.

E. Molecular phylogeny is greatly complicated by another characteristic of microbes: Throughout the history of life, single-celled organisms have tended to swap sections of DNA with other species. The idealized picture of an unbroken branching history of DNA evolution from parent to daughter is invalid. A modern view of the tree of life encompasses countless perplexing cross-links and no one "last common ancestor."

F. Despite these uncertainties, molecular phylogeny points to a few shared biochemical mechanisms in all cells. Two observations stand out.

   1. Every cell uses RNA, both to carry genetic information and to assemble proteins. This common heritage suggests that there once was an *RNA World* as a key step in life's emergence (see Lectures Twenty-One and Twenty-Two).

   2. A simple metabolic strategy called the *citric acid cycle* lies at the heart of every cell. This cycle consists of a sequence of simple chemical reactions essential to survival. Any successful model of life's origins must incorporate the early emergence of this cycle (see Lectures Nineteen and Twenty).

G. What, then, does molecular phylogeny tell us? The earliest cells on Earth weren't so different from those of today. They probably relied on similar, if simpler, biochemical pathways. Those shared chemical traits suggest an experimental approach to studies of emergent biochemistry.

V. No matter how abundantly biomolecules were made, the primitive ocean, with its estimated volume greater than 320 cubic miles, must have been hopelessly dilute. The chances of having just the right combination of molecules bump into each other and assemble into anything useful were practically zero. A surface must have been involved in life's emergence.

A. Perhaps lipid molecules concentrated to form a primordial "oil slick." Evaporating tidal pools provide other interesting surfaces, where water contacts rock, and intriguing cycles of evaporation might have concentrated chemicals. Deep within the crust or in hydrothermal volcanic zones, a variety of mineral surfaces might have selected, concentrated, and organized molecules on their periodic, crystalline surfaces (see Lectures Seventeen and Eighteen).

B. What if carbon-based molecules could organize themselves without the help of a mineral? In the next lecture, we'll see that modern cells do indeed organize themselves so as to provide their own surfaces.

Essential Reading:

Hazen, *Genesis: The Scientific Quest for Life's Origin*, chap. 10.

Lehninger et al., *The Principles of Biochemistry*.

Supplemental Reading:

Knoll, *Life on a Young Planet*, chap. 2.

Lahav, *Biogenesis: Theories of Life's Origin*, part III.

Questions to Consider:

1. Is the fact that all known cells use a similar limited selection of small molecular building blocks evidence for or against a common origin for all life-forms? What if different groups of cells used very different suites of organic molecules?
2. Phylogenetic approaches only work if there is some mechanism by which small changes are introduced from one generation to the next. What are some of the mechanisms for such changes in manuscripts of Chaucer's *Canterbury Tales*? What are mechanisms for changes in cells?

Lecture Fifteen—Transcript
Macromolecules and the Tree of Life

By now I'm sure that you're getting the sense that to understand the origin of life is a daunting task. One thing that helps us is that we have a pretty good idea of both the first and the last step of life's emergence on Earth. In the first step, perhaps 4 billion years ago, we have the emergence of life's basic molecular building blocks. Molecules formed by the Miller-Urey processes and in the deep ocean, and even earlier in nebular clouds in deep space. All sorts of environments produced amino acids, sugars, hydrocarbons, and more. The last half century of synthesis experiments leads us to the inevitable conclusion that prospective biomolecules must have littered the ancient Earth.

We can also imagine the last step of life's origins. However it happened, biology took over from chemistry with the appearance of a relatively simple, encapsulated precursor to modern microbial life. That entity must have possessed all of life's characteristics, including the abilities to grow, to reproduce, and to evolve. We've seen, from various top-down studies of the fossil record, that such cellular life may have been firmly established not much less than 4 billion years ago.

So, the great mystery of life's origins—the missing chemical link, if you will—now really lies in the gap—short in time, but huge in complexity—between simple organic molecules, on the one hand, and primitive cells on the other. It's fascinating and reassuring to know that ancient Earth had an ocean full of promising organic molecules. But those isolated molecules were something like piles of bricks and stacks of lumber at a building supplier. A lot of assembly is required.

Living cells depend on the assembly of just the right combination of small organic molecules into much larger collections, called biopolymers, or bio-macromolecules, or just macromolecules. A macromolecule, though the exact definition is somewhat arbitrary, is a big molecule that is assembled from dozens, or hundreds, or even thousands of smaller molecules. Each macromolecule has a specific cellular function: to provide protective barriers, to promote chemical reactions, to store energy, and more.

Making big macromolecules from lots of little molecules is not straightforward. For every potentially valuable, small molecule in the prebiotic soup, there must have been dozens of other molecules with no ultimate use. As we saw in Lecture Seven, life as we know it is incredibly picky about its molecular building blocks and a survey of known biomolecules reveals all kinds of idiosyncrasies. The great majority of organic molecules in all of the known, or suspected, prebiotic synthesis processes are biological dead ends. That's why many origin-of-life researchers have turned their focus to the emergent steps by which just the right combination of carbon-based molecules might have been selected from the soup, concentrated at a surface, and assembled into the essential structures of life.

In this lecture we're going to begin our consideration of this second great emergent step in the path from geochemistry to biochemistry, the emergence of macromolecules. We'll review the macromolecules of current life and then we'll turn to consider findings from a top-down effort to map the tree of life, findings that suggest that early life may have used a more diverse, and a more interactive, set of building blocks than does familiar life today.

All of us realize that life is much more than a chemical phenomenon. Living cells display all kinds of complex behaviors that are much more than chemical reactions. Nevertheless, we are all chemical beings. Every living creature, from the most primitive microbe to the most advanced multicellular plant and animal, is formed from millions of intricate molecular components and those components interact by an exquisitely balanced cascade of chemical reactions. At the heart of every cell is the chemical machinery necessary to grow, to develop, to reproduce, and to respond to changes in external environment.

There is a paradox related to life's chemical complexity. On the one hand, even the most primitive single-celled organisms are amazingly intricate in their structures and functions. Indeed, chemical complexity seems to be tied up in the very definition of life. Yet, on the other hand, living cells are built from relatively simple molecular parts.

One of the transforming biochemical discoveries of the last two centuries is that all known life forms rely on just a few basic types of chemical reactions, and these chemical reactions result in just a handful of different molecular building blocks. Life's construction materials are almost exclusively carbon-based organic molecules that bond together to form layered membranes or chainlike polymers. Just a few kinds of small molecules are able to assemble into a tremendous variety of larger structures.

Two centuries ago, many chemists had concluded that the carbon-based chemicals of life formed by their own mysterious rules, perhaps governed by a "life force." At that time, the world of chemistry applied only to the inorganic domains of rocks, minerals, and metals. This conventional wisdom changed in 1828, with the discoveries of a young German chemist, mineral collector, and gynecologist by the name of Friedrich Wöhler. Wöhler, who lived from 1800 to 1882, was able to demonstrate that the molecules of life are no different, in principle, from any other chemicals.

In a classic experiment, he took a common laboratory chemical called cyanic acid—it's just the elements HOCN—mixed the cyanic acid with ammonia and wound up producing urea, which is the principal component of urine. Wöhler's publication reporting his important discovery contained an unusual sense of humor. He wrote, "I can no longer, as it were, hold back my chemical urine, and I have to let out that I can make urea without needing a kidney, whether of man or dog." In spite of this humorous tone, Wöhler had accomplished a significant advance. He was able to convince his colleagues that organic chemistry is rather ordinary, because he employed standard lab techniques to make a chemical substance previously known only from life.

So, is there anything special about the molecules of living organisms? One distinctive trait of the molecules of life is their modular design. You've undoubtedly run into this familiar strategy in the context of your home. Modern architects rely, for the most part, on standard building materials. They use all sorts of mass-produced items, bricks and beams, doors and windows, lighting fixtures, ceiling tiles and so on, to assemble an almost infinite variety of commercial and private buildings.

Now, it isn't absolutely required that you build in that way. I know a multimillionaire who designed and had built the most extraordinary mansion. Almost every square foot of that house was hand crafted, and let me tell you, it's an absolutely amazing house. It has secret passageways, unexpected nooks, and hidden closets sort of like Chinese puzzles. Everything is lovingly made of fine woods, rare stones, and other equally extravagant materials. It's wonderful to see such craftsmanship, but you can imagine how inordinately expensive it all was.

That's why most of us have to choose a more economical solution. We rely on a few standardized construction modules, so buildings are much faster and much cheaper to build. This really isn't such a sacrifice. Modularity doesn't imply uniformity or poor quality. You can design a unique building of almost any shape or size from the kind of basic materials available at any hardware store.

Life adopts exactly the same modular principle in its design and construction of carbon-based molecules. Rather than use a different molecule for every purpose, the majority of life's structures are built from just four key types of molecules—sugars, amino acids, nucleic acids and hydrocarbons.

We've already met some of these molecules in Lecture Ten, but let's take a closer look at how each one combines to form macromolecules. Sugars are the basic building blocks of carbohydrates, which include what are, by far, Earth's most abundant macromolecules. Our diets include a number of common sugars, including the fructose of honey, the sucrose of cane sugar, and the glucose of fruits, but also ribose and deoxyribose, which are part of every cell's complement of genetic material. These are all small molecules that incorporate, at most, a few dozen atoms.

All of these energy-rich sugars have a characteristic ratio of carbon to hydrogen to oxygen of about 1-to-2-to-1. So, for example, glucose is $C_6H_{12}O_6$. But most of life's sugars are not found as isolated molecules. Instead, they're typically locked into macromolecules in which millions of individual sugar molecules link together into giant polymers. The two most familiar examples of sugar polymers are the fibrous cellulose of plant stems or the bulky starch of potatoes. Both types of polymers are formed from glucose molecules. They're just linked in different ways.

Amino acids, which we met in Lectures Nine and Ten, form a second key group of biomolecules. Amino acids are the small molecules that Stanley Miller made in his 1953 experiment. They link together, end to end with peptide bonds to form proteins and proteins serve in many different capacities. They've been called the chemical workhorses of life, because of their myriad vital functions. I've already mentioned how proteins help to strengthen the mineral constituents of bone. They act as hormones, like insulin to control glandular functions. Hemoglobin is a protein that carries oxygen in the blood. Proteins also clot blood, digest food, and promote the thousands of chemical operations essential to life. Every one of the tens of thousands of different proteins in your body forms from long chains of amino-acid molecules. Hundreds to thousands of amino acids just line up like beads on a string. These chains fold into all sorts of different shapes and each protein folds in a way that accomplishes a specific chemical task.

A third type of essential biomolecule is the nucleotide, which combines end to end in long polymers to form the familiar genetic molecules DNA and RNA. Nucleotides are a little more complicated than sugars or amino acids. Each nucleotide is built from three small molecules. First there's a 5-carbon sugar, either ribose in RNA, or its close relative deoxyribose in DNA. Then a so-called "base", that's one of five closely related ring-shaped molecules. Finally there's a phosphate group, that's a tiny cluster made up of a phosphorus atom surrounded by four oxygen atoms. RNA is a long chain of single nucleotides, while DNA has two chains that link together in a ladder-like structure and twist into the famed double-helix structure.

Finally we have lipids, the fourth and final type of common biomolecule. We'll focus on lipids in the next lecture but they are elongated hydrocarbon molecules that include a wide variety of fats and oils. Lipids organize themselves in every cell to form membranes, store energy, and perform other critical functions.

So, we have four common types of biomolecules—sugars, amino acids, nucleic acids and lipids. The most appealing and convincing aspect of Stanley Miller's experiment and the discoveries of subsequent prebiotic chemists, was that they were able to synthesize examples, or at least close relatives, of all four groups. In fact, the distribution of molecules in Miller's original experiments was remarkably similar to that of biology. So it's little wonder that so many scientists and science watchers were optimistic that an understanding of life's emergence was just around the corner.

But, as I've suggested, the story isn't quite as simple as that. It's easy to catalog all the biological molecules that were synthesized in the early experiments, but this catalog of success ignores another equally important part of the prebiotic synthesis story. Because, for every useful molecule that was produced, there were many other molecular species that have no obvious biological use. Let's just take sugar molecules as one example. Every living cell relies on two different kinds of 5-carbon sugar molecules, ribose and deoxyribose. Sure enough, chemists have identified plausible prebiotic synthesis pathways that yield small amounts of these sugars. But for every molecule of ribose or deoxyribose produced, you'll find many other 5-carbon sugar molecules as well. Xylose, arabinose and lyxose, for example, are always formed. There is even more chemical complexity. Sugar synthesis typically produces a morass of more than a hundred 3-carbon, 4-carbon, 6-carbon, and 7-carbon sugars that occur in chains, in branching structures and in rings, but life uses only a small handful of these sugars.

Then, as if that weren't enough, life turns out to be even choosier in its sugar molecules. Look at this. Many of the most common biological molecules, including the 5-carbon ribose and deoxyribose sugars, come in pairs that are related by mirror symmetry. In most respects, these left- and right-handed varieties are chemically identical. They have exactly the same chemical formula and identical physical properties. So they have the same melting and boiling points, for example. But these pairs of molecules differ in their shapes in the same way your left and right hands differ. Here's a real mystery. Laboratory synthesis almost always produces equal amounts of left- and right-handed sugars, but life is highly selective. Life only employs the right-handed sugars, never the left.

So this chemical selectivity of modern life raises a knotty problem. We generally think of the first life form as being chemically simple compared to the modern scene. But, given the rich variety of prebiotic molecules that must have been available, is it possible that we're fooling ourselves? Might the earliest life forms have used a much more diverse suite of organic molecules, and then, over time, might much of this diversity have been stripped away? Is it possible that ancient life developed with a completely different set of biochemical pathways?

Well, to understand the biochemistry of ancient living cells, the answer may lie in yet another kind of top-down investigation. We can turn to the remarkable field of molecular phylogeny.

So what is molecular phylogeny? Every chemical reaction in a cell is recorded in its "genome." That's the encoded sequence of millions to billions of DNA letters. Recall that there are four different bases in the DNA alphabet—A, G, T, and C. Those letters stand for the purines, adenine and guanine, and the pyrimidines, thymine and cytosine. In DNA, matched pairs of these letters, either A-T or C-G, form the steps of the DNA ladder. These four letters are sufficient to hold all of the genetic information for any organism. What's more, every cell has the exact same mechanism for converting the genetic instructions in DNA into all the various proteins that play so many chemical and structural roles. I don't have time to describe that process here, but I go into some detail in Lectures Forty-Nine through Fifty-Two of my Teaching Company course, *The Joy of Science*.

The point here is that the genetic mechanisms that convert the information of DNA into functional proteins must be ancient. All known organisms share them. That means we should be able to compare the genetic information in different cells and tease out the common threads that point to the most primitive characteristics of a last common ancestor.

The central assumption of molecular phylogeny is that genomes in modern cells evolved over billions of years from ancestral cells. The DNA of the last common ancestor gradually mutated and branched out to achieve the diversity of life on Earth today. So we can resort to comparative studies of lots of different genomes and look for shared traits.

Differences among DNA sequences suggest the timing of evolutionary branching. If two DNA sequences are quite similar, then they must have branched only recently, but if two sequences differ by a great deal, they are likely to have split from a common ancestor long ago.

Here's the catch. There are some vital chemical functions in every cell that you just can't change all that much. Every cell has to make ATP for energy. Every cell needs to convert the information of DNA to proteins. You can't mutate those instructions very much. That means, if you can find DNA sequences that show relatively little variation among many diverse species—these are known as highly conserved sequences—then they are likely to represent the most ancient and essential biochemical traits.

Let me give you an example of how molecular phylogeny works. I want to tell you about a recent study in English literature, of all things. A group of scientists and literary scholars from the University of Cambridge used this approach to analyze 58 different 15th-century manuscript copies of the "Wife of Bath's Prologue," from *The Canterbury Tales* by Geoffrey Chaucer. I was surprised to learn that there's not a single known copy of *The Canterbury Tales* from the 14th century, when Chaucer wrote it. Every manuscript is a copy, and these 58 manuscripts differ widely in the text. So is there any way to recover Chaucer's original version?

Here's how they use the phylogenetic approach. The British team employed the same computer techniques used by evolutionary biologists, but instead of a sequence of DNA letters, they entered each of the Prologue's 850 lines, from all 58 versions, into the computer. That's a huge job, so I imagine they had graduate students do the typing. Then they let the computer search for textual similarities and differences. So, any lines of text that are common to most of the sources presumably reflected the original text, while any variations that might have arisen from copying errors or from deliberate changes were used to construct a kind of genealogy of the manuscripts.

The results are really intriguing. It turns out that 44 manuscripts fall neatly into five groups. These groups represent five different copying sources. But one group of 11 manuscripts proved crucial. Those are the ones they labeled "O," for original. These 11 copies appear to lie much closer to the presumed original manuscript than any of the others, but it turns out that these 11 texts had received much less study from literary scholars. So it's possible that study of these manuscripts will give us a new view of Chaucer's intentions.

What does that have to do with the origin of life? It turns out that the molecular phylogenists who try to figure out the nature of the earliest life form use exactly the same analytical approach. In this case, the "texts" are sequences of genetic molecules, either DNA or RNA. New sequencing techniques have allowed us to determine numerous complete genomes, including more than a hundred genomes of different kinds of microbes. Then the phylogenists use a powerful computer to compare all these genetic texts.

The key to success is finding some gene that is present in every cell, one that is highly conserved. Many researchers have zeroed in on a sequence called "16S-ribosomal RNA". That's a sequence of RNA letters that's part of the ribosome, which assembles proteins. This structure is so critical to all life that it is one of the most highly conserved regions in genomes. In fact, the 16S-ribosomal RNA sequence in the simple bacterium *E. coli* overlaps yours by about 50 percent. That's an amazing match after billions of years of evolution.

The pioneer who is universally acknowledged as the master of molecular phylogeny is University of Illinois geneticist, Carl Woese. He's the one who first started to compare 16S-ribosomal RNA sequences back in the 1970s. Before the so-called "Woesian Revolution," life was divided into several kingdoms, including plants, animals, fungi, and various single-celled groups. What Woese found was that all the plants, animals, and fungi are genetically very similar and thus constitute a single kingdom, or "domain of life," called the *Eukarya*. Most single-celled creatures, on the other hand, display astonishing chemical diversity and fall into two additional domains. He called them Bacteria and Archaea.

In Woese's new and astonishing phylogeny shown here, all the plants and animals and fungi are in one tiny region, way up in one corner of the diagram. All the rest of the *Eukarya* are single-cell organisms, like amoebas. Then there is a vast range of bacteria, which are well known for their role in causing a wide range of disease. Then there's another entire domain of life called the Archaea, which was an entirely unrecognized kingdom before Woese. This is a really radical shift in our view of life, and it took almost two decades to gain general acceptance.

It's really not all that surprising that the Archaea weren't discovered until the late 1970s. Most species in these two microbial domains are sort of nondescript, microscopic spheres or rods. It's really hard to tell them apart without sophisticated chemical and physiological tests, so previous researchers tended to lump them together. Only with the advent of molecular phylogeny was it possible to tell them apart.

The original motivation of Carl Woese was to figure out the evolutionary tree of life, just like a branching tree of Chaucer manuscripts. I think you can see how such a top-down effort can lead to insights regarding the nature of the last common ancestor. Woese was able to figure out, for example, that Archaea and Bacteria arose long before *Eukarya.*

Woese noticed other primitive features. For one thing, many of the most primitive microbes are "extremophiles" that live at high temperatures. That discovery encouraged advocates of the idea of a deep, hot emergence of life. Of course, other scientists took issue with that conclusion because life could have first formed at the surface, then radiated into deep regions. Then those deep microbes might have been the only survivors following a massive asteroid impact, and so they became the last common ancestors of life by default.

Then molecular phylogeny points to other seemingly primitive traits. For example, the most primitive members of the Archaea usually turn out to be autotrophs; they synthesize their own biomolecules. That result is at odds with the view of many origin specialists, including Stanley Miller, who have long assumed that the first cells must have been heterotrophs. Heterotrophs would have scavenged molecules from the prebiotic soup. Once again, you can't resolve the controversy with phylogenetic studies.

It's always possible that the deep-dwelling autotrophs evolved from surface heterotrophs and the autotrophs were the sole survivors of a globe-sterilizing event.

The entire approach of molecular phylogeny is greatly complicated by an additional characteristic of microbes. It turns out that single-celled organisms have a tendency to swap sections of DNA. Throughout the history of life, cells have picked up segments of DNA from other species. It's as if a new copy of *The Canterbury Tales* was cobbled together with different stories from different manuscripts. So the idealized picture of an unbroken branching history of DNA evolution from parent to daughter turns out to be invalid. In fact, in the modern view of the tree of life there are countless perplexing cross links that really confuse the story. In this scenario, there is no one "last common ancestor."

In spite of all these uncertainties, molecular phylogeny points to a few shared biochemical mechanisms in all cells and that information informs origins research. I think that two observations really stand out. First, it's clear that every cell employs RNA, both to carry genetic information and to assemble proteins. This common heritage suggests that there once was an "RNA World" as a key step in life's emergence. I'll tell you a lot more about the RNA World in Lectures Twenty-One and Twenty-Two.

A second conclusion is that a simple metabolic strategy called the citric acid cycle lies at the heart of every cell. As we'll see in Lectures Nineteen and Twenty, this cycle consists of a sequence of simple chemical reactions that are essential to survival. Any successful model of life's origins has to incorporate the early emergence of this cycle.

So what does molecular phylogeny tell us? I think the most important conclusion is that the earliest cells on Earth weren't so different from those of today. Those cells probably relied on very similar, though simpler, biochemical pathways. It is those shared chemical traits that suggest an experimental approach to studies of emergent biochemistry.

I began this lecture by suggesting that the greatest challenge in understanding life's emergence, lies in identifying the processes by which just the right assortment of smaller molecules was selected, concentrated, and organized into key macromolecules, including RNA, and into self-replicating cycles of molecules, especially the citric acid cycle. But no matter how abundantly biomolecules were made, the primitive ocean, with its estimated volume greater than 320 cubic miles, must have been a hopelessly dilute soup. The chance of having just the right combination of molecules bump into each other and assemble into anything useful was practically zero. Recall, from Lecture Eight, that complex emergent systems require a minimum concentration of interacting agents.

The only logical answer is an answer arrived at independently by many scientists that life's molecules must have concentrated on some kind of surface. Interesting chemistry often occurs on surfaces, where two different kinds of materials meet. Molecules often congregate at such surfaces. Think about the surface of the ocean, for example, where water contacts air. Perhaps lipid molecules concentrated to form a primordial "oil slick." Evaporating tidal pools provide other interesting surfaces, where water contacts rock. These environments also feature intriguing cycles of evaporation that might have concentrated chemicals. Deep within the crust, or in hydrothermal volcanic zones, a variety of mineral surfaces might have selected, concentrated, and organized molecules on their periodic, crystalline surfaces.

These ideas have become conventional wisdom within the origins-of-life community. In some way or another, a surface must have been involved in life's emergence. We will also examine the possible contributions of mineral surfaces in Lectures Seventeen and Eighteen. But what if a collection of carbon-based molecules could somehow organize themselves, without the help of a mineral?

In the next lecture, we'll see that modern cells do indeed organize themselves so as to provide their own surfaces.

Lecture Sixteen

Lipids and Membrane Self-Organization

Scope: Lipid self-organization seems to have been an essential step in the emergence of life. Several remarkable processes occur by which lipid molecules form and organize themselves into tiny cell-like spheres. In these processes, a membrane, for the first time, isolated the insides from the outsides of protolife.

Outline

I. Almost all prebiotic synthesis processes wind up in the ocean—in the prebiotic soup. All cells are primarily water on the inside, and most cells are enveloped by water on the outside. That environment poses a problem.

A. Water is one of the most well known solvents (see Lecture Twelve). It is a polar molecule, with a positively charged side and a negatively charged side. This configuration helps to pull apart other molecules. Thus, life had to develop some kind of protective membrane that isn't soluble in water. What chemical fits the bill?

B. There are two possibilities, both of which involve fatty molecules called *lipids*, distinguished by their insolubility in water.

1. Because cells are mostly water on the inside, a lipid that merely repelled water wouldn't allow cells to exist. The lipid barrier against water on the outside of the membrane must be compatible with water on the inside.
2. Phosphorus plus a lipid results in the kind of phospholipid molecule that makes most modern cell membranes. The molecules are shaped something like a tiny bobby pin. They have two long hydrocarbon chains attached to a rounded end.
3. On one end of this molecule are two long hydrocarbon chains with exposed tips that are *hydrophobic*—that is, "water hating." These parts of the elongated molecule act like molecular water-repellent.

A PHOSPHOLIPID MOLECULE

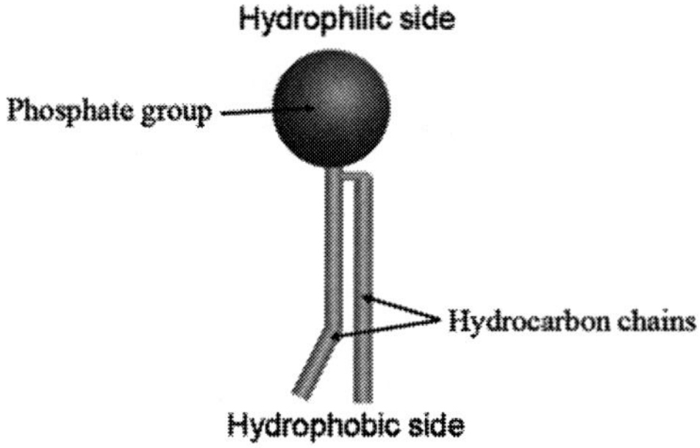

4. Behaving just the opposite is a rounded end that incorporates a phosphate group, which is *hydrophilic*, or "water loving."
5. The phosphate end attracts water, whereas the rest of the molecule repels water. Such a molecule is called an *amphiphile*.

C. Every natural system tends to rearrange itself to reduce its total energy. In the case of amphiphilic molecules, that means that only hydrophilic parts of the molecule touch water; the hydrophobic parts do not.
1. How do phospholipids accomplish this? In the early 1960s, Alec Bangham, a biophysicist from the University of Cambridge in England, discovered that lipid molecules put in water undergo a spontaneous reorganization. They automatically arrange themselves into tiny spheres—structures now known as *vesicles*.
2. At first, the lipids float around. The hydrophobic parts want to escape from the water, so they clump together.
3. Thousands or millions of individual amphiphilic molecules line up side by side and form a *double* layer of lipids—a *lipid bilayer*. In this spherical vesicle, all the hydrophobic parts are away from water, in the middle of the structure.

A LIPID BILAYER

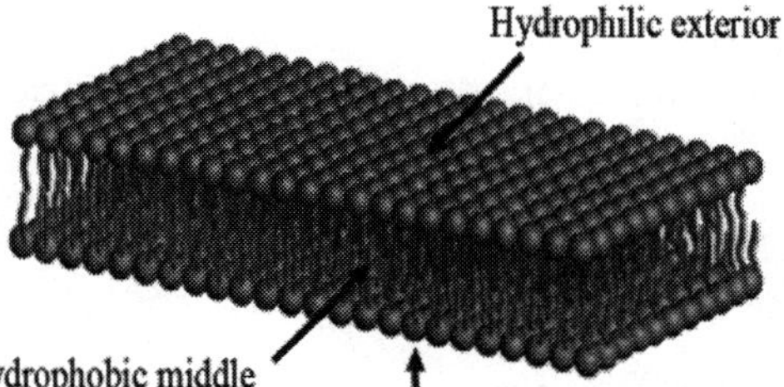

4. At the same time, all the hydrophilic ends wind up on the outside of the cell, facing the wet external environment, or on the inside, facing the water-filled interior of the vesicle.
5. This arrangement is used by virtually all cells. The cell membrane holds the cell together while separating the inside from the outside.

D. What about the prebiotic soup? Could a similar self-organizing process have arisen spontaneously, before life?

1. Biochemist Pier Luigi Luisi and his colleagues have shown that under the right conditions, vesicles can grow larger by gradually incorporating new lipid molecules from solution. In some instances, these vesicles display the ability to trigger the formation of more vesicles—they are *autocatalytic*. Under just the right circumstances, they split in two, a kind of self-replication.
2. After observing this self-assembly, growth, and division—Luisi proposed the *Lipid World* scenario: Prebiotic lipids were synthesized abundantly on the early Earth and in space, and they entered the primordial soup, where they self-organized into cell-like vesicles. From time to time, those vesicles captured a primitive genetic molecule, perhaps a strand of RNA that could copy itself. If a self-replicating molecule were incorporated into a self-replicating vesicle, the result could have the essential properties of life.

E. The Lipid World hypothesis requires a primordial source of lipids. However, lipids are one group of molecules that are not well represented in the Miller-Urey experiment. Is there another possible source of lipids? The answer is an emphatic yes.

II. David Deamer's discovery of lipid self-organization in molecules from space has changed the way we think about the origins of life.

A. A professor in the University of California system, Deamer became interested in the origins of life in 1975 when he went to Cambridge, England, to study lipids with Alec Bangham, who first identified vesicles. Together, they found that the size and stability of vesicles depends on a variety of factors, including the size and shape of the dissolved lipid molecules. Big molecules, for example, form vesicles at lower concentrations than small ones. These researchers thought that vesicles might have provided an ideal sheltering environment for the first life, but did lipids exist in the early oceans?

B. Later, Deamer began a study of molecules extracted from the carbonaceous Murchison meteorite, which held a wealth of exactly those molecules that must have accumulated on the prebiotic Earth.

1. Deamer and his colleagues ran a sample of the Murchison meteorite through a series of chemical steps that would break apart the dense black mass into a series of chemically distinct fractions. Those steps were designed to imitate the exposure to weathering that must have occurred on the early Earth. Whatever molecules they extracted from the Murchison meteorite were very likely to have occurred in abundance on the prebiotic Earth, as well.

2. The first step was to grind the rock into a fine black powder while it was submerged in a mixture of three liquids—water, alcohol, and chloroform, which dissolve different suites of interesting organic molecules. Amino acids and sugars concentrate in the water-plus-alcohol fraction, whereas lipids concentrate in the chloroform.

3. Deamer poured everything into a thick-walled test tube, which he placed into a tabletop centrifuge. The centrifuge separated the solution into three layers. All the pulverized rock fragments settled to the bottom of the tube, because they were the densest fraction.

4. The real treasure lay in the middle layer of chloroform, an ideal solvent for any lipids the meteorite might have held. Indeed, the chloroform had dissolved more than a tenth of a percent of the meteorite's mass. That's a remarkably high concentration of tantalizing organic species. So what are they?

5. Deamer used chromatography to further separate his compounds. The dried glass plate was pretty dull in daylight, but when Deamer turned on his black light, he saw a colorful display in a broad arc.
6. Several distinct fluorescent regions were visible. Each of these regions had a different, unknown suite of ancient cosmic organic molecules. Deamer's next step was to outline each of these areas by scratching the soft, powdery white surface. Then he scraped off the powder from each of those areas into different test tubes.
7. Because lipid molecules are soluble in chloroform, he washed each powdered sample in chloroform. The chloroform thus re-concentrated the lipids. That's all that was required to recover the different suites of Murchison molecules, and any one of those fractions might contain lipid molecules that form vesicles.
8. One-by-one, Deamer placed the chemical fractions into water and made observations. When he reached the molecules that had been concentrated in an elongated area on the chromatography plate, he watched the familiar magical transformation.
9. The invisible molecules spontaneously arranged themselves into tiny spheres no more than a hundredth of an inch across. Each of these structures had an inside and an outside. The molecules had self-organized into lipid bilayers, just like a cell membrane.
10. This result, published in 1989, was a breakthrough for origins researchers. Deamer had discovered that these ancient lipid molecules, which were synthesized at some unknowable distant place in space and delivered to Earth, spontaneously form tiny enclosed structures. And those vesicles are remarkably similar to the membranes that encase living cells.

III. David Deamer's great discovery taught us that one of life's most basic requirements—the separation of the inside from the outside—appears to be an integral part of the fabric of the universe. When new lipid-rich samples came along, Deamer repeated the process using those samples.

A. In their experiments simulating deep space, Lou Allamandola and his NASA Ames research team had synthesized a variety of organic molecules under conditions of ultracold vacuum with ultraviolet radiation. Some of their experiments produced a residue, just like the Murchison meteorite. The highest yields occurred when they irradiated ice made principally of water and alcohol, with a bit of ammonia and carbon monoxide. That recipe produced a residue of fluorescent material. Other molecules might have an amphiphilic character.

1. Allamandola went to Dave Deamer's lab to check for amphiphiles.
2. It took Deamer less than a day to verify what the NASA researchers had suspected. He used thin-layer chromatography to isolate the most promising fraction of fluorescent molecules, and beautiful vesicles appeared spontaneously in water
3. The implications are clear: Even before the formation of planets, even before the Miller-Urey process got started, essential raw materials for life were abundant in the deep freeze of space. The universe is littered with lipid molecules that are poised to organize spontaneously into cell-like structures.

B. Our group at the Carnegie Institution began studying the high-pressure and high-temperature reactions of pyruvate in 1996 (see Lecture Twelve). That work had been sitting on hold for years, because we did not have a lot of information on the specific compounds, and the relevance of the experiments to life's origins was not clear. Deamer's work pointed to another angle of interest.

1. Perhaps the yellow, oily "goo" that oozed out of our gold capsules included a fraction of self-organizing molecules. We had started with pyruvate, one of the core metabolic molecules. Perhaps we could show a simple chemical pathway from primitive metabolism to membranes.
2. In early 2001, I made a new batch of pyruvate-plus-water capsules, ran them at our standard conditions, and flew them out to Dave Deamer's specially equipped laboratory.
3. I opened the capsules, and they responded with the familiar *bang!* and intense oily foaming. We went through the step-by-

step process of chloroform extraction, concentration, and preparation for chromatography.

4. Under Deamer's ultraviolet fluorescent lamp, the results were a gorgeous pattern of yellow, blue, and purple, with several distinctly fluorescing areas—amazingly close to the Murchison experimental results. Deamer assumed that the correspondence was not just a coincidence, so we zeroed in on a blue fluorescing area that closely matched the area of his Murchison plate that had produced the most vesicles. We found that we had made self-organizing bilayer membranes starting from nothing more than pyruvate and water.

IV. All sorts of puzzles are associated with the formation of lipid vesicles.

A. Recent work by Deamer's group shows that lipid self-organization is inhibited by the presence of dissolved calcium and magnesium. Yet these salts are inevitable ingredients of seawater. Does that mean that life can only begin in fresh water? Or are there some unidentified varieties of lipid molecules that do self-organize in saltwater?

B. Many biologists point out that lipid vesicles are a far cry from cell membranes. A living cell must have ways to bring food in and get waste out. That's done in modern cells by an amazing array of *protein receptors* that regulate the flow of molecules and chemical energy into and out of the cell. What served as receptors before proteins?

C. Such questions will occupy researchers for many decades to come. For now, the emergence of cell-like vesicles that form spontaneously from simple molecules is one of the best-understood features of life's origin.

V. Clever new ideas keep popping up about origins-of-life research, and ideas about the emergence of self-organized molecular systems are no exception.

A. One of the most intriguing recent proposals comes from chemist Christopher Dobson of Oxford University and his collaborators at the National Oceanic and Atmospheric Administration in Boulder, Colorado. In 2000, they proposed a hypothesis about how lipids might organize themselves in the atmosphere as the result of ocean spray.

B. Here's the crux of Dobson's idea: Organic molecules could have accumulated at the ocean's surface like an oil slick. The turbulence of the ocean would have produced a continuous fine mist of aerosol particles. Countless trillions of tiny droplets would have been propelled from the oily surface into the atmosphere.

1. Each water droplet would have contained amphiphile molecules that formed a layer around the water-filled interior—a *single* layer, not a double layer, because there is only water on the inside.
2. Larger droplets were heavy enough to fall back into the sea, but Dobson's calculations show that smaller aerosol particles are quite stable and could have remained suspended in the atmosphere for months or even years.

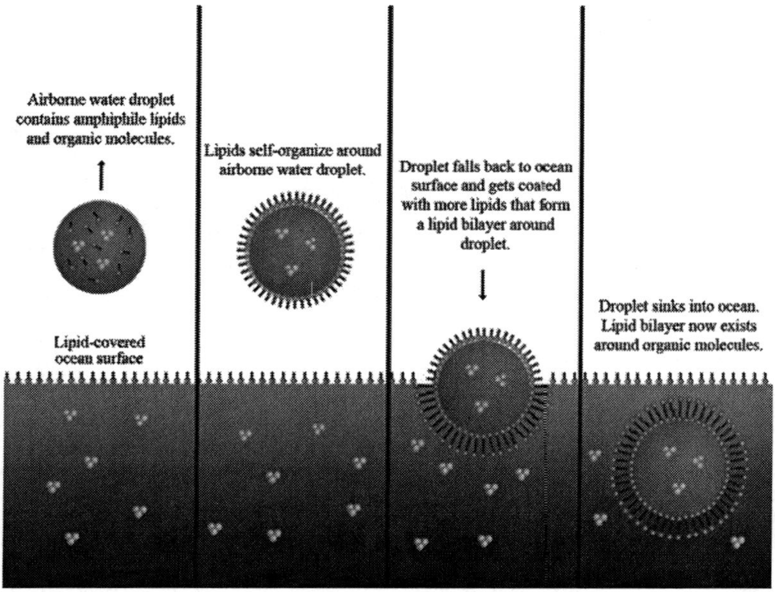

C. Dobson and his coworkers speculate that the lipid molecules in each aerosol particle form a sphere with the hydrophobic ends facing the outside atmosphere and the hydrophilic ends facing the water-filled interior. Many of these aerosol particles must have incorporated reactive, water-soluble organic molecules. As the spheres ascended to the highest levels of the atmosphere, those reactive molecules might have undergone additional chemical reactions in sunlight.

1. Think about the vast number of different origin experiments represented by these aerosol particles. Every one would have had weeks or months to undergo energetic chemical reactions. Each particle would have been, in effect, a tiny chemical experiment. And this process went on for hundreds of millions of years, as astronomical numbers of aerosols drifted into the skies.
2. Consider what happened when an aerosol particle finally descended to the ocean. While in the air, each aerosol particle had a hydrophobic exterior, but upon contact with water, more lipid molecules at the ocean's surface would have spontaneously coated the particle. Thus, the lipids formed a bilayer structure, leaving countless trillions of tiny cell-like objects in the primitive ocean.

VI. Lipid self-organization seems to have been an essential step in life's emergence. A membrane, for the first time, isolated the insides from the outsides of protolife. Yet a membrane by itself is not life. Life also required essential biomolecules, including proteins, carbohydrates, and nucleic acids, all of which had to be assembled from the prebiotic soup. All the building blocks of these macromolecules are water-soluble, and the molecules just cannot self-assemble in water. As we'll see in the next lecture, the best way to assemble life's molecules in water is to call in the rocks.

Essential Reading:

Hazen, *Genesis: The Scientific Quest for Life's Origin*, chap. 11.

Supplemental Reading:

Fry, *The Emergence of Life on Earth*, chap. 12.

Lahav, *Biogenesis: Theories of Life's Origin*, chap. 19.

Wills and Bada, *The Spark of Life: Darwin and the Primeval Soup*, chap. 7.

Questions to Consider:

1. Why is David Deamer's experiment on the organic molecules in the Murchison meteorite considered to be important?

2. If scientists are able to synthesize lipid vesicles that grow and divide as more lipids are added to the solution, will this system represent a primitive life-form? Why or why not?

Lecture Sixteen—Transcript

Lipids and Membrane Self-Organization

We've been exploring one of the central questions in origins research. By what process did life learn to use just the right combination of molecules? How were those molecules selected, concentrated, and organized into the macromolecular structures that form living cells?

We saw that this is a knotty problem, in part because prebiotic synthesis processes produce such a wealth of different kinds of organic molecules. Most of those molecules aren't useful; at least they aren't useful in modern organisms.

But there's another difficulty I haven't yet mentioned. Almost all prebiotic synthesis processes eventually wind up in the ocean, in the prebiotic soup. This makes sense because water provides the universal medium for life. All cells are primarily water on the inside, and the majority of cells are enveloped by water on the outside as well, but that environment poses a problem. Recall from Lecture Twelve that water is one of the best-known solvents. It's a polar molecule, with a positively charged side and a negatively charged side. This configuration helps to pull apart other molecules. But you certainly don't want your body to dissolve every time you take a shower, so life had to develop some kind of protective membrane that isn't soluble in water. But what chemical fits the bill? In this lecture we'll explore two possible answers to this question, both of which involve fatty molecules called "lipids".

It turns out that lipid molecules provide perfect protection. Recall that lipids molecules feature hydrocarbon chains, that is, a row of carbon atoms surrounded by hydrogen atoms. Lipids, which include the full range of fats, oils and waxes, are distinguished by their insolubility in water. As the old saying goes, oil and water don't mix. This suggests a strategy for protecting cells.

But wait. Cells are mostly water on the inside, so a lipid that merely repelled water wouldn't allow cells to exist. The lipid barrier against water on the outside of the membrane needs a way to be compatible with water on the inside. It turns out that phosphorus can do the trick. Phosphorus plus a lipid gives you the kind of phospholipid molecule that makes most modern cell membranes. The molecules are shaped something like a tiny bobby pin. They have two long hydrocarbon chains that are attached to a kind of rounded end. Now, there's something quite remarkable about this molecule. On one end, there are two long, hydrocarbon chains with exposed tips that are hydrophobic, that is, the chains and tips are literally "water hating." So these parts of the elongated molecule act like molecular water repellent.

Behaving just the opposite, there is also a rounded end that incorporates a phosphate group, which is just a phosphorus atom and four oxygen atoms. That end is hydrophilic, or "water loving." So the phosphate end attracts water, while the rest of the molecule repels water. Such a molecule, which is of two minds when it comes to water, is called an amphiphile.

Well, so what? Of what possible use is an amphiphilic molecule that can't decide whether it loves water or hates it? It is also true that just a few amphiphilic molecules aren't very useful. But if you place large numbers of these molecules in solution, they wind up dealing with their love-hate relationship in a truly remarkable way.

Every natural system has a tendency to rearrange itself in such a way as to reduce its total energy. You can observe this behavior all the time: A tightly stretched elastic band lowers its energy by snapping; a precariously perched boulder lowers its energy by tumbling down the mountainside; a firecracker lowers its energy by exploding. In the exact same way, a large collection of lipid molecules in solution always strives for a state of lower energy; and in the case of amphiphilic molecules that means that only hydrophilic parts of the molecule touch water, while the hydrophobic parts are kept away from water.

So how do phospholipids accomplish this trick? In the early 1960s, a biophysicist from the University of Cambridge in England, named Alec Bangham, discovered that lipid molecules extracted from egg yolk and put in water undergo a spontaneous reorganization. They automatically arrange themselves into tiny sphere-shaped structures that are now known as "vesicles."

At first the lipids are just floating around, but the hydrophobic parts really want to get away from the water, so they clump together. Thousands or millions of individual amphiphilic molecules line up side-by-side. What's more, they wind up forming a double layer, what's called a lipid bilayer. This resilient lipid bilayer provides a simple and elegant solution to the phospholipid's dilemma. Look at this. In this spherical vesicle, all of the hydrophobic parts are away from water, in the middle of the structure. At the same time, all of the hydrophilic ends wind up on the outside of the cell facing the wet external environment, or on the inside facing the water-filled interior of the vesicle. This is the exact kind of arrangement used by virtually all cells. The cell membrane holds the cell together while it separates the inside from the outside.

The spontaneous self-assembly of a lipid bilayer from lots of individual molecules is almost magical. University of California chemist David Deamer, whom we'll soon meet as a major player in this game, puts it this way, "The self-assembly process seems to defy our intuitive expectation from the laws of physics that everything on average becomes more disordered."

Modern life has gotten very skillful at this task of separating cellular insides from outsides. What about the prebiotic soup? Could a similar self-organizing process have arisen spontaneously, before life? Once again, laboratory experiments provide answers. Some of the most amazing experiments in prebiotic self-organization have been done by the Italian biochemist Pier Luigi Luisi, working in Switzerland. He is a pioneer in this field and has spent decades studying lipid self-organization.

The bottom line is that Luisi and his colleagues are able to form vesicles with ease. What's more, they have shown that, under the right conditions, these structures can grow larger by gradually incorporating new lipid molecules from solution. In some instances these vesicles display the ability to trigger the formation of more vesicles; they are autocatalytic. Even more significant, under just the right circumstances—at an adequate concentration of lipids and the correct acidity—Luisi's vesicles have the ability to split into two. This is an awful lot like a kind of self-replication.

After observing these intriguing, life-like behaviors—self-assembly, growth, and division—Luisi proposed the "Lipid World" scenario for life's origins. By the way, this is one of many examples of calling an origins theory a "world," such as "RNA world," or "thioester world." I'll be telling you about a lot of these "worlds" during the next few lectures.

Anyway, in Luisi's appealing Lipid World model, prebiotic lipids were synthesized abundantly on the early Earth and in space. These lipids naturally entered the primordial soup, where they self-organized into cell-like vesicles. From time to time, those vesicles captured some kind of primitive genetic molecule, perhaps a strand of RNA that could copy itself.

So, if a self-replicating molecule was incorporated into a self-replicating vesicle, one can imagine an object that has the essential properties of life.

The Swiss team set its sights on encapsulating this kind of self-replicating RNA into self-replicating vesicles. If successful, they claim, they will have created the first synthetic life-form. This advance hasn't happened yet, but the chemical pieces are certainly close to falling into place.

If the Lipid World hypothesis is part of the origins story then, naturally, there had to be a primordial source of lipids. But it turns out that lipids are one group of molecules that are not well represented in the Miller-Urey experiment. So is there another possible source of lipids? The answer is an emphatic yes.

In the past half century of origins research, there have been a few experiments that are widely acknowledged as classics. Of course, Louis Pasteur's refutation of spontaneous generation is one. Stanley Miller's electric-spark synthesis experiments also have achieved that status. These results have changed the way we think about the origins of life. I think it is also safe to say that a similar high regard has been accorded to David Deamer's discovery of lipid self-organization in molecules from space.

David Deamer has been an extremely popular professor in the University of California system for three decades. His specialty is biochemistry, though he teaches a wide range of courses, from molecular biology to bioethics. One of his hallmarks is a wonderfully gentle and reassuring voice, and a low-key lecture delivery. He speaks a lot like Mr. Rogers, if you remember that children's show.

Anyway, Deamer became interested in the origins of life in 1975. That's when he took a sabbatical leave from the University of California at Davis and went to Cambridge, England to study lipids with Alec Bangham. Recall that Bangham is the researcher who first identified vesicles. Together, Deamer and Bangham found that the size and stability of vesicles depends on a whole variety of factors, including the size and shape of the dissolved lipid molecules. So big molecules, for example, form vesicles at lower concentrations than small ones. In the course of this research it occurred to them that vesicles might have provided an ideal sheltering environment for the first life. So, the natural question to ask was whether lipids existed in the early oceans.

When Deamer returned to California, he resolved to continue this line of research. He realized that one possible source of prebiotic lipids was carbon-rich meteorites, so he began a systematic study of molecules extracted from the carbonaceous Murchison meteorite.

The Murchison meteorite is arguably the most celebrated of all carbon-rich meteorites, largely because of the nature of its discovery. The meteorite was actually observed to fall to Earth in a cow field outside the small town of Murchison, which is about 100 miles north of Melbourne, Australia. That was on September 28, 1969. Meteorites hit Earth all the time, but the Murchison fall was unusual in several respects. For one important aspect, it was quite big, several kilograms of rock all told. Another unusual aspect was that it was relatively uncontaminated. In fact, several large pieces were collected while they were still warm from the fall. Added to these factors, the Murchison contained a whopping 3.5 percent by weight of organic molecules, making it a rare type of meteorite known as carbonaceous chondrite. The organic molecules in the meteorite, including amino acids, formed a black, resinous substance, a residue that must have formed billions of years ago in dense molecular clouds and the protoplanetary disk that predated our Solar System. In short, the Murchison meteorite held a wealth of exactly those molecules that must have accumulated on the prebiotic Earth.

Well, as you can imagine, ever since it crashed to Earth in 1969, origin-of-life scientists around the world wanted to study the prize. In the mid 1980's, Deamer received a precious Murchison fragment that weighed about 90 grams, roughly the size of a walnut. Right from the start, Deamer was focused on any possible lipids that might be in the rock. Maybe, he thought, the carbonaceous meteorites provided the essential raw materials for life's membranes.

Let me describe the experiments that Deamer and his colleagues ran. The strategy was to run the rock sample through a series of chemical steps that would break apart the dense black meteorite mass into a series of chemically distinct fractions. Those steps were designed to imitate the kind of exposure to weathering that must have occurred on the early Earth. So, whatever molecules they extracted from the Murchison meteorite were very likely to have occurred in abundance on the prebiotic Earth, as well.

The first step was to grind the rock into a very fine black powder, which contained both organic molecules and lots of rock fragments. They simply used a mortar and pestle to grind the rock while it was submerged in a mixture of three different liquids: water, alcohol, and chloroform. The reason they used three solvents was clever. First, these solvents don't affect the crystalline minerals that form the bulk of the Murchison at all. The mineral fragments just accumulate at the bottom of the mortar. However, each of these solvents does dissolve different suites of interesting organic molecules. Amino acids and sugars concentrate in the water-plus-alcohol fraction, whereas lipids concentrate in the chloroform. Well, after several minutes of grinding, Deamer poured everything into a special thick-walled test tube. He placed the test tube into a tabletop centrifuge and let it spin.

The idea of the centrifuge is simply to increase gravity. That speeds up the separation of chemicals according to their density. In this case, the centrifuge separates the solution into three distinct layers. All the pulverized rock fragments settled to the bottom of the tube because they were the densest fraction. These mineral separates were set aside just in case more studies were required. The top layer in the test tube consisted of the water-alcohol solution, which had dissolved and concentrated amino acids, sugars and a variety of other water-soluble organic species. Deamer also set aside this fraction. The real treasure lay in the middle layer of chloroform, which is an ideal solvent for any lipids the meteorite might have held. Indeed, the chloroform had dissolved more than a tenth of a percent of the meteorite's mass. That's a remarkably high concentration of tantalizing organic species. But what are they?

To find out, Deamer used chromatography to further separate his compounds. He followed the same protocols that Stanley Miller had used decades earlier to separate his amino acids. He evaporated a portion of the chloroform sample until it was concentrated down to a yellowish-brown solution. Deamer placed a small droplet of this concentrate on the corner of a 4- by 4-inch glass plate that was coated with a soft, porous, white powder. Remember how we do this? Use two different solvents in a glass tray. First he used ether, which is a colorless, strong-smelling solvent. So, ether stretched the dried dot into a streak. Then he rotated the glass plate a quarter turn and used chloroform to spread the streak into a distinctive two-dimensional array of compounds.

That dried glass plate was pretty dull in daylight. All you could see was the original brownish spot and a few faint yellowish areas nearby. But many organic molecules that are invisible in daylight fluoresce brightly under an ultraviolet "black light." So when Deamer darkened the room and turned on his black light, he saw this colorful display in a broad arc.

In this, Deamer saw several different distinct fluorescent regions. Each of these regions had a different, unknown suite of ancient cosmic organic molecules. Deamer's next step was to outline each of these areas by scratching the soft, powdery, white surface. Then he scraped off the powder from each of those areas into different test tubes. Now, because lipid molecules are soluble in chloroform, he just washed each powdered sample in chloroform. The chloroform thus reconcentrated the lipids. That is all that it took to recover the different suites of Murchison molecules and any one of those fractions might contain lipid molecules that form vesicles.

I think you can imagine Dave Deamer's anticipation. One by one he placed the chemical fractions into water and watched to see if anything interesting happened. Then he got to the molecules that had been concentrated in an elongated area on the chromatography plate. Deamer watched the familiar magical transformation. The invisible molecules spontaneously arranged themselves into tiny spheres, no more than a hundredth of an inch across. These weren't just little droplets of oil or fat floating in water. Each of these little structures had an inside and an outside. The molecules had self-organized themselves into lipid bilayers, just like a cell membrane. Dave Deamer watched emergence at work. He knew that the exact same kind of process must have occurred in the primitive oceans, billions of years ago.

This result, published in 1989, was a breakthrough moment for origins researchers. Deamer had discovered that these ancient lipid molecules, which were synthesized at some unknowable distant place in space and then delivered to Earth, spontaneously form tiny enclosed structures, and those vesicles are remarkably similar to the membranes that encase living cells today. David Deamer's great discovery taught us that one of life's most basic requirements—the separation of the inside from the outside—appears to be an integral part of the fabric of the universe.

People admire Deamer's Murchison experiments because they were conceptually simple and beautifully executed. So it is not surprising that when new lipid-rich samples came along, Dave Deamer repeated the process using those new samples.

For example, in Lecture Fourteen, I told you about experiments simulating deep space by Lou Allamandola and his NASA Ames research team. Recall that they synthesized a variety of organic molecules under conditions of ultra-cold vacuum, with ultraviolet radiation. Now, some of their experiments produced a significant amount of yellow, oily residue, just like the Murchison meteorite. The highest yields occurred when they irradiated ice made principally of two components, water and alcohol, with a bit of ammonia and carbon monoxide as well. That recipe produced a residue of fluorescent material. A lot of that material was just multi-ringed hydrocarbons charred by heat. These are the PAHs, or polycyclic molecules, that we saw in Lecture Four that initially raised such excitement about the possibility of life on Mars. But there were other molecules, as well, and they might have an amphiphilic character. So Allamandola went to Dave Deamer's lab to check for amphiphiles.

It took Deamer less than a day to verify what the NASA researchers had suspected. He used thin-layer chromatography to isolate the most promising fraction of fluorescent molecules and, bingo, beautiful vesicles appeared spontaneously in water. This discovery actually made bigger headlines than the original 1989 article on the Murchison meteorite. NASA has a very active public relations group and the press is primed for discoveries about the origins of life. So a colorful photograph of the tiny spheres appeared on the front page of the *Washington Post* with the headline, "IN SPACE; CLUES TO THE SEEDS OF LIFE" That was on January 30, 2001.

Unlike the 1996 headlines about life on Mars, self-organizing lipids have been generally accepted. The implications are clear. Even before the formation of planets and even before the Miller-Urey process got going, essential raw materials for life were abundant in the deep freeze of space. The universe is littered with lipid molecules that are poised to organize spontaneously into cell-like structures.

Here's a second example of Deamer's meteorite experiments, again applied to something other than a meteorite. I personally had the chance to do this kind of experiment with Dave Deamer, following a conversation in April of 2000, at one of the first NASA astrobiology meetings. In Lecture Twelve, I told you about how our group at the Carnegie Institution began studying the high pressure and high-temperature reactions of pyruvate. That was way back in 1996. Well, that work had been sitting on hold for years. It wasn't that the results were uninteresting. We knew that we'd made a lot of interesting organic molecules by heating and squeezing pyruvate. But other than our messy chromatograms of humpane, we didn't have a lot of information on the specific compounds. In addition, the relevance of these experiments to life's origins wasn't really clear.

Deamer's work pointed to another angle of interest. Perhaps the yellow, oily goo that oozed out of our gold capsules included a fraction of self-organizing molecules. That result would be interesting because we had started with pyruvate, one of the core metabolic molecules. Perhaps we could show a simple chemical pathway from primitive metabolism to membranes.

In early 2001, I made a new batch of pyruvate-plus-water capsules. I ran them at our standard conditions of two hours at 2,000 atmospheres and 250°C. Then I flew out West, carrying the unopened capsules to Dave Deamer's specially equipped laboratory on the beautiful campus at UC Santa Cruz.

I opened the capsules and they responded with the familiar *bang* and the intense, oily foaming, after which we went through the step-by-step process of chloroform extraction, concentration, and preparation of a 10- x 10-inch glass plate for chromatography. I had read his 1989 paper several times, so it was fun to duplicate that work with my own samples under his expert supervision.

There are these special moments in an experiment when you're about to find out a big result or if an idea actually worked. That's the way I felt as the lights went out and Dave Deamer turned on his ultraviolet fluorescent lamp. I must tell you, the results were a gorgeous pattern of yellow, blue and purple, with several distinctly fluorescing areas. It was amazing how similar this plate was to the Murchison sample. Dave assumed that this correspondence wasn't just a coincidence, so we zeroed in on a blue fluorescing area that closely matched the area of his Murchison plate that produced the most vesicles.

So we followed all of those procedures from the late 1980s, and the result was fantastic. In this case we had made self-organizing bilayer membranes, starting from nothing more than pyruvate and water. I should keep this result in perspective. The pyruvate experiment was, at best, a footnote to the self-organization of lipids from the Murchison meteorite and the deep space simulations by the NASA Ames team. If nothing else, the pyruvate experiment underscores what seems to be the inevitable emergence of self-organizing molecular systems in the path to life.

There are all sorts of puzzles associated with the formation of lipid vesicles. For one thing, recent work by Deamer's group shows that lipid self-organization is inhibited by the presence of dissolved calcium and magnesium. But these salts are inevitable ingredients of seawater. Does that mean that life can only begin in fresh water? Or were there some, as yet, unidentified varieties of lipid molecules that do self-organize in salt water? Another thing, many biologists have been quick to point out that lipid vesicles are a far cry from actual cell membranes. After all, a living cell has to have ways to bring food in and get waste out. That's done in modern cells by an amazing array of so-called "protein receptors" that regulate the flow of molecules and chemical energy into and out of the cell. What served as receptors before proteins?

You can be sure that such questions will occupy researchers for many decades to come and there will probably be more headlines about lipids in your local newspaper. Anyway, for now, the emergence of cell-like vesicles that form spontaneously from simple molecules is one of the best-understood features of life's origins.

One of the things I like most about origin-of-life research is that clever new ideas keep popping up and ideas about the emergence of self-organized molecular systems are no exception. One of the most intriguing recent proposals comes from chemist Christopher Dobson of Oxford University and his collaborators at laboratories of the National Oceanic and Atmospheric Administration in Boulder, Colorado. In 2000, they proposed a fascinating hypothesis on how lipids might organize themselves in the atmosphere as the result of ocean spray.

Here's the crux of Dobson's ideas: Lots of organic molecules could have accumulated at the ocean's surface like an oil slick. That's especially true for lipid molecules like the ones Dave Deamer isolated from the Murchison meteorite. Four billion years ago, wind would have kicked up whitecaps and waves would have crashed onto the shore, just as today. Such turbulence would have produced a continuous fine mist of aerosol particles. Countless trillions of tiny droplets, many smaller than a thousandth of an inch across, would have been propelled from the oily surface into the atmosphere.

Here's the trick. Each water droplet would have contained amphiphile molecules that quickly formed a layer around the water-filled interior. Those amphiphiles would have self-assembled into a *single* layer, not a double layer, because there's only water on the inside. Larger droplets were heavy enough to fall back into the sea, but Dobson's calculations show that smaller aerosol particles are quite stable and could have remained suspended in the atmosphere for months or even years. They just ride wind currents like microscopic hang gliders. They can even reach high into the stratosphere.

Dobson and his coworkers speculate that the lipid molecules in each aerosol particle form a sphere, with the hydrophobic ends facing the outside atmosphere and the hydrophilic ends facing the water-filled interior. What's more, many of these aerosol particles must have incorporated lots of reactive, water-soluble organic molecules. As the spheres ascended to the highest levels of the atmosphere, those reactive molecules might have undergone additional chemical reactions in sunlight.

Think about the vast number of different origin experiments represented by these aerosol particles. Every one would have had weeks or months to undergo these kinds of energetic chemical reactions. Each particle would have been, in effect, a tiny chemical experiment. This process went on for hundreds of millions of years, as astronomical numbers of aerosols drifted into the skies. Who knows what emergent steps might have occurred?

Now, consider what happened when an aerosol particle finally descended to the ocean. While in the air, each aerosol particle had a hydrophobic exterior. However, on contact with water, more lipid molecules at the ocean's surface would have spontaneously coated the particle. So the lipids wind up forming a familiar bilayer structure. So now we have countless trillions of tiny cell-like objects in the primitive ocean.

Christopher Dobson summarizes the significance of the model in this way, "Organic aerosols offer more than freedom from the tyranny of the tidal pool or Darwin's 'warm little pond.' They offer a possible mechanism for the precursory production and the subsequent evolution of populations of cells."

To summarize this lecture, we've seen several remarkable processes by which lipid molecules form and organize themselves into tiny cell-like spheres. We have the windblown aerosols of Dobson and the water-bound vesicles of Deamer. In either case, lipid self-organization seems to have been an essential step in the emergence of life, a membrane, for the first time, isolated the insides from the outsides of protolife.

But a membrane, by itself, is not life. Life also required a variety of essential biomolecules, including proteins, carbohydrates and nucleic acids. All those molecules had to be assembled from the prebiotic soup. But, as we've seen, all of the building blocks of these macromolecules, including the amino acids, the sugars and the bases, are water soluble and these molecules just cannot self-assemble in water.

So, what to do? As we'll see in the next lecture, the best way to assemble life's molecules in water is to call in the rocks.

Lecture Seventeen

Life on Clay, Clay as Life

Scope: Many earth scientists like the idea that clays played a central role in the origin of life. There is a satisfying unity in integrating the solid, liquid, and gaseous parts of our planet to make life. The behavior of clays is fascinating, but we don't know what's going on at the atomic scale. And if we don't understand the chemistry behind our experiments, it is harder to say whether they're relevant to the beginnings of life or not.

Outline

I. Rocks and minerals provide attractive surfaces for the concentration and assembly of macromolecules in the emergence of life. Rocks can act as a kind of scaffolding for the organization of proto-life. Several geologists first proposed this idea more than half-a-century ago, even before the Miller-Urey experiments.

II. Let's look at some of the ways that minerals might have played a role in selecting and organizing biomolecules. We'll focus on the ubiquitous group of minerals called *clays*.

- **A.** One of the earliest advocates of mineral-assisted life was British biophysicist John Desmond Bernal (1901–1971). Bernal's fascination with clay minerals lies in their crystal structure.
  1. All clay minerals feature sheets of atoms arranged like a sandwich.
  2. The outer sheet in every clay is made of silicon and oxygen. Four oxygen atoms surround each silicon atom to form a tetrahedron.
  3. Then the layer is made up of interconnected rings of six tetrahedra. Think of this silicate tetrahedral sheet as one piece of bread in the sandwich.
  4. The silicate sheet is linked to a second layer with magnesium, aluminum, iron, and other elements.
  5. Six oxygen atoms surround each of these atoms to form an octahedron. A number of octahedra link edge to edge to form a layer. Think of the octahedral sheet as the meat of the sandwich.

6. Some clays, including the *kaolinite group*, have a structure similar to a stack of open-faced sandwiches. A unit of one tetrahedral sheet joined to one octahedral sheet is the basic layer, and billions of these layers stack on top of each other.
7. Other clays, including talc and montmorillonite, are like closed-face sandwiches. These minerals have a unit of one octahedral sheet sandwiched between two tetrahedral sheets to form a three-layer unit. Again, billions of these units stack on top of each other.

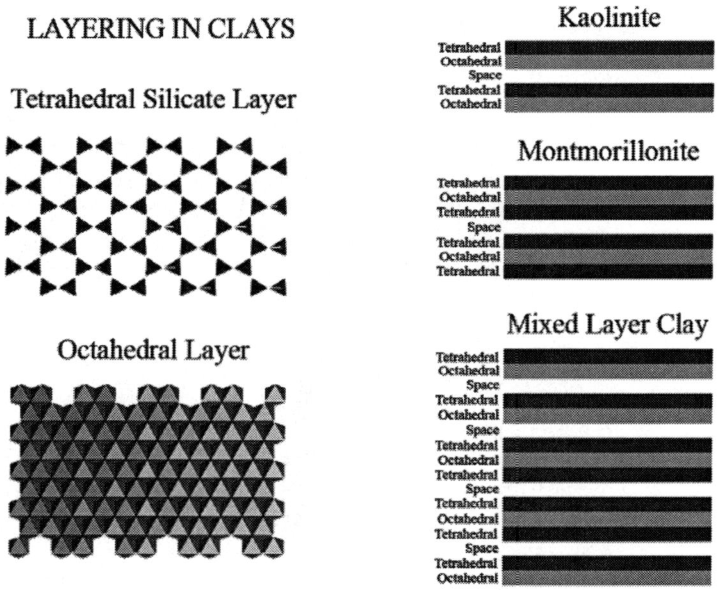

B. The layered structure of clay is the key to many of its distinctive properties. The layer units themselves are quite strong, but the forces between the layers, in those spaces, are extremely weak. That's why clays are so soft, smooth, and slippery when wet.

C. The weak bonding between layers allows foreign atoms and molecules to enter and concentrate in the regions between layers. These impurity molecules, including water and all sorts of small organic molecules, can react with each other to form larger molecules.

D. Yet another advantage to clays is that they often have an electrically charged surface. This kind of surface has the ability to attract and bind molecules out of solution—just the kind of selection and concentration we need for life's origin.

III. Bernal and others recognized that clays could have provided the ideal solid surface for the origins of life. Clays are the products of weathering rocks and are everywhere on Earth's surface. They're exceptionally fine grained and, thus, have an immense surface area. A small pile of ordinary clay can have a reactive surface area greater than 1,000 square feet. What's more, these charged, chemically active surfaces attract interesting biomolecules and can even catalyze their reactions.

A. Subsequent experiments have tended to bolster Bernal's idea that clay can host biomolecules. In 1978, Israeli biochemist Noam Lahav and his colleagues discovered that amino acids concentrate and polymerize on clay surfaces to form small, protein-like molecules. They induced such reactions by repeatedly wetting clays with a solution containing amino acids, then evaporating the solution—similar to what might happen in a shallow pond or tidal pool that dries up in the Sun.

B. Since the mid-1980s, a NASA-sponsored research team at Rensselaer Polytechnic Institute led by James Ferris has demonstrated that clay minerals can adsorb and assemble organic molecules. Clays have also been shown to act as scaffolds in the formation of RNA, the polymer that carries the genetic message and mediates protein synthesis.

1. Ferris's team prepared a solution of RNA nucleotides, each consisting of a ribose sugar bonded to a phosphate and a base.
2. Ferris then "activated" the nucleotides by attaching a reactive, nitrogen-bearing molecule called *imidazole* that promotes bonding between the nucleotides.
3. Adding a small amount of a suitable clay mineral, such as montmorillonite, into this solution causes the RNA pieces to link up. In a matter of hours, the individual pieces combine into strands as long as 10 nucleotides. In two weeks, this simple process yields RNA strands more than 50 nucleotides in length.

C. The dramatic success of James Ferris has inspired other researchers to try similar experiments to make other biopolymers. Leslie Orgel at the Salk Institute for Biological Studies formed a variety of protein-like chains of amino acids, some up to several dozen molecules long.

1. Orgel and his students found that different kinds of minerals preferentially select and polymerize different kinds of molecules from solution. They learned to form a variety of polymers by selecting the right combination of mineral and molecule.
2. These experiments also led Orgel to develop a more general theory of how minerals might have promoted polymerization on the ancient Earth—what he calls *polymerization on the rocks*.
3. In a series of publications, Orgel pointed out the advantages and problems of mineral surfaces. Layer-structured minerals, such as clays, certainly can adsorb interesting biomolecules, including the RNA nucleotides and amino acids essential to life. Orgel also showed how these surfaces promote bonding between two of these molecules once they're adsorbed close to each other.
4. However, there is a problem. First one molecule sticks to the surface, then another and another. They begin to form bonds between themselves—to make a polymer. However, as more of these molecules are added to the elongating chain, the polymer becomes more strongly bound to the mineral surface. It is hard to see how a polymer can contribute to life if it is stuck tight to a rock.
5. One possible solution is to incorporate tiny bits of clay in the first cells. This idea gained credibility after experiments at Harvard University in the laboratory of geneticist Jack Szostak.
6. In an early experiment, Szostak's students mixed together finely powdered clays, RNA nucleotides, and lipids in the same solution. The clays not only adsorbed RNA, as in the Ferris experiments, but they also triggered the formation of lipid vesicles. The vesicles formed around the RNA-decorated clay particles. The cell-like membranes emerged by the self-

organization of lipids, and in the process, the membrane encapsulated RNA.

IV. The *clay life theory* of Scottish geologist Graham Cairns-Smith has captured the imagination of people both inside and outside the world of science.

A. Graham Cairns-Smith proposes that fine-grained crystals of clay might have been the first life-forms on Earth. Virtually all other authorities assume that life emerged by a gradual and continuous chemical process that relied exclusively on carbon-based chemical compounds.

1. Today's biochemistry, according to most experts in the field, is a highly evolved successor of what must have been ancient carbon geochemistry. The operative word here is *carbon.* All life is, and always has been, based on the element carbon.
2. Cairns-Smith disagrees. He says, "… the unity of biochemistry is of no direct help. Evolution did not start with the organic molecules that have now become universal to life: indeed I doubt whether the first organisms…had any organic molecules in them at all."
3. That's an extraordinary statement, but he points to more than five decades of failed attempts to synthesize some of the most fundamental macromolecules, including RNA and proteins. No plausible prebiotic scenario leads to those essential components of life. In Cairns-Smith's view, nucleic acids and proteins are too complicated to build without help from the world of minerals.

B. Cairns-Smith's popular book on the clay life hypothesis is entitled *Seven Clues to the Origin of Life* (1985). Let's look at his hypothesis.

1. Cairns-Smith notes that nature may have used some sort of scaffolding in constructing life. The parts that now lean together surely used to lean on something else—something low tech. Cairns-Smith proposes that the scaffolding was a clay mineral.
2. He lists the many advantages of clays: their ubiquity, their negative surface charge, their affinity for organic molecules, and their huge chemically reactive surface area.

3. Cairns-Smith also assumes that clays selected and concentrated organic molecules from their environment.
4. He adds a twist that reveals a great deal of mineralogical sophistication: We know that clay crystal layers contain a variety of internal variations that have at least the character of an information storage system. The layered structure of clay can carry—and, possibly, pass on—something similar to "genetic information" in the form of a reproducible sequence of layer orientations, of crystal defects, or of variable chemical compositions.
5. If they can grow and pass on this information from one generation to the next, much the way that modern genes do, then might not clays be thought of as living?
6. Genetic information in all modern cells is stored in DNA and RNA molecules. These polymers are constructed from sequences of four nucleotides, which provide, in effect, a four-letter genetic alphabet. With a sequence of four different letters, any conceivable message can be conveyed.
7. The alphabet of clay minerals may lie in one of three possibilities related to the orientation of rows of atoms, which occur in every layer. The first possibility has to do with the orientation of adjacent sandwich-like layers. These layers occur in stacks, like a stack of paper. Each of these layers can adopt any one of three different orientations relative to its neighbors—at 0°, 120°, or 240°. The sequence is analogous to a three-letter alphabet, with A, B, and C representing the three different orientations.

LAYER ORIENTATION IN CLAYS

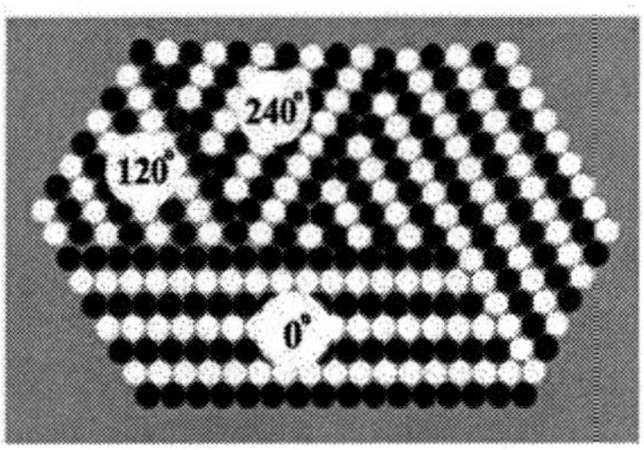

8. A second kind of information can be stored in the form of a stack of different kinds of layers. It is possible, for example, to

mix two- and three-layer sandwiches at random in the same clay. In this case, the information is stored in a sequence of 2s and 3s.

9. In addition to variations *across* layers—whether in the sequence of angle orientations or double versus triple layers—variations in clay can also occur *within* any given layer. Each sandwich layer can display a feature called *twinning*, which means that any given layer can incorporate small regions in all three orientations.
10. What's more, many clays have extremely complex chemical compositions, especially in their octahedral sheets. Thus, varying mixtures of aluminum, magnesium, iron, and other elements can be present, as well as vacancies where certain atoms are missing.
11. The surface of a clay particle contains an octahedral site about every 10-billionths of an inch. These sites display a more or less random sequence of iron, aluminum, magnesium, and vacancies—yet another way that clays might carry information.

V. Clay minerals certainly have the potential to store information in a variety of ways, but how does that translate into life-like behavior?

A. A clay would have to accomplish two critical tasks to be considered alive. First, it would have to be able to make copies of the information.

1. Cairns-Smith suggests that if certain sequences of layers grow from a seed crystal, then simple cleaving of flakes would be an adequate copying mechanism. He speculates that clay crystals grow in such a way as to copy any given arrangement of layers, or of elements within layers, over and over again, then flake apart in what amounts to an act of self-replication.
2. In this scenario, identical clay particles crystallize repeatedly. Particularly stable sequences of layers or elements ultimately win out in a mineralogical version of evolution by natural selection. That seems possible, though no experiments have verified the idea.

B. The second step is that at some point in history, the clay scaffolding had to pass its information on to a purely carbon-based life-form.

1. According to Cairns-Smith's model, different exposed layers or elements at clay surfaces might have affinities for different kinds of organic molecules. A given clay layer sequence, therefore, would tend to attract a given pattern of organic molecules. In this way, a self-replicating clay could have acted as scaffolding for the organic world. Cairns-Smith calls this process *genetic takeover*.
2. The result is a macromolecule like a protein or RNA strand that might possess exactly the sort of information-rich structure needed to jumpstart life.

VI. One potentially testable feature of the clay-life hypothesis is the central requirement that clays evolve.

A. In a provocative 1988 article, Cairns-Smith proposed that different clay sequences may be more or less stable in a specific environment. These sequences will, in effect, compete for the limited resources of more atoms as the clays continuously form and dissolve.

B. Perhaps clays can evolve by selective copying of favorable patterns, combined with the selective dissolution of unfavorable patterns. In this model, the evolution of clays is closely analogous to that of cells.

1. Research in a similar process is done frequently in studies of microbial evolution, but that work depends on genetic equipment to collect, refine, and sequence DNA. No such technology exists to sequence clays.
2. Do sequences replicate accurately enough through growth? Underlying that question is the technical barrier of determining the exact elemental or layer sequence of a clay particle. For clays, no such sequencing methods exist.
3. Characterizing the layer arrangement in a *single* clay particle is beyond any current technology; thus, measuring the sequence of the thousands of separate clay particles necessary for an adequate population survey is inconceivable, at least for the time being.

VII. The behavior of clays is fascinating: It often seems that chemists just sprinkle in a little "magic" clay powder to make an experiment work. Nucleotides do nothing in a beaker until clay is sprinkled in, then polymerization takes place to form RNA. Lipids do nothing until clay is added, then suddenly, self-assembly of vesicles takes place.

A. Clays are so variable in their composition, so poorly defined in their surface properties, and so fine-grained that it is virtually impossible to be sure what molecule is binding to what surface in these experiments.

B. Mineralogists have another approach that uses large, faceted, single crystals. In the next lecture, we'll explore this alternative approach to the selection and concentration of organic molecules.

Essential Reading:

Cairns-Smith, *Seven Clues to the Origin of Life.*

Hazen, *Genesis: The Scientific Quest for Life's Origin*, chap. 12.

Supplemental Reading:

Fry, *The Emergence of Life on Earth*, chap. 10.

Lahav, *Biogenesis: Theories of Life's Origin*, chap. 20.

Wills and Bada, *The Spark of Life: Darwin and the Primeval Soup*, chap. 4.

Questions to Consider:

1. Is a self-replicating clay mineral that has the ability to evolve and to pass on its distinctive set of crystal defects alive? Why or why not?
2. One of the principal objections to the clay life hypothesis is that clays play no known role in modern organisms. How does Cairns-Smith answer this objection?

Lecture Seventeen—Transcript

Life on Clay, Clay as Life

The origin of life is viewed as a sequence of chemical steps, each of which adds a bit of complexity to the world. As we have seen, we have a pretty good idea about the first steps, which must have been the synthesis of all kinds of small organic molecules.

A bigger hurdle in the transition from non-life to life is the emergence of macromolecules. Even if making the small building-block molecules, such as sugars and amino acids, seems to be reasonably well understood, assembling those small pieces into big functional molecules is much more difficult.

We do know that the production of macromolecules requires two concerted steps. First, the right building-block molecules must be concentrated from the prebiotic soup. Second, these pieces must be organized into the correct structure. Now, in the last lecture, we saw how these two tasks occur virtually simultaneously and spontaneously in the case of lipid vesicles. Lipid molecules in water can separate and self-organize into a bilayer structure. However, that's not the case for other important biological macromolecules, especially carbohydrates and proteins. Water breaks carbohydrates into sugars and proteins into amino acids. Moreover, sugars and amino acids are soluble in water. So in water, carbohydrates and proteins tend to break down, not self-organize.

What to do? In the last two lectures, I talked about the role of surfaces. One very promising way to assemble molecules from a dilute solution is to select and concentrate them on a surface. For most of the last 50 years the prevailing view has been that life's molecules assembled right at the ocean-atmosphere interface. So you can picture the surface of a calm tidal pool, or perhaps a primitive oil slick on the ocean itself, where molecules accumulate. Then, there's also the problem of lightning storms and ultraviolet radiation, which can disintegrate useful molecules.

That's why many origins scientists have turned to rocks and minerals, which provide all sorts of attractive surfaces for the concentration and assembly of macromolecules. Rocks can act as scaffolding for the organization of protolife. Several geologists first proposed this idea more than half a century ago, even before the Miller-Urey experiments. Minerals have always been lurking in the wings of origins research, especially when more traditional chemical approaches fail.

Therefore, in this lecture we're going to look at some of the ways that minerals might have played a role in selecting and organizing biomolecules. In particular, we will focus on the ubiquitous group of minerals called clays.

One of the earliest advocates of mineral-assisted life was not a geologist, but rather British biophysicist John Desmond Bernal, who lived from 1901 to 1971. Clays have regularly layered atomic structures and they are everywhere. Pick up any handful of soil or mud and much of what you hold is made of clay minerals. If you do any gardening, then you'll know how much more nutrient-rich clay soils can be compared to sandy soils.

The reason for Bernal's fascination with clay minerals lies in their crystal structure. So, let us spend a minute or two considering the beautiful crystal structures of clays. All clay minerals feature regular sheets of atoms that are arranged something like a sandwich. The outer sheet in all clay is made of silicon and oxygen. Four oxygen atoms surround each silicon atom to form a tetrahedron. Then the layer is made up of interconnected rings like this, of six tetrahedra. You can think of this silicate tetrahedral sheet as one piece of bread in the sandwich.

The silicate sheet is always linked to a second layer with magnesium, aluminum, iron and other elements. Six oxygen atoms surround each of these atoms to form an octahedron, and lots of octahedra link, edge to edge, to form a layer. So you can think of the octahedral sheet as the meat of the sandwich.

Some clays, including the so-called "kaolinite group," have a structure that is a lot like a stack of open-faced sandwiches. A unit of one tetrahedral sheet joined to one octahedral sheet is the basic layer and countless billions of these layers stack on top of each other. So, we have a tetrahedral sheet, then an octahedral sheet, then a space and then repeat that sequence.

Other clays, including talc and montmorillonite, are like closed-face sandwiches. These minerals have one octahedral sheet sandwiched between two tetrahedral sheets to form a three-layer unit. Again, countless billions of these units stack on top of each other. So, we have a tetrahedral sheet, octahedral sheet, tetrahedral sheet, then a space, and that sequence repeats.

The layered structure of clay is the key to many of its distinctive properties. For one thing, the layer units, themselves, are quite strong but the forces between the layers, in those spaces, are extremely weak. That is why clays are so soft, smooth, and slippery when wet. Think about a potter shaping clay on a wheel. The layers just glide over each other. Talcum powder, or baby powder, displays the same kind of smoothness—all the result of atomic layers sliding past each other.

The weak bonding between layers also allows lots of foreign atoms and molecules to enter and concentrate in the regions between layers. These impurity molecules, including water and all sorts of small organic molecules, can react with each other to form larger molecules.

Yet another advantage to clays is that they often have an electrically charged surface. This kind of surface has the ability to attract and bind molecules out of solution—just the kind of selection and concentration we need for life's origins.

Bernal, and others, quickly recognized that clays could have provided the ideal solid surface for the origins of life. Clays are the products of weathering rocks, so they're everywhere on Earth's surface. They're exceptionally fine grained and have an immense surface area. In fact, a small pile of ordinary clay, a pile that would fit in the palm of your hand, can have a reactive surface area greater than 1,000 square feet. What's more, these charged, chemically active surfaces attract interesting biomolecules and can even catalyze their reactions.

Subsequent experiments have tended to bolster Bernal's ideas that clay can host biomolecules. One pioneering study took place in 1978, when Israeli biochemist Noam Lahav and his colleagues discovered that amino acids concentrate and polymerize on clay surfaces to form small, protein-like molecules. They induced such reactions by repeatedly wetting clays with a solution containing amino acids and then evaporating that solution. That's a situation similar to what might happen in a shallow pond or tidal pool that dries up in the sun. I find this experiment especially appealing because it relies on cycles of heating and evaporation. Recall, from Lecture Eight, that cycles of energy are one of the key factors in the emergence of complexity. Lahav and his colleagues proposed that patterns of daily and seasonal changes resulted in the emergence of new molecular structures.

Clays have continued to be important in origins-of-life research. Probably the most newsworthy efforts have been by a NASA-sponsored research team at Rensselaer Polytechnic Institute in Troy, New York. Chemistry professor James Ferris leads that group. Since the mid-1980s, Ferris and coworkers have demonstrated that clay minerals can adsorb and assemble a variety of organic molecules. Of special note, clays have been shown to act as scaffolds in the formation of RNA. That's the polymer that carries the genetic message and that mediates protein synthesis.

The procedures used by Ferris were simple and elegant. The first step was to prepare a solution of RNA nucleotide. Recall that nucleotides combine to form RNA, each nucleotide consisting of a ribose sugar bonded to a phosphate, and a base. Ferris then "activated" his nucleotides by attaching a reactive, nitrogen-bearing molecule called imidazole, which promotes, or "activates", bonding between the nucleotides.

If you leave such a solution sitting on a lab bench it will do almost nothing, even after weeks. However, if you sprinkle in a small amount of a suitable clay mineral such as montmorillonite, then the RNA pieces start to link up. In a matter of a few hours, the individual pieces combine into strands as long as 10 nucleotides. If you wait a full two weeks, this simple process yields RNA strands more than 50 nucleotides in length. No one knows exactly how the process happens but somehow the fine-grained clay particles cause the RNA nucleotides to polymerize.

The dramatic success of James Ferris has inspired other origin-of-life researchers to try similar kinds of experiments to make other biopolymers. Leslie Orgel, who is a research professor at the Salk Institute for Biological Studies in San Diego and a pioneer in origins research in his own right, was able to form a variety of protein-like chains of amino acids, some of which were up to several dozen molecules long. Orgel and his students found that different kinds of minerals preferentially select and polymerize different kinds of molecules from solution. They learned to form a variety of different polymers by selecting the right combination of mineral and molecule.

These experiments also led Leslie Orgel to develop a more general theory of how minerals might have promoted polymerization on the ancient Earth—what he calls, "polymerization on the rocks." In a series of publications, he pointed out both the advantages and the problems with mineral surfaces. He reviewed how layer-structure minerals like clays certainly have the ability to adsorb interesting biomolecules, including the RNA nucleotides and amino acids so essential to life. He also showed how these surfaces promote bonding between two of these molecules once they're adsorbed close to each other.

But there's a problem. Think about it this way. First one molecule sticks to the surface, then another, and another. They begin to form bonds between themselves, to make a polymer. However, as more and more of these molecules are added to the elongating chain, the polymer becomes increasingly strongly bound to the mineral surface. It is hard to see how a polymer can contribute to life if it's stuck tight to a rock.

One possible solution to this problem is to incorporate tiny bits of clay in the first cells. This idea gained a lot of credibility after experiments at Harvard University, in the laboratory of geneticist Jack Szostak. In an early experiment, Jack's students mixed together finely powdered clays, RNA nucleotides and lipids in the same solution. In this case, the clays did something quite magical. They not only adsorbed RNA, as in the Ferris experiments, but they also triggered the formation of lipid vesicles. The vesicles actually formed around the RNA-decorated clay particles. The cell-like membranes emerged by the self-organization of lipids, and in the process that membrane encapsulates RNA. This experiment is admittedly a long, long way from synthesizing life but it is a tremendous accomplishment to combine both a membrane and a genetic molecule in a single step. I will tell you about Szostak's further work with RNA in Lecture Twenty-Three.

The idea that clay played a central role in the origins of life seems to have a deep resonance –beyond any scientific reasoning. Part of the reason may lie in mythology. The "golem" in Jewish legend, for example, was a magical humanoid creature formed from clay. Somehow clay isn't like other minerals. You can mold clay, shape it and fire it into vessels of almost any form.

I think that's why the so-called "clay-life" theory of Scottish geologist Graham Cairns-Smith has so captured the imagination of people both inside and outside the world of science. Indeed, the theory that I'm about to describe actually plays a role in a novel by John Updike. Let me read you a passage from *Roger's Version*, which Updike wrote in 1986. Updike puts this description in the mouth of a naïve graduate student who is trying, rather unsuccessfully, to grasp the meaning of life:

But I happen to know exactly how life arose; it's brand-new news, at least to the average layman like yourself. Clay. Clay is the answer. Crystal formation in fine clays provided the template, the scaffolding, for the organic compounds and the primitive forms of life. All life did, you see, was take over the phenotype that crystalline clays had evolved on their own.

Well that, in a nutshell, is a pretty good summary of Graham Cairns-Smith's theory of life's origin.

We've seen that quite a number of researchers looked to minerals to help out in the origins of life. Minerals have been proposed as protective containers, as organizing surfaces, as catalysts for prebiotic synthesis and more. Graham Cairns-Smith has carried these ideas quite a bit further. In a landmark 1966 article, and in more than a dozen subsequent books and articles, he proposes that fine-grained crystals of clay might, all by themselves, have been the very first life forms on Earth.

This view is provocative and out of the mainstream, because virtually all other authorities assume that life emerged by a gradual and continuous chemical process that relied exclusively on carbon-based chemical compounds. Today's biochemistry, according to virtually all experts in the field, is consequently a highly evolved successor of what must have been ancient carbon geochemistry. The operative word here is "carbon". All life is, and always has been, based on the element carbon.

As, I think, you've guessed, Cairns-Smith disagrees. He says, "I believe the unity of biochemistry is of no direct help. Evolution did not start with the organic molecules that have now become universal to life. Indeed, I doubt whether the first organisms ... had any organic molecules in them at all."

That is an extraordinary statement, but he points to more than five decades of failed attempts to synthesize some of the most fundamental macromolecules, including RNA and proteins. No plausible prebiotic scenario leads to those essential components of life. In Cairns-Smith's view, nucleic acids and proteins are too complicated to build without help from the world of minerals.

Given the fact that this view flies in the face of conventional wisdom, how is it that anyone takes Graham Cairns-Smith seriously? If he's so out of sync with most people in the field, why does anyone listen? I think the answer to that question reveals something interesting and unusual about the origins-of-life research field. In the first lecture of this series, I talked a bit about the sociology of the origins-of-life community. Before going into the details of Cairns-Smith's hypothesis, therefore, I want to revisit this topic.

At least two factors work in Cairns-Smith's favor. First, origins researchers appreciate creative, original ideas if they're well argued, scientifically plausible and experimentally testable. Cairns-Smith's ideas might at first seem fanciful, but he's a persuasive scientist, a lucid writer, and an eloquent lecturer. He's never too pushy with his ideas, but he has a knack for gentle persuasion. He draws others into his way of thinking.

A second reason for Cairns-Smith's scientific success is his exceptional ability as a writer, coupled with the long and distinguished tradition of the book as a primary medium for proposing theories of life's origin and evolution. Ever since the 1859 publication of Darwin's *Origin of Species*, and indeed long before that, important new ideas about the origins of life have often appeared as books that are accessible to scientists and nonscientists, alike. In Lecture One, I mentioned some of the scientific periodicals that feature origin-of-life research, but I didn't mention the inordinate number of books on the subject.

This reliance on books is not the norm in most areas of science. Perhaps only studies of complexity, including environmental science and consciousness, have shown a somewhat similar reliance on books. Original contributions in particle physics, or genetics, or plate tectonics are almost always in the form of relatively short technical articles in specialized journals. But repeatedly, new ideas in origins research have appeared as book-length works.

Graham Cairns-Smith embraced this tradition. His writings for both academic peers and general audiences avoid jargon. They present simple, vivid illustrations. His best-known popular book on the clay life hypothesis is entitled *Seven Clues to the Origin of Life*. It was published in 1985—which, by the way, was just in time for John Updike to incorporate the ideas in his 1986 novel. It really is a tour de force that's presented like a mystery novel, complete with catchy dialog, false leads and lots of quotations from the Sherlock Holmes stories of Arthur Conan Doyle. Indeed, Holmes's words of wisdom perfectly compliment the speculative scientific tract, "I suspect myself...of coming to conclusions too rapidly," he says. "It is a capital mistake to theorize in advance of the facts." My favorite quote is, "There should be no combination of events for which the wit of man cannot conceive an explanation." Whether you are persuaded or not, you have to admire Cairns-Smith's style.

So, let us take a close look at Cairns-Smith's hypothesis. The crux of the argument rests on a simple analogy. Cairns-Smith compares life's origins to the assembly of a stone archway. The archway is built from a series of carefully fitted blocks, including the crucial central keystone that locks the whole structure in place. However, you cannot make an arch simply by piling one stone on top of the next. The structure would collapse before you got close to fitting in the keystone.

Nature must have resorted to the same kind of strategy that builders have used throughout the ages; a simple support structure aids the construction and is then removed. Here is another quote from his book, "The answer," he says, "is with a scaffolding of some kind. I think this must have been the way our amazingly 'arched' biochemistry was built in the first place. The parts that now lean together surely used to lean on something else, something low tech." What was that low-tech "something"? He proposes it was a clay mineral.

Just as J. D. Bernal had done back in the 1940s, Cairns-Smith lists the many advantages of clay—their ubiquity, their negative surface charge, their affinity for organic molecules and their huge chemically reactive surface area. Like James Ferris, on nucleotides, and Noam Lahav, on amino acids, Cairns-Smith assumes that clays selected and concentrated organic molecules from their environment. This is all pretty familiar territory and broadly accepted.

However, Cairns-Smith embellishes this story with a surprising twist that reveals a great deal of mineralogical sophistication. We know that clay crystal layers contain a variety of internal variations that have at least the character of an information storage system. The layered structure of clay can carry, and possibly pass on, something similar to "genetic information" in the form of a reproducible sequence of layer orientations, of crystal defects, or of variable chemical compositions. If they can grow, and if they can pass on this information from one generation to the next, much the way that modern genes do, then might not clays be thought of as being alive? Now, I like minerals but even the most dedicated mineralogist would tend to balk at Cairns-Smith's concept of "clay life" and "crystal genes." But let's take a closer look at this idea.

Genetic information in all modern cells is stored in DNA and RNA molecules. These polymers are constructed from sequences of four different nucleotides, which provide, in effect, a four-letter genetic alphabet. With a sequence of four different letters, you can convey any conceivable message. In fact, it really only takes two letters, like the zeroes and ones of binary computer code.

So what is the alphabet of clay minerals? It turns out that there are three possibilities that are related to the orientation of rows of atoms, which occur in every layer. Look at this. The first possibility has to do with the orientation of adjacent sandwich-like layers. These layers occur in stacks, like a stack of paper. Each of these layers can adopt any one of three different orientations relative to its neighbors—that's at 0°, or 120°, or 240°. The sequence of clay layer orientations is analogous to a three-letter alphabet, with A, B, and C representing the three different orientations.

A second kind of information can be stored in the form of a stack of different kinds of layers. It's possible, for example, to mix two- and three-layer sandwiches at random in the same clay. In this case, the information is stored in a sequence of twos and threes.

In addition to variations across layers, whether in the sequence of angle orientations or double versus triple layers, variations in clay can also occur within any given layer. Each sandwich layer can display a feature called "twinning", which means that any given layer can incorporate small regions in all three different orientations. What's more, many types of clay have extremely complex chemical compositions, especially in their octahedral sheets. Therefore, you can have varying mixtures of aluminum, magnesium, iron, and other elements, as well as vacancies where the atom is missing. If you could shrink down to the size of an atom and move along the surface of the clay particle, you'd find an octahedral site every 10 billionths of an inch or so, just as regular as clockwork. But these sites would display a more-or-less random sequence of iron, aluminum, magnesium, vacancies, and so forth. So that kind of sequence is yet another way that clays might carry information.

Clay minerals certainly have the potential to store information in a variety of ways. But how does that translate into life-like behavior? Clay would have to accomplish two critical tasks to be considered alive. First, it would have to be able to make copies of the information. Cairns-Smith suggests that if certain sequences of layers grow from a seed crystal, then simple cleaving of flakes would be an adequate copying mechanism. Look at this sequence. He speculates that clay crystals grow in such a way as to copy any given arrangement of layers, or of elements within layers, over and over again, then the flake breaks apart in what amounts to an act of self-replication. In this scenario, identical clay particles crystallize repeatedly. Particularly stable sequences of layers or elements ultimately win out in a mineralogical version of evolution by natural selection. That seems possible, though no experiments have verified the idea.

The second step is that at some point in history, the clay scaffolding had to pass its information on to a purely carbon-based life form. Here is a possibility. According to Cairns-Smith's speculative model, different exposed layers or elements at clay surfaces might have affinities for different kinds of organic molecules. A given clay layer sequence, therefore, would tend to attract a given pattern of organic molecules. In this way, self-replicating clay could have acted as scaffolding for the organic world. Cairns-Smith calls this process "genetic takeover". The result is a macromolecule like a protein or RNA strand that might possess exactly the sort of information-rich structure that you need to get life going.

It's all well and good to propose speculative models for life's origins, but any acceptable scientific theory has to make testable predictions. If a model makes no predictions, then it just isn't useful to science. Graham Cairns-Smith is perfectly aware of this requirement and he's embellished his model with several unambiguous predictions.

One potentially testable feature of the clay-life hypothesis is the central requirement that clays evolve. In 1988 Cairns-Smith explored this point in a provocative article entitled "The Chemistry of Materials for Artificial Darwinian Systems". In this paper he proposed that different clay sequences might be more or less stable in a specific environment. These sequences will, in effect, compete for the limited resources of more atoms as the clays continuously form and dissolve. Perhaps, he suggested, clays can evolve by selecting, copying of favorable patterns, combined with the selective dissolution of unfavorable patterns. In this model, the evolution of clays is closely analogous to that of cells. "The first step is to grow up a batch of the organisms," Cairns-Smith writes. "Then a small sample of this is used to seed a new batch, which is in turn grown up—and so on. In such circumstances those types which reproduce fastest will be most likely to be carried on between the batches and will eventually become the only types present."

It is a fascinating idea. But is there any way to test this idea in a laboratory experiment? Similar experiments are done all the time in studies of microbial evolution, but that work depends on an arsenal of genetic equipment to collect, refine and sequence DNA. No such technology exists to sequence clays. Cairns-Smith knows the challenges. He once wrote this to me, "Can the material be synthesized in the laboratory on a practical time scale? Can we find conditions that will allow crystals to grow?" Then the $64,000 question, "Do sequences replicate accurately enough through growth?"

Underlying all of these questions is the technical barrier of determining the exact elemental or layer sequence of a clay particle. For clays, no such sequencing methods exist. Characterizing the layer arrangement in a single clay particle is beyond any current technology. So measuring the sequence of the thousands of separate clay particles necessary for an adequate population survey is inconceivable, at least for the time being. But this situation is hardly new. Scientific discoveries often have to wait for new technology to come along. Meanwhile, the clay-life hypothesis of Graham Cairns-Smith will remain an intriguing conjecture in our quest for life's origins.

It's only natural that many earth scientists like the idea that clays played a central role in the origin of life. It is true that there is a satisfying unity in integrating the solid, liquid, and gaseous parts of our planet to make life. Even so, I have to confess that I'm not entirely comfortable with the reliance on clays. That primarily is due to the way clays are often used in experiments. So often, it seems that chemists just sprinkle in a little clay powder to make the experiment work. Nucleotides just sit there in a beaker until you sprinkle in clay, and bingo, you get polymerization to form RNA. Lipids just sit there until you sprinkle in clay and suddenly you get self-assembly of vesicles.

The behavior of clays is fascinating but we really don't know what's going on at the atomic scale. Clays are so variable in their composition, so poorly defined in their surface properties, and so fine grained that it's virtually impossible to be sure what molecule is binding to what surface. It's those "magic powder" aspects of clay that can make it hard to determine just what happens in those experiments. In addition, if we don't understand the chemistry behind these experiments, it's harder to say whether they're relevant to the beginnings of life or not.

But mineralogists aren't limited to powders. We have another experimental choice in using big, faceted, single crystals. In the next lecture, we'll explore an alternative approach to the selection and concentration of organic molecules, an approach that not only uses big crystals, but that also exploits the curious fact of life's molecular handedness.

Lecture Eighteen
Life's Curious Handedness

Scope: The origin of biochemical one-handedness, known as *homochirality*, is still one of the mysteries of life's emergence. It is possible that life's molecules self-select for handedness, or perhaps, chiral mineral surfaces did the selecting. What we can say for sure is that mineral surfaces have the ability to select, concentrate, and organize organic compounds in their transition to life. Thanks to many minerals, including quartz, calcite, and a growing list of other mineral substrates, we seem to be closer to understanding how organized molecular systems emerged.

Outline

I. One of the early stages of life's emergence must have been the incorporation of *handedness*.

A. Many of life's molecules occur in mirror-image pairs (see Lecture Eight). Handedness is a pervasive feature of life, from simple amino acids and simple sugars to the DNA double helix and complex proteins.

B. Each of these molecules has a mirror image, and these mirror-image pairs are known as *chiral molecules*. They have the exact same composition and structure and many of the same physical properties, but a chiral molecule always looks different in a mirror. It has a mirror image that cannot be exactly superimposed on itself.

C. At the molecular level, life is one-handed, even though nonlife shows no preference. Prebiotic, nonliving synthesis pathways make chiral molecules in 50:50 mixtures. Living cells almost always choose right-handed sugars over left (see ribose diagram in Lecture Fifteen, p. 16) and left-handed amino acids over right (see alanine diagram below).

1. Without chiral purity, most proteins would not function properly. Also, the uniform spiral of the DNA double helix would be a mess, curving this way and that.
2. In fact, the spiral of a DNA helix is disrupted if even one of its deoxyribose sugars has the wrong chirality.

D. Molecular chirality occurs because carbon atoms tend to form four chemical bonds to four *different* groups of atoms.

1. The amino acid alanine, for example, has a central carbon atom linked to one NH_2 molecule (amino group), one COOH molecule (carboxyl group), one CH_3 molecule (methyl group, indicated below with the letter R), and one hydrogen atom. The four groups form a triangular-based pyramid, or *tetrahedron*, around the carbon atom.
2. Orienting the hydrogen atom up and looking down on the molecule, there are two mirror-image ways to arrange the remaining three groups. So-called *right-handed* or D-alanine has the clockwise arrangement of carboxyl–amine–methyl groups. *Left-handed* or L-alanine features a different clockwise arrangement: amine–carboxyl–methyl.

CHIRALITY IN ALANINE

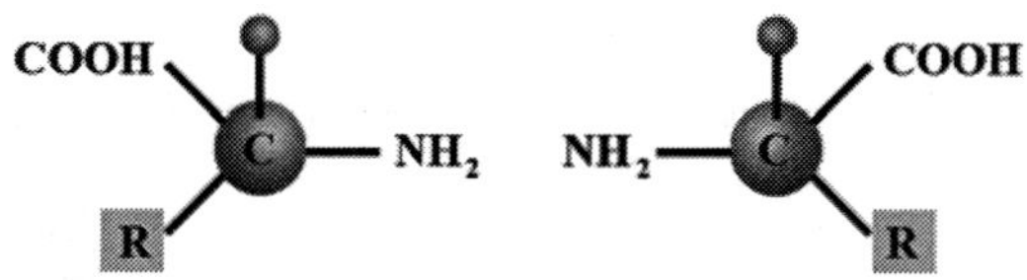

Right-handed **Left-handed**

3. Life almost always selects L-amino acids and D-sugars, rather than D-amino acids and L-sugars, which has important consequences. Cells often respond differently to molecules of different chirality.

II. Chirality is a diagnostic characteristic of life. The mystery is how this selectivity emerged from the random prebiotic world.

A. Some researchers think that prebiotic synthesis was inherently asymmetrical. They say that there must have been a global excess of L- over D-amino acids, for example. How could this be possible if all synthesis processes yield 50:50 mixtures?

1. One possible solution is based on a discovery by Louis Pasteur, who found that left- and right-handed crystals cause polarized light to be rotated in opposite directions. If shining polarized light on a chemical-rich environment were to result in disintegration of those molecules, a selective breakdown of D-amino acids might be possible. One optical isomer, as chiral

molecules are sometimes called in this context, would be favored over the other. This process might conceivably have occurred in deep space in the vicinity of a rapidly rotating star. Such a situation could have produced polarized light of that kind.

2. Physicists have proposed other explanations for a global excess of one handedness based on the physics of the weak force, which operates over distances smaller than a proton. Almost every known physical process, such as gravity and electromagnetism, treats left- and right-handed molecules in the same way. However, a process involving the weak force, radioactive beta decay, for example, violates mirror symmetry. Beta decay produces polarized radiation of only one handedness. It is possible that this chiral radiation could have selected molecules of one handedness over the other.
3. However, calculations predict that such asymmetric processes should produce only the most trivial chiral excess—generally less than a minuscule fraction of 1 percent.
4. Similarly, researchers have found that chiral molecules may have an almost immeasurably tiny difference in energy of their two forms, suggesting that the lower-energy forms favored by life might be ever-so-slightly more stable than their nonlife mirror images.
5. However, it is hard to imagine how such minute effects could tip the global balance toward L-amino acids or D-sugars. Such miniscule effects would have to be amplified many, many times over by something else.

B. Many origins-of-life researchers argue that chiral selectivity is most likely the result of a local physical environment on Earth, not a global or universal effect. Accordingly, the emergence of life consisted of a series of chemical events at a specific place and time. It is possible, even likely, that those emergent steps were repeated countless millions of times across the globe, but each individual emergent step was local. Thus, we need to understand *local* chiral environments, where chiral molecules might have emerged.

C. The origins of life may be likened to the formation of a crystal. In life, as in crystals, two essential and largely independent steps are required.

1. The first step in crystal formation is called *nucleation*, the precise assembly of a relatively small number of atoms or molecules into a "seed crystal." Given many local chiral environments, this seed might purely by chance have been either D or L.
2. Nucleation is followed by the second stage of crystal formation, called *growth.*
3. In life, as in crystal formation, these two distinct stages usually occur at very different speeds.
4. Assume that the nucleation of life—that is, the assembly of molecules into some kind of replicating entity—was a relatively rare event. Once that self-replicating set of molecules formed, it is likely that it grew rapidly. The first successful life-form must have consumed most of the available molecular food. In the process, it frustrated further origin events.
5. If that life-form was by chance chiral, either D or L, then that handedness would have been passed on to subsequent generations.
6. Every chiral molecule is *itself* a tiny local chiral environment—either left-handed or right-handed. Some of these chiral molecules may have selected other molecules of the same handedness. Still, synthesis experiments don't back up this idea. D-L links between amino acids seem to form just as easily as D-D or L-L links.

III. Another more promising, answer to the problem of chiral selection is chiral *mineral* surfaces.

A. Minerals can promote polymerization by concentrating and aligning desirable molecules. Every mineral also can provide chiral surfaces. Indeed, left- and right-handed mineral surfaces are potentially the perfect approach for concentrating and separating a 50:50 mixture of L and D molecules.

B. A few minerals occur in both right-handed and left-handed structural variants. The common mineral quartz is one example of a chiral mineral. Even minerals that don't come in chiral pairs commonly feature pairs of crystal faces whose surface structures are chiral. The structures are mirror images of each other. These chiral surfaces have atomic arrangements that are perfectly suited to select and concentrate L versus D molecules, such as amino acids or sugars.

C. Natural left- and right-handed surfaces occur in roughly equal numbers. *On average*, the Earth's surface would have had equal proportions of both. However, specific places would have varied.

1. The first living entity arose as a bundle of self-replicating chemicals *at a specific place and time.* That living system adopted whatever chirality was imposed by the local environment. Once that chiral system began to make copies of itself, the handedness of life was fixed.

2. It is conceivable that primitive systems of opposite handedness emerged in different places on Earth at different times. If so, however, one of those systems ultimately won the competitive battle for resources, and that's the handedness we see today.

D. The problem of life's *homochirality*—the selection of one handedness over the other—is a puzzling aspect of the much more general question of molecular selection. Prebiotic synthesis processes produce huge numbers of different molecules of which life uses relatively few.

IV. Scientists have long recognized the ability of minerals to sort out molecules. Some mineral surfaces attract some molecules more strongly than others. If two molecules differ from each other in size or shape, then such selection is easy to understand. Selection is just a matter of which molecule fits best, but selection between two chiral molecules is more difficult to explain.

A. Because chiral pairs of molecules are chemically identical, they form exactly the same kinds of bonds with the mineral surface. In fact, the only way for chiral selection to occur is for a molecule to form *three* separate points of attachment, and those three bonds cannot be in a straight line.

B. One way to study the interactions between chiral minerals and chiral organic molecules is to expose the mineral to both left-handed and right-handed organic molecules and see if they react differently. Experiments of this type have been tried since the 1930s.

C. The basic idea of these experiments is sound. Differently handed molecules have the potential to attach selectively to different chiral surfaces. Still, such experiments on quartz had a flaw. In every case, researchers ground the quartz crystals to a fine powder to maximize the surface area of interaction. The process destroyed all the flat crystal surfaces, the only surfaces where we can be sure of what's going on.

D. In 2000, I tried a different approach using calcite, which is calcium carbonate, $CaCO_3$.

1. This mineral offers fist-sized crystals with large, flat faces. Big crystals make it easier to figure out the atomic-scale interactions of crystals with organic molecules.
2. I also needed crystals that have pairs of faces with surface structures that are mirror images of each other. The only way to select chiral molecules on a surface is with a chiral surface. Calcite is not only an abundant mineral, but its most common crystal faces are chiral.

E. How to design an experiment to test the hypothesis that minerals can selectively concentrate left-handed and right-handed molecules on different faces? In particular, would D- or L-amino acids attach themselves preferentially to left- or right-handed calcite faces? The problem is that our modern environment is already contaminated with the excess L-amino acids and D-sugars of living cells.

1. I mixed a 50:50 solution of D- and L-aspartic acid and exposed left- and right-handed calcite faces to both to see which molecules stuck to which faces. If crystal faces have the ability to concentrate chiral molecules, then the left- and right-handed calcite faces must attract *equal and opposite* excesses of L- and D-molecules.
2. Calcite specimens commonly have cleavage planes. These flat surfaces along which crystals easily break are *not* chiral. In addition to the equal and opposite effects that we should observe on left- and right-handed chiral faces, we should see no preference for D- or L-aspartic acid adsorbed onto the cleavage surfaces.
3. After cleaning the crystal surfaces, I poured a 50:50 aspartic-acid solution into glass beakers and allowed the calcite crystals to soak in the amino acid mix for 24 hours. I then removed the crystals and washed them quickly in pure water. Any molecules adsorbed to the calcite surface were still

present. If the experiment worked, each crystal face would have an excess of D- or L-aspartic acid, whereas the cleavage surfaces would show the 50:50 mix.

4. It was fairly easy to extract attached molecules from the calcite crystal surfaces, because calcite dissolves easily in dilute hydrochloric acid. Using a dropper, I covered one crystal face with a thin puddle of acid, waited about 20 seconds, then sucked the acid back up into the dropper and emptied it into a glass vial. Any attached molecules were removed from the crystal, along with the outer layers of calcite.
5. I applied the acid wash to each face on each of four crystals, a total of 23 surfaces in all, with a mix of left-handed, right-handed, and cleavage surfaces. Three day's work resulted in 69 vials of aspartic-acid solution washed from the crystal faces.
6. I needed extremely accurate analyses of the D-to-L ratio of aspartic acid, for which I turned to Glenn Goodfriend at George Washington University.
7. Amino acids have a surprising property. Over many thousands of years, they tend to flip back and forth between D and L configurations. When an organism dies, it has virtually all L-amino acids, but over time, the ratio of D-to-L rises to 50:50. Consequently, the D- to-L ratio can be used to determine the relative ages of shells or bones. The ratio also varies according to other environmental conditions, including temperature and salinity.
8. In processing the samples, Glenn and I both knew what chiral effects we hoped to see. Certain crystal faces should select L-molecules and others, D-molecules, whereas the cleavage surfaces should display no preference.
9. After randomly numbering the samples and running the analyses in triplicate, a clear and compelling story emerged from the data. The left-handed calcite faces almost always displayed D-to-L ratios a few percent less than 1.00. These faces preferentially retained L-aspartic acid compared with the D form. By contrast, the right-handed calcite crystal faces all displayed an equal and opposite affinity for D-aspartic acid.

All the cleavage surfaces produced D-to-L ratios indistinguishable from 1.00.

10. These results held a plausible scenario for the origin of key macromolecules. We had found that chiral crystal faces of calcite selectively adsorb D- and L-amino acids. If the adsorbed amino acids line up side by side on the crystal surface, then they might be poised to link to each other to form a protein-like polymer made entirely of L- or D-amino acids. Perhaps this is the way that mineral surfaces selected, organized, and assembled the first homochiral biomolecules.

Essential Reading:

Hazen, *Genesis: The Scientific Quest for Life's Origin*, chap. 13.

Supplemental Reading:

Lahav, *Biogenesis: Theories of Life's Origin*, chap. 20.

Questions to Consider:

1. Some authorities state that the universe is biased toward life with left-handed amino acids and right-handed sugars, while others think the choice is random. How might we discover which point of view is correct?
2. Some authorities state that if we find a chiral excess of any sort on Mars, then that will be proof of life. Is that a reasonable conclusion?

Lecture Eighteen—Transcript
Life's Curious Handedness

In the last three lectures, we have explored how just the right subset of biomolecules might have been selected and assembled from the prebiotic soup. We have seen that prebiotic processes yield a bewildering variety of molecules. While some of those molecules played essential roles as the starting materials of life, most of that molecular jumble played no role whatsoever. The selection of just the right mix remains a central puzzle in origins-of-life research.

Some aspects of the problem are at least comprehensible. We have seen from David Deamer's work how lipid molecules, if present in sufficient concentrations, will spontaneously separate and self-organize. Other key molecules might have concentrated on the surfaces of clays or other minerals. However, I've been saving the most confounding problem in the prebiotic selection of molecules and that's the origin of the curious handedness of life.

One of the early stages of life's emergence must have been the incorporation of handedness. In Lecture Eight I introduced the idea that many of life's molecules occur in mirror-image pairs, like your left and right hands. Chemists call these enantiomers. Your hands have the exact same structure but they cannot be exactly superimposed. In fact, handedness turns out to be a pervasive feature of life, from simple amino acids and simple sugars to the DNA double helix and extremely complex proteins. Each of these molecules has a mirror image and these mirror-image pairs are known as chiral molecules. They have the exact same composition and structure and many of the same physical properties, but they are mirror images of each other. In fact, you can tell a chiral molecule from a non-chiral molecule because a chiral molecule always looks different in the mirror. It has a mirror image that cannot be exactly superimposed on its own self; just as your left hand doesn't exactly superimpose on your right hand.

Yet, at the molecular level, life is curiously one-handed, even though non-life shows no preference. All of the prebiotic, non-living synthesis pathways I have told you about make chiral molecules in a 50:50 mixture. There is just no obvious reason why left or right should be preferred. Yet, living cells are highly selective. They almost invariably choose right-handed sugars over left and left-handed amino acids over right. There is a good reason for this selection, too. Without chiral purity, most proteins would not function properly. Also, the uniform spiral of the DNA double helix would be a mess, curving this way and that. In fact, the spiral of a DNA helix is disrupted if even one of its deoxyribose sugars has the wrong chirality. This distinctive characteristic of life's molecules has been known for a century and a half but the origin of this biochemical one-handedness, known as "homochirality," is still one of the great mysteries of life's emergence.

In this lecture, the eighteenth of our series, I want to explore some of the many creative ideas that have been developed to address the question of life's molecular handedness. Researchers have investigated the possible contributions of chiral crystals, such as quartz, and I want to show you one of my own experiments with a big crystal called calcite to give you an idea how some scientists approach this question of chirality.

So what exactly is molecular chirality and what makes it possible? Molecular chirality occurs because of the common tendency of carbon atoms to form four chemical bonds to four different groups of atoms. Let's consider the amino acid alanine, for example. It has a central carbon atom linked to one NH_2 molecule, that's an amino group; one COOH molecule, that's a carboxyl group; one CH_3 molecule, that's a methyl group; and one single hydrogen atom. Those four groups form a little triangular-base pyramid, or tetrahedron, around the carbon atom.

Now, if you orient the hydrogen atom up and look down on the molecule, there are two, mirror-image ways to arrange the remaining three groups. Again, your two hands are mirror images in this sense. So-called "right-handed," or D-alanine, has the clockwise arrangement of carboxyl–amine–methyl groups. Chemists usually employ the letter "D," which stands for the Greek *dextro* for "right," though that designation of molecular "right" versus "left" is completely arbitrary. Then "left-handed," or L-alanine, features a different clockwise arrangement, amine–carboxyl–methyl. In this case here, the "L" stands for the Greek *levo*, for "left."

Well, the big mystery lies in the fact that life usually selects L-amino acids and D-sugars, rather than D-amino acids and L-sugars. That might, at first, seem a trivial distinction but there's an important consequence of this selection. Our cells often respond very differently to molecules of different chirality. One benign example is a familiar flavoring called limonene. This molecule tastes like lemons in its right-handed form but like oranges in the left-handed variant. The reason for this difference has to do with the molecules' shapes. Different shapes trigger different taste receptors.

The infamous drug thalidomide, shown here, is a much more sinister example. The left-handed form of this three-ring molecule relieves the symptoms of morning sickness, but the right-handed form induces birth defects. Many women who took the right-handed form wound up delivering babies with serious deformities. Today, the Food and Drug Administration requires chiral purity in many pharmaceuticals. This is a difficult chemical processing step, adding to our drug costs every year. Moreover, for some substances such as thalidomide, utilizing chiral purity remains difficult as the human body can convert the pure form of thalidomide back into a mixture.

So chirality is important and it's a diagnostic characteristic of life. The mystery is how this selectivity emerged from the random prebiotic world. There have been many theories but all the ideas fall into one of two general categories.

On the one hand, some researchers think that prebiotic synthesis was inherently asymmetrical. They say that there must have been a global excess of L- over D-amino acids, for example, but how could this be if all synthesis processes yield 50:50 mixtures?

One possible solution is based on a discovery by Louis Pasteur using polarized light more than a century and a half ago. Polarized light is light whose electro-magnetic vibrations have been filtered to align along one common axis. Pasteur found that left- and right-handed crystals cause polarized light to be rotated in opposite directions. So you can flip this around. If you shine polarized light on a chemical-rich environment, and if that light tends to disintegrate those molecules, then it is possible that you could get a selective breakdown of D-amino acids. One optical isomer, as chiral molecules are sometimes called in this context, would be favored over the other. This kind of chiral-selective process might conceivably have occurred in deep space in the vicinity of a rapidly rotating star. Such a situation could have produced polarized light of that kind. It is speculative, but no one knows what hundreds of millions of years of such processing might yield.

Physicists have proposed other ways to get a global excess of one handedness based on the physics of the weak force, which operates over distances smaller than a proton. Almost every known physical process looks the same in a mirror. Gravity, electromagnetism and the laws of thermodynamics all treat left- and right-handed molecules the same, but a process involving the weak force, radioactive beta-decay, for example, does violate mirror symmetry. Beta-decay produces polarized radiation of only one handedness. Therefore, it is possible that this chiral radiation could have selected molecules of one handedness over the other. The trouble is that calculations say such asymmetric processes should produce only the most trivial chiral excess, generally less than a minuscule fraction of one percent. Similarly, researchers have found that chiral molecules may have an almost immeasurably tiny difference in energy of their two forms, suggesting that the lower-energy forms favored by life might be ever-so slightly more stable than their non-life mirror images. However, it is hard to imagine how such minute effects could tip the global balance toward L-amino acids or D-sugars. Such miniscule effects would have to be amplified many, many times over by something else.

The second general type of answer involves looking for factors specific to Earth. Many origins-of-life researchers argue that chiral selectivity is most likely the result of a local physical environment on Earth, not a global or universal effect. According to this view, the emergence of life consisted of a series of chemical events. Each of these events occurred at a specific place and a specific time. It is possible, even likely, that those emergent steps were repeated countless millions of times across the globe but each individual, emergent step was local. So what we need is to understand local chiral environments. That is where the reactions leading to life occurred. That is where chiral molecules might have emerged.

Think about the origin of life as being like the formation of a crystal. In life, as in crystals, you need two essential and largely independent steps. The first step in crystal formation is called nucleation. That is the precise assembly of a relatively small number of atoms or molecules into what's known as a "seed crystal". Given lots of local chiral environments, this seed might, purely by chance, have been either D or L.

Nucleation is followed by the second stage of crystal formation called growth. The seed crystal provides a template for the addition of more atoms and molecules. However, that growth will just mimic the original seed, whether it is left handed or right handed. It seems logical to assume that each step in the chemical origin of life must also have required nucleation followed by growth.

Now in life, as in crystal formation, these two distinct stages usually occur at very different speeds. In some materials, such as in frosted glass, nucleation takes place quickly, but growth is very slow. Such a situation leads to countless tiny crystallites. That's what gives colorful agates and frosted glass that distinctive translucent appearance. If, on the other hand, nucleation is very slow but growth is rapid, then you typically get a single large crystal.

I think it's safe to say that life is much more complex than any possible crystal, so it's reasonable to assume that the nucleation of life—that is the assembly of molecules into some kind of replicating entity—is a relatively rare event. Maybe it only happened once in the history of Earth, or maybe it happened a bunch of times, but it was certainly not happening everywhere all the time. However, once that self-replicating set of molecules formed, then it is likely that it grew rapidly. The first successful life form must have consumed most of the available molecular food. In the process, it frustrated further origin events. If that life form was, by chance, chiral, either D or L, that handedness would have been passed on to subsequent generations. In this model, all it takes is an initial local chiral environment.

Such local chiral environments are everywhere on Earth. Think about it. For one thing, every chiral molecule is itself, a tiny local chiral environment—either left-handed or right-handed. It is very possible that some of these chiral molecules selected other molecules of the same handedness. In fact, that is exactly what happened in the pioneering chirality studies of Louis Pasteur. When he evaporated a 50:50 mixture of D- and L-tartaric acid, he got pure D and pure L crystals. That is because D-D or L-L pairs of tartaric acid molecules happen to fit together more easily than D-L pairs. It is possible that the exact same kind of selection will have happened with amino acids of RNA nucleotides. Maybe some of these macromolecules have to grow with chiral purity. If so, that would solve the problem. Still, synthesis experiments do not back up this idea. D-L links between amino acids seem to form just as easily as D-D or L-L links.

Another, potentially more promising, answer to the problem of chiral selection is chiral mineral surfaces. We've seen how lots of scientists have resorted to mineral surfaces. James Ferris used clay to synthesize RNA, Leslie Orgel developed a theory of "polymerization on the rocks" and Graham Cairns-Smith proposed that self-replicating clay may have preceded life based on DNA. Minerals can promote polymerization by concentrating and aligning desirable molecules. It also turns out that some mineral surfaces have an added advantage. Every mineral, every grain of sand or particle of silt also has the potential to provide chiral surfaces. Indeed, left- and right-handed mineral surfaces are potentially the perfect approach for concentrating and separating a 50:50 mixture of L and D molecules.

So just how common are chiral mineral faces? A few minerals occur in both right-handed and left-handed structural variants. The common mineral quartz, which makes up most of the grains of white beach sand, is one example of a chiral mineral. The quartz structure has helices of silicon and oxygen atoms that in some crystals spiral up to the left, and in other crystals up to the right. So every grain of beach sand is potentially a chiral environment.

Even minerals that don't come in chiral pairs commonly feature pairs of crystal faces whose surface structures are chiral. The structures are mirror images of each other. These chiral surfaces have atomic arrangements that are perfectly suited to select and concentrate L versus D molecules, such as amino acids or sugars.

It is important to stress that natural left- and right-handed surfaces occur in roughly equal numbers. So we're not talking here about chiral selection occurring on a global scale. On average, the Earth's surface would have had equal proportions of both. However, specific places would have varied. The first living entity arose as a bundle of self-replicating chemicals at a specific place and time. That living system adopted whatever chirality was imposed by the local environment. Once that chiral system began to make copies of its own self, the handedness of life was fixed. Of course, it is conceivable that primitive systems of opposite handedness emerged in different places on Earth at different times. If so, however, one of those systems ultimately won the competitive battle for resources. This is the handedness we see today.

The problem of life's homochirality, the selection of one handedness over the other, is a particularly puzzling aspect of the much more general question of molecular selection. Prebiotic synthesis processes produce huge numbers of different molecules, and as we've seen, life uses relatively few of these. While it's virtually impossible to study the full range of prebiotic products in one experiment, it is possible to investigate the behavior of chiral molecules. It's relatively easy to design an experiment that starts with a 50:50 mix of D- and L-molecules—this is called a racemic mixture—and then looks for environments that are able to separate left from right; this is called resolving a racemate. If we can discover a process by which handed molecules were selected and concentrated, which is a difficult problem, then there is more hope that we can solve the more general selection problem as well.

Scientists have long recognized the ability of minerals to sort out molecules. Some mineral surfaces attract some molecules more strongly than others. If two molecules differ from each in size or shape, then such selection is pretty easy to understand. Selection is just a matter of which molecule fits best. But selection between two chiral molecules is more difficult to explain. Chiral pairs of molecules are chemically identical, so they form exactly the same kinds of bonds with the mineral surface. In fact, the only way for chiral selection to occur is for a molecule to form three, separate points of attachment and those three bonds cannot be in a straight line. You may have experienced this kind of selection process if you have gone ten-pin bowling. The big bowling ball typically has three holes, one for your thumb and two for your index and middle fingers. The three holes are not in a line, so a right-handed bowler cannot use a left-handed ball and vice versa.

So, one way to study the interactions between chiral minerals and chiral organic molecules is to expose the mineral to both left-handed and right-handed organic molecules and see if they react differently. Experiments of this type have been tried ever since the 1930s, when scientists in Greece and in Japan tried separating left- and right-handed molecules by pouring D- and L-solutions over left- or right-handed quartz. Over the past 70 years, at least a dozen similar experiments have been attempted. The basic idea of these experiments is sound.

Differently handed molecules do have the potential to attach selectively to different chiral surfaces. Still, these experiments on quartz had a serious flaw. That is because, in every case, the researchers ground the quartz crystals to a fine powder in an effort to maximize the surface area of interaction. You might think that increasing the surface area would magnify the desired effects. After all, powdering can increase the surface area of crystal a thousand times or more. But at the same time, it completely destroys all the flat crystal surfaces and those are the only surfaces where you can be sure of what's going on.

Think about it. A powder will feature every possible crystal surface simultaneously because you are creating lots of tiny grains with lots of surfaces. While some of these surfaces may very well select L-amino acids with great efficiency, other surfaces with a different atomic structure will just as likely select D-amino acids. Even if an experiment displays a slight preference for L or D molecules, there is absolutely no hope at all of determining which crystal surface is doing the selecting.

In the year 2000, I decided to try a different approach. I wanted big, fist-sized crystals with large, flat faces. A layer of molecules, adsorbed onto a surface an inch square, weighs at most a few billionths of a gram and it's a daunting analytical challenge to detect that tiny amount. Big crystals make it easier to figure out the atomic-scale interactions of crystals with organic molecules. Big crystals pose a different problem because they tend to be astronomically expensive, thanks to the competition among mineral collectors. In my experiments, I needed to find a common rock-forming mineral that collectors do not covet. That not only saves money, the common inexpensive minerals are also likely to be the most relevant to origins-of-life scenarios.

There is another critical aspect to these experiments. We need crystals that have chiral faces; that is, pairs of faces with surface structures that are mirror images of each other. The only way to select chiral molecules on a surface is with a chiral surface. To find the best crystals, I picked up a favorite mineralogy book. It's this old, frayed, dog-eared copy of Edward Dana's *A Textbook of Mineralogy* that I purchased at the American Museum of Natural History when I was just 14 years old. Almost every page has line drawings of crystal forms, so I just leafed through the volume. It was quickly obvious that the majority of crystal faces are not chiral. Nevertheless, there are a few classic chiral mineral surfaces.

The mineral that really caught my attention was calcite. Calcite is calcium carbonate, $CaCO_3$. It's not only an abundant mineral, but it's the one mineral most closely associated with life. Calcite is the mineral that makes clamshells and snail shells. It forms pearls and coral. What's exciting is that its most common crystal faces are chiral. Calcite crystals usually take the shape of a six-sided, pointed form. The technical name for this crystal shape is scalenohedron, but most mineralogists just call it a dogtooth.

If you look on eBay on any given day, you will find several calcite crystals for sale. You can often pick up a nice, big dogtooth crystal for $20 or $30.

So I was quickly able to accumulate a sizable collection of calcite crystals for my experiments.

So, how do you design an experiment to test the hypothesis that minerals can selectively concentrate left- and right-handed molecules on different faces? In particular, would D- or L-amino acids attach themselves preferentially to left- or right-handed calcite faces? I had to be extremely careful because there had been decades of suspect results and non-reproducible results. The biggest problem is that every nook and cranny of our modern environment is already contaminated with the excess L-amino acids and D-sugars of living cells. So if you look for a chiral effect, you're probably going to find a chiral effect. It is easy to be fooled.

My approach was to mix up a 50:50 solution of D- and L-aspartic acid. That's an amino acid that's known to have special affinity for calcite. Aspartic acid is a key component in proteins that add strength and resiliency to all sorts of shells. My idea was to expose both left- and right-handed calcite faces to both D- and L-aspartic acid. Then, we'd see which molecules stick to which faces. If crystal faces have the ability to concentrate chiral molecules, then the left- and right-handed calcite faces must attract equal and opposite excesses of L- and D-molecules. So, for example, if the left-handed calcite faces adsorb a 5% excess of L-aspartic acid, then the right-handed calcite face has to adsorb a 5% excess of D-aspartic acid; otherwise, the results are ambiguous.

Calcite has another advantage as a surface. In addition to the chiral crystal faces, calcite specimens commonly have what are known as cleavage planes. These are flat surfaces along which crystals easily break. But these cleavage surfaces are not chiral. So in addition to the equal and opposite effects that we should observe on left- and right-handed chiral faces, we should see absolutely no preference for D- or L-aspartic acid adsorbed onto the cleavage surfaces.

So here is how the experiment is done. The first step is to clean the crystal surfaces. Many people, including the miners, the mineral dealers, various collectors, and, of course I, had handled my calcite crystals. Contamination is everywhere and any little contamination, even a fingerprint or a breath, could completely skew the results.

Once the crystals were cleaned, I poured a 50:50 aspartic-acid solution into glass beakers and just dunked the fist-sized calcite crystals into the liquid. I let the crystals soak in the amino acid mix for 24 hours, came back the next day, removed the crystals and washed them off. So now, most of the excess solution was rinsed off but any molecules that were adsorbed onto the surface were still there. If the experiment worked, then each crystal face would have an excess of D- or L-aspartic acid, while the cleavage surface would show the 50:50 mix.

It turns out to be easy to extract attached molecules from calcite crystal surfaces because calcite dissolves easily in dilute hydrochloric acid. So, I held one of the crystals in my gloved left hand and oriented one crystal face parallel to the bench top. With my right hand, I used a dropper to cover that face with a thin puddle of acid. I let that acid sit there about 20 seconds and then I sucked it back up into the dropper and then deposited that liquid into a vial. So any attached molecules were removed from the crystal along with the outer layers of calcite.

I applied the acid wash to each face on each of four crystals. That was a total of 23 surfaces in all, with a mix of left-handed, right-handed and cleavage surfaces. Three days' work resulted in 69 vials of aspartic-acid solution washed from various crystal faces, plus several vials with individual samples of each day's aspartic acid solution, of the pure water and of the acid wash as well. In all, more than 75 sample vials needed to be processed and each had to be analyzed at least three times.

Many laboratories are able to analyze amino acid samples, but I needed extremely accurate analyses of the D-to-L ratio of aspartic acid. Two-hundred-plus amino-acid analyses are a huge, specialized job and the Geophysical Lab facilities just weren't suitable. Luckily, one of the world's experts, geochemist Glenn Goodfriend, worked nearby at George Washington University.

Glenn Goodfriend was a master at amino acid analysis. Glenn's technique of choice is gas chromatography, just as I described in Lecture Twelve when George Cody analyzed our gold capsule results to discover "humpane." But Glenn Goodfriend had devised meticulous, specialized procedures to get precise and accurate ratios of D-to-L amino acids. The reason that's an interesting thing to know has to do with a surprising property of amino acids. Over long periods of time, typically many thousands of years, amino acids tend to flip back and forth between D and L configurations. When an organism dies, it has virtually all L-amino acids but gradually, over time, the ratio of D-to-L rises back to 50:50.

Consequently, the D-to-L ratio can be used to determine the relative ages of shells or bones. The ratio also varies according to other environmental conditions, including temperature and salinity. Consequently, Glenn was in constant demand as a collaborator, and he had ongoing projects to study the salinity of Venetian lagoons, and climate change of the Baja Peninsula, and the evolution of land snails from the Bahamas. The latter project with Stephen Jay Gould, in particular, had attracted a lot of attention. So it was a big deal for me to gain Glenn Goodfriend's help.

Glenn and I developed a routine for processing samples. I did the chemical preparations, under his strict supervision. Glenn proved to be one of the most exacting, finicky experimentalists I'd ever met. He prepared amino acid samples for analysis exactly the same way every time. He heated them at exactly 100°C for exactly one hour in a small oven. He instructed me to open the oven door quickly and place the tray of vials on the shelf in one swift gesture. I had to close the door within four seconds to keep the temperature at the proper level. If the temperature dropped as little 0.2°C, he recorded it in his lab notebook. I had to remove the samples from the oven with a similar smooth motion exactly an hour later. If I was off by 10 or 15 seconds, he recorded it in his lab notebook.

Once the samples were prepared, they were run automatically overnight. Glenn and I were paranoid about the potential for unconscious bias in these experiments. We both knew exactly what chiral effects we hoped to see. Certain crystal faces should select L-molecules and others D-molecules, while the cleavage surfaces should display no preference at all. So, to avoid any chance of bias, I randomly renumbered all my samples and then Glenn renumbered them again in his own notebook. That way we obtained all the analyses first and neither of us had any idea which sample came from which face. Only after we had completed all the analyses and compared numbers would we know the results. It took several days to prepare all the samples and even longer to run everything in triplicate.

As soon as we'd sorted out which analysis went with which face, a clear and compelling story emerged from the data. The left-handed calcite faces usually displayed D-to-L ratios a few percent less than 1.00. So these faces preferentially retained L-aspartic acid compared to the D form. By contrast, the right-handed calcite crystal faces all displayed an equal and opposite affinity for D-aspartic acid. Just as important, all of the cleavage surfaces, which served as our experiment's internal control, produced D-to-L ratios indistinguishable from 1.00.

We wrote up those results as quickly as possible and submitted a short manuscript to the *Proceedings of the National Academy of Sciences*. Hat Yoder served as the sponsoring Academy member, which put us on a fast track for publication.

We realized that these results held a plausible scenario for the origin of key macromolecules. We had found that chiral crystal faces of calcite selectively adsorb D- and L-amino acids. If the adsorbed amino acids line up side-by-side on the crystal surface, then they might be poised to link to each other to form a protein-like polymer made entirely of L- or D-amino acids. Perhaps this is the way that mineral surfaces selected, organized and assembled the first homochiral biomolecules.

Glenn and I had been pretty sure that some workers in the origins-of-life community would notice our results. What we hadn't anticipated was the significant interest that came from chemical engineers, who are engaged in the design and purification of chiral pharmaceuticals. So, one consequence of this study is a new search for commercial applications of chiral mineral surfaces.

At the time, in late 2001, Glenn Goodfriend and I saw the aspartic-acid study as the beginning of what, we hoped, would be a long and fruitful collaboration. We had a whole series of follow-up experiments planned, with different amino acids and different chiral minerals.

But in late 2001 Glenn got a strange, drug-resistant infection in his jaw. It gradually spread, eventually attacking his brain. He died in October of 2002 at the age of 51. We've subsequently carried on this research and have lots of new results, but it was tremendously sad to watch his decline and his loss has been a real blow to science.

In summary, scientists have not yet learned how life became homochiral. Indeed, we may not know for sure how L-amino acids and D-sugars were selected from the random prebiotic soup. It's possible that life's molecules self-select for handedness. Or perhaps chiral mineral surfaces did all of the selecting.

What we can say for sure is that mineral surfaces have the ability to select, concentrate and organize organic compounds in their transition to life. Thanks to many minerals, including quartz, calcite and a growing list of other mineral substrates, we seem to be a lot closer to understanding how organized molecular systems emerged.

Nevertheless, even if we knew every chemical step along the path to macromolecules, a huge gap would remain in our understanding of life's origins. We'd still need to discover the process by which a collection of molecules begins to grow and make copies of itself. So, in the next four lectures, we'll turn to the processes of molecular metabolism and reproduction.

Lecture Nineteen

Self-Replicating Molecular Systems

Scope: Most researchers agree that life can be defined as a chemical process possessing three key abilities: to grow, to evolve, and to reproduce. At the dawn of life, growth and evolution were not impressive, but the third attribute, reproduction, must have progressed swiftly and set the early world of life apart from the prebiotic era. To invent a self-replicating chemical system in the laboratory, researchers face a challenging task, but one worth the attempt.

Outline

I. For many origins-of-life researchers, the experimental Holy Grail has become the creation of a self-replicating molecular system in a test tube.

- **A.** Self-replicating molecular systems are not necessarily alive, but they do involve something similar to metabolism, which brings us much closer to our goal of understanding the origin of life.
- **B.** The simplest self-replicating molecular system consists of just one kind of molecule that makes copies of itself.
  - **1.** In the right chemical environment, with the right combination of smaller chemical compounds to act as food, a single molecule will increase in a geometrical expansion.
  - **2.** Chemists call this kind of molecule *autocatalytic*—it literally brings about its own transformation. Autocatalytic molecules assemble themselves: They act as templates that latch onto and assemble their own chemical components from a solution.
- **C.** All self-replicating molecules must be *self-complementary*.
  - **1.** Complementary molecules fit snugly together because of their matching shapes and the arrangement of their chemical bonds.
  - **2.** Many of the proteins in our bodies, for example, are able to function because they're complementary to other vital molecules: food, for example, or toxins.
- **D.** Only a small handful of molecules are self-complementary; even fewer are able to self-replicate.

1. In 1989, chemist Julius Rebek, Jr., created one of the first examples to do both. His group formed a self-replicating molecule by linking three smaller molecules. When joined end to end, the resulting adenine-napthalene-imide molecule is self-complementary.
2. If *one* of these combined molecules is immersed in a solution containing many of these three component molecules, it will make copies of itself. The combination is not a likely biological molecule, but it does demonstrate the concept.
3. In 1996, chemist Reza Ghadiri and his group designed the first self-replicating peptide, a sequence of 32 amino acids. For self-replication to occur, however, this peptide required a steady supply of two specialized reactive fragments, one of them 15 amino acids long and the other 17 amino acids long.

E. The base pairs of DNA are probably the most famous complementary pairs of molecules.

SELF-COMPLEMENTARY DNA STRANDS

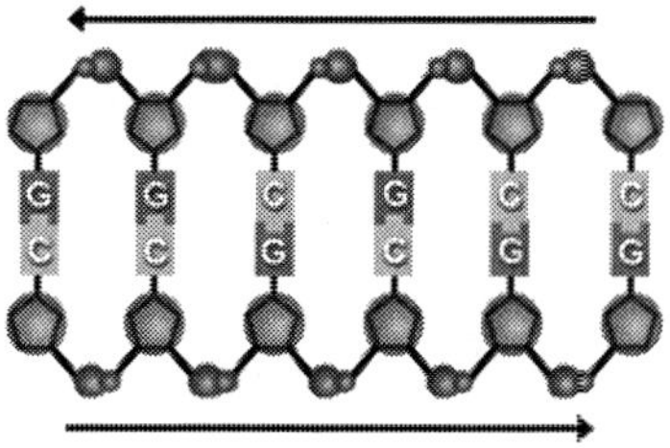

1. The double-helix structure of the DNA molecule is like a long, twisted ladder, the rungs of which form from complementary pairs of the molecules called *bases*. The base cytosine (C) is complementary to guanine (G), and the base adenine (A) is complementary to thymine (T). Not only are the resulting base pairs, C-G and A-T, complementary, but so also are the two halves of every DNA double-helix strand.
2. The beauty of this molecule lies in its two-fold ability to both carry and duplicate information. When the double helix splits apart, each half can make a copy of the complete double helix.
3. The two halves of most double-helix strands have complementary yet different base sequences. Still, given the symmetry of DNA, it is an easy step, conceptually, to design self-complementary strands.

4. In 1986, chemist Guenter von Kiedrowski and his coworkers produced the first such strand—the six-nucleotide sequence CCGCGG, which makes exact copies of itself as long as enough fresh C and G nucleotides are provided.

F. These few known types of self-replicating molecules still don't meet the minimum requirements for life.

1. Such molecular systems need a constant supply of small, highly specialized molecular building blocks, but there is no evidence that these building-block molecules could arise spontaneously in a natural environment.
2. Self-replicating molecules never change or evolve.

II. Research is underway to increase the molecular complexity to systems with two or more molecules that form a self-replicating cycle or network.

A. The simplest systems are *cross-catalytic*; that is, two molecules, which we can call AA and BB, are made from smaller feedstock molecules, A and B. If AA catalyzes the formation of BB from many individual B molecules, and if BB does the same to form AA from individual A molecules, a self-replicating system results. This system will continue to make copies as long as a supply of the A and B molecules is provided.

A CROSS-CATALYTIC SYSTEM

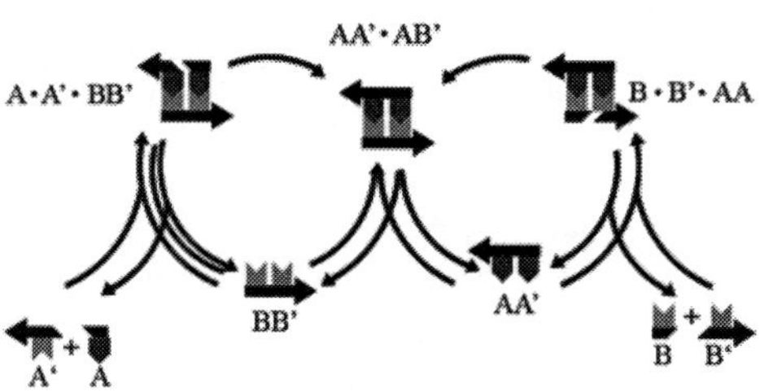

B. A system with a dozen or more different molecules, each of which promotes the production of another chemical compound in the system, is an alternative model. Stuart Kauffman of the Santa Fe Institute argues that such *autocatalytic networks* are the most likely form of protolife.

AN AUTOCATALYTIC NETWORK SYSTEM

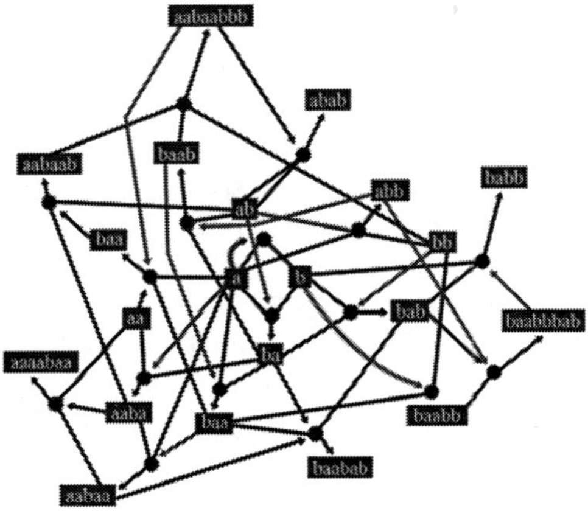

C. Kaufmann proposes that the molecular repertoire of that first living autocatalytic network might have been vast—hundreds or thousands of different interacting molecular species.

D. Kauffman's theory provides an appealing model for the emergence of primordial life. The various molecules in the network catalyze each other; thus, complexity is present and a genetic mechanism is unneeded. Nevertheless, the system is likely to vary and evolve gradually. Thus, the network would have the ability to grow, to reproduce, and to evolve.

E. A simple metabolic cycle can be designed in the lab to test Kaufmann's theory. Metabolism is life's preferred cycle of chemical self-replication.

1. Every cycle requires two kinds of input: energy (in experiments, it is best to rely on chemical energy) and feedstocks of building-block molecules, such as carbon dioxide, ammonia, and water.
2. The goal in metabolism is to capture as much energy as possible and use it to make new molecules that can reinforce the cycle.

III. What was a primitive metabolic cycle like? Several origins-of-life researchers have proposed metabolic models, which reflect the chemist's job to invent a cycle of chemical reactions that obey three basic rules.

- **A.** The first rule of metabolism is that the cycle must consist of a sequence of progressively larger molecules. Each molecule in the cycle combines with carbon dioxide or water or some other small molecule to make the next molecule in the cycle. The largest should be able to split into two smaller molecules, *both* of which are also in the cycle. By constantly building up to a larger molecule that splits in two, the number of cycles keeps doubling, and the system grows.
- **B.** The second rule is to identify a plausible prebiotic environment to supply all the necessary raw materials, including small molecules made up of carbon, nitrogen, sulfur and other essential elements, plus a steady source of energy. The chemical energy of unstable minerals is preferable to harsh sunlight or lightning. A*ll* molecules that make up the cycle must survive long enough to keep the cycle going.
- **C.** The third rule is that an unbroken chemical history must connect Earth's past with the present. All the molecules and energy sources must lie close to the basic metabolism of modern cells.
- **D.** Every chemical step in a metabolic cycle is a potential experiment, and every step must be shown to work. Two of the most influential models of primitive metabolism involve baked amino acids and sulfur-carbon bonds, respectively.

IV. Chemist Sidney Fox proposed the first comprehensive model of primitive metabolism, the *Protenoid World*, which arose out of his study of the role of proteins in life's origin.

- **A.** Fox imagined a scenario in which amino acids from the primordial soup dried and baked on cooling volcanic rocks. He tried to duplicate these conditions in his lab by drying amino acids on a surface at 170°C. His chemicals polymerized into a lumpy substance that he called *proteinoid* or *protenoid*, implying a relationship to biological proteins.
- **B.** Protenoids displayed fascinating behavior, and Fox did many follow-up experiments in his Florida lab.
  1. Some protenoids acted like catalysts, though less so than protein catalysts in modern cells.
  2. Protenoids self-organized, not unlike lipids; when placed in hot water, they formed microbe-sized spheres.

3. Fox claimed that his *microspheres* had a bilayer membrane of sorts. Sometimes these spheres appeared to divide. These intriguing cell-like spheres absorbed more protenoids, and they could grow and divide to yield something like a second generation of microspheres.
4. The detailed structure of these microspheres and the mechanisms for self-replication were not obvious, but these objects, which grew from nutrients in solution, represented a true metabolic model for life's origin.

C. At first glance, Fox's model seems to score rather well in the game of metabolism. The metabolic cycle is made with widely available amino acids. The volcanic environment near a tidal zone is certainly plausible for the early Earth. And protenoids seem to be consistent with the principle of biological continuity, because proteins are central to modern biochemistry.

D. By the mid-1960s, Fox's claims for the Protenoid World model became more extreme: The origin-of-life problem had been solved. Protenoids were "alive in some primitive way."

E. Stanley Miller and Harold Urey raised numerous objections. They disagreed that the Protenoid World could have arisen before any sign of a genetic mechanism to pass information from one generation to the next. How could DNA and RNA arise in such a world?

F. Today, the majority opinion is that Sydney Fox went too far in his uncritical advocacy of the Protenoid World. Still, his model was influential because it was the first comprehensive origin model in which metabolism came before genetics, and because he introduced the philosophical position that origin processes might be non-random and deterministic, a view later embraced by numerous origins workers.

V. Nobel Prize-winning chemist Christian de Duve devised a different model of primitive metabolism. His central hypothesis is that lots of energy-rich thioester molecules were available on the primordial Earth—hence, his model has been called the *Thioester World.*

A. De Duve assumes that life began in some early "volcanic setting" that featured hydrogen sulfide gas and iron sulfide minerals, along with a diverse suite of organic molecules. Of special importance was the production of a group of sulfur-containing molecules called *thioesters*. These molecules feature a strong bond between one sulfur atom and one carbon atom. If broken, this bond releases significant energy.

B. Essential in the metabolism of all modern cells, thioesters transfer energy to promote metabolic reactions by breaking the carbon-sulfur bond. That's how thioesters promote construction of larger molecules from smaller ones. A steady supply of thioesters represents the chemical energy needed to drive protometabolism in the Thioester World.

C. Thioesters also tend to form bonds with the amino acids that were surely available in the prebiotic soup. When they do, amino acids become more reactive.

D. De Duve's Thioester World scores well in the game of metabolism. Because the model relies on a reasonable prebiotic environment, rich in plausible organic molecules, it has been generally well received by the origins community. Miller and his allies were quite sympathetic.

VI. Many origins-of-life researchers would agree that Christian de Duve's Thioester World scenario is attractive in its broad outlines, but it lacks many specific chemical details on which to commence an experimental program. Because details matter, the next lecture will consider the remarkable model of Günter Wächtershäuser—a model known as the *Iron-Sulfur World.*

Essential Reading:

De Duve, *Vital Dust: Life as a Cosmic Imperative.*

Hazen, *Genesis: The Scientific Quest for Life's Origin*, chap. 14.

Supplemental Reading:

Dick and Strick, *The Living Universe: NASA and the Development of Astrobiology*, chap. 2.

Fry, *The Emergence of Life on Earth*, chap. 7.

Lahav, *Biogenesis: Theories of Life's Origin*, chap. 21.

Wills and Bada, *The Spark of Life: Darwin and the Primeval Soup*, chap. 2.

Questions to Consider:

1. Why was Sidney Fox's work on the Protenoid World discredited by most scientists in the origin-of-life field? Why do you suppose philosophers of science take him more seriously?
2. Why was Christian de Duve's work on the Thioester World respected by most scientists in the origin-of-life field?

Lecture Nineteen—Transcript
Self-Replicating Molecular Systems

We are now three-quarters of the way through our investigation of life's chemical origins. We do not know exactly how life emerged, but our strategy has been to close the gap between the top-down study of the oldest fossil evidence, on the one hand, and the bottom-up study of increasing chemical complexity in the prebiotic world, on the other.

We have seen top-down research has suggested that traces of life remain from almost four billion years ago on an Earth that was still violent and alien. Massive volcanoes poured out rivers of lava and spewed noxious gases into the atmosphere. Asteroids and comets blasted the surface in a brutal bombardment. The Earth's surface was also subject to violent weather and giant tides.

Next, we traced the emergence of life through two initial steps. First, the ancient Earth had accumulated vast amounts of useful organic molecules that had emerged from all sorts of chemical reactions—in the sunlit oceans and atmosphere, in the deep crust and in the far reaches of space. Second, some of these molecules became concentrated and self-assembled into vesicles. Others were selected on mineral surfaces where they were assembled into polymers and other macromolecules of biological interest.

To consider what else happened, let's revisit the question, "What is life?" Most researchers agree that life, at a minimum, can be defined as a chemical process possessing three key attributes—the ability to grow, the ability to evolve, and the ability to reproduce. At the dawn of life, growth and evolution were not very impressive. Individual cell growth occurred on an almost invisible scale for most of Earth's history. For at least three billion years, from about 3.8 billion years ago to about 700 million years ago, all living organisms on Earth were microscopic single-celled objects. You would have been hard pressed to watch growth. The evolution of primitive cellular life was even slower. It took billions of years for cells to develop such basic biochemical tasks as using sunlight for energy or oxygen for respiration.

However, the third attribute, reproduction, must have progressed with an unstoppable swiftness. Rapidly reproducing populations of cells doubled and redoubled, and spread throughout Earth's oceans. In a geological blink of an eye, life colonized and transformed the globe, perhaps as fast as a few centuries. So, I think it is fair to say that reproduction, more than any other characteristic of life, set that early world of life apart from the prebiotic era.

For many origin-of-life researchers, the experimental Holy Grail has become the creation of a self-replicating molecular system in a test tube. As you might expect, it is a lot more difficult to achieve chemical self-replication than to study the earlier stages of life's emergence, including the synthesis of biomolecules and the selective concentration of those molecules into membranes and polymers. In a self-replicating system, it is not enough to make a few interesting chemicals. In a reproducing chemical system, one group of molecules has to multiply repeatedly, while other molecules serve as food. That is an extremely challenging chemical task but it is worth the attempt. In fact, the first scientist to invent a self-replicating chemical system in the laboratory will have a legitimate claim to creating a life-like system.

In this lecture and the next, I want to consider the topic of self-replicating molecular systems. Such systems are not necessarily alive but they do have something like metabolism. The emergence of metabolism brings us much closer to our goal of understanding the origins of life.

The simplest self-replicating molecular system that you can imagine consists of just one kind of molecule that makes copies of itself. If you put such a molecule in the right chemical environment, with just the right combination of smaller chemical compounds to act as food, then a single molecule becomes two, then four, then eight, and so forth. The concentration of molecules will increase in a geometrical expansion.

Chemists call this kind of molecule "autocatalytic". Literally, it brings about its own transformation. Autocatalytic molecules assemble themselves; they act as templates that latch onto and assemble their own chemical components from a solution. You can imagine that a single, self-replicating molecule is a rather complex object but organic chemists have managed to synthesize several different kinds.

One important attribute of all self-replicating molecules is that they must be "self-complementary". Now complementary molecules are common in nature. These are pairs of molecules that happen to fit snugly together because of their matching shapes and the arrangement of their chemical bonds. Many of the proteins in our bodies, for example, are able to function because they are complementary to other vital molecules: food, for example, or toxins. It is because of this biological activity that the synthesis of complementary molecules has become a huge business in the world of flavors, fragrances, and pharmaceuticals. There is a whole field of computer-aided drug design that tries to create molecules of a specific shape. After all, if you can design a molecule that fits precisely into pain receptors and thus reduce pain, then you can make a lot of money. So, complementary molecules are common and new ones are being made all the time.

Nevertheless, only a small handful of molecules are known to be self-complementary and even fewer are able to self-replicate. Hungarian-born chemist Julius Rebek Jr. created one of the first examples to do both. Rebek, who is now at the Scripps Research Institute in La Jolla, California, has been designing self-replicating molecules since the 1980s. In 1989, Rebek's group managed to form a self-replicating molecule by linking together three smaller molecules—the one-ring molecule adenine; the two-ring molecule napthalene; and imide, which is a nitrogen atom, bonded to hydrogen and two carbon monoxides. When joined end to end, the resulting adenine-napthalene-imide molecule is self-complementary. If you immerse one of these combined molecules in a solution containing lots of these three-component molecules, it will make copies of itself. Now, this combination is not a likely biological molecule but it does demonstrate the concept.

Chemist Reza Ghadiri, who studies self-replicating peptides, leads another active research group at the Scripps Institute. Peptides are arguably of great relevance to the origin of life. Peptides are just like proteins. They're chainlike molecules built from long sequences of amino acids and their building blocks are drawn from the same set of twenty amino acids that make up proteins. In fact, the only real difference between peptides and proteins is their length. Peptides are typically up to a few tens of amino acids long, whereas proteins may be hundreds or thousands of amino acids long.

Ghadiri's group achieved a major breakthrough in 1996, when they designed the first self-replicating peptide. This peptide was a sequence of 32 amino acids. However, there is a catch. For self-replication to occur, this peptide required a steady supply of two specialized reactive fragments, one of them 15 amino acids long and the other 17 amino acids long. Only in the presence of a steady supply of these two exotic precursor fragments could new peptides form spontaneously.

Base pairs of DNA are probably the most famous complementary pairs of molecules. As James Watson and Francis Crick so famously announced in 1953, the double-helix structure of the DNA molecule is like a long, twisted ladder. The vertical supports are chains of alternating phosphate and sugar molecules, while the rungs of the ladder are formed from complementary pairs of the molecules called bases. The base cytosine, which is C, is complementary to guanine, G; and the base adenine, A, is complementary to thymine, T. Not only are the resulting base pairs, C-G and A-T, complementary, but also are the two halves of every DNA double helix strand. So, for example, a single DNA strand with the base sequence ACGTA is complementary to a second single strand TGCAT. The beauty of this molecule lies in its two-fold ability to both carry and duplicate information. When the double helix splits apart, each half can make a copy of the complete double helix. In their 1953 landmark paper, Watson and Crick remarked, "It has not escaped our attention that the specific pairing we have postulated immediately suggests a possible copying mechanism for the genetic material." This is a very famous quote in the history of science.

The vast majority of DNA strands are complementary, yet not self-complementary. That is, the two halves of most double helix strands have complementary, yet different, base sequences. Still, given the elegant symmetry of DNA, it is an easy step, conceptually, to design self-complementary strands. In 1986, the German chemist Guenter von Kiedrowski and his coworkers produced the first self-complementary DNA strand, the six-nucleotide sequence CCGCGG, which makes exact copies of itself as long as enough fresh C and G nucleotides are provided.

The ability of these few known types of self-replicating molecules to make copies of themselves seems almost magical. Nevertheless, these molecules do not come close to meeting the minimum requirements for life for at least two reasons. For one thing, such molecular systems need a constant supply of small, highly specialized molecular building blocks. Those odd pieces have to be synthesized and supplied by the researchers. There is no evidence that these idiosyncratic building block molecules could ever arise spontaneously in a natural environment. In addition, these self-replicating molecules cannot be alive because they can never change or evolve, any more than a Xerox copy can evolve from an original. Living systems require that added level of complexity.

Now, let's up the ante and increase the molecular complexity to systems with two or more molecules that, together, form a self-replicating cycle or network. Intense effort is being focused on this research and a variety of strategies for molecular self-replication has been discovered. The simplest systems are "cross-catalytic". That is two molecules, which we can call AA and BB, are made from smaller feedstock molecules A and B. So, if AA catalyzes the formation of BB from lots of individual B molecules and if BB, in turn, catalyzes the formation of AA from lots of individual A molecules, then you have a self-replicating system. This system will keep making copies as long as a steady supply of the food molecules A and B are provided.

One familiar example of this process is DNA, every strand of which is cross-catalytic. If you split the double helix down the middle, then each half makes copies of the other half, provided sources of the four DNA nucleotides, A, C, G, and T, are available.

It's easy for theorists to imagine more elaborate, self-replicating networks of molecules. Rather than just two cross-catalytic molecules, think of a system with a dozen or more different molecules, each of which promotes the production of another chemical compound in the system. A leading proponent of this type of model is Santa Fe Institute theorist Stuart Kauffman, who argues that such "autocatalytic networks" are the most likely form of protolife. Given the bewildering complexity of living cells, Kaufmann proposes that the molecular repertoire of that first living autocatalytic network might have been vast, perhaps hundreds or thousands of different interacting molecular species. To make his point, Kauffman draws complex spaghetti-like illustrations of hypothetical reaction cycles, with each molecule playing a role in the entire integrated molecular network.

You can see that we are quickly getting into complex territory and that has an important advantage in thinking about life's origins. A single self-replicating molecule does not evolve, nor does a cross-catalytic pair of molecules. But vast networks of reactive molecules like the ones Kaufmann advocates do have the potential to increase in complexity and efficiency; that is, to evolve through numerous different interactions among molecules. Almost inevitably, the large number and variety of molecular interactions fosters the emergence of new metabolic pathways and leads to nested cycles of synthesis. Kauffman's networks are just like living metabolic cycles in that they incorporate a certain degree of variability. Such inherent variations enable the system to change. The collection of molecules finds new, more efficient, reaction pathways and, in the process, it displays one of the defining traits of life, the ability to evolve.

In many ways, Stuart Kauffman's theory of autocatalytic networks of molecules provides an appealing model for the emergence of primordial life. You see, in Kaufmann's model, the various molecules in the network catalyze each other, so there is lots of complexity but you do not need a genetic mechanism. Replication of the entire network is intrinsic to the system. Nevertheless, the system is likely to vary and gradually evolve. So, the autocatalytic network would appear to meet the minimum criteria for life. It has the ability to grow, to reproduce and to evolve.

On paper, Kauffman's model just makes good chemical sense. However, there is a huge gap between a plausible theory on paper and an actual experiment with real chemicals. What exactly is the chemical "A" in Kaufmann's spaghetti diagrams? What is "B"? Which molecule catalyzes which? This is a job for the chemists.

The best way to test Stuart Kaufmann's theory of autocatalytic networks is to design a simple metabolic cycle in the lab. Metabolism is life's preferred cycle of chemical self-replication. Every metabolic cycle requires two kinds of input. First, metabolism requires a source of energy. In experiments, it is best to rely on chemical energy, since that follows what all simple living cells seem to use. Of course, there are many photosynthetic cells that use sunlight but photosynthesis requires layers of chemical complexity far beyond what is found in ancient, simple cells. Therefore, it is better to stick to chemical energy. Secondly, metabolism requires reliable feedstocks of small building block molecules. Ideally, those molecules should not be too big or exotic. The most convincing primitive metabolic cycles are those that rely on the most abundant small molecules, including carbon dioxide, ammonia and water.

A metabolic cycle is sometimes compared to fire, but that is actually not a very good analogy. In burning, oxygen and some kind of carbon-rich fuel, such as coal or natural gas, react and release energy to the surroundings. That energy quickly dissipates. However, the objective in metabolism is to capture as much of that energy as possible and use it to make new, useful molecules that can reinforce the cycle. Therefore, unlike burning, metabolism requires a sequence of chemical reactions that work together in concert.

What was a primitive metabolic cycle like? Several origin-of-life researchers have proposed metabolic models and I tend to think of this effort as a kind of a game. It's the chemist's job to invent a cycle of chemical reactions that obey three basic rules.

The first rule of the game of metabolism is that the cycle has to consist of a sequence of progressively larger molecules. Each molecule in the cycle combines with carbon dioxide or water or some other small molecule to make the next molecule in the cycle. Then the largest molecule in the cycle should be able to split into two smaller molecules, both of which are also in the cycle. You see, by constantly building up to a larger molecule that splits in two, you keep doubling the number of cycles and the system grows. One molecule becomes two, two become four and so on. The cycle keeps reproducing itself.

The second rule of the game of metabolism is to identify a plausible prebiotic environment. That environment has to provide a reliable supply of all the necessary raw materials, including small molecules made up of carbon, nitrogen, sulfur and other essential elements, plus a steady source of energy. Also, as I've indicated, the chemical energy of unstable minerals is preferable to harsh sunlight or lightning. Here is a key point; *all* of the different types of molecules that make up the cycle have to be able to survive in that environment, at least long enough to keep the cycle going. If any one molecule in the cycle is unstable, then the whole cycle is unstable.

The third rule of the game of metabolism is that an unbroken chemical history must connect Earth's past with the present. The earliest metabolic cycle has to form a chemical continuum with the metabolism of today's cells. Once life began, there was no opportunity to start over with an entirely new chemical repertoire. Therefore, deep within the metabolism of our cells, we're likely to find "fossil" biochemical pathways—primitive chemical steps that point to life's simpler beginnings. Any plausible model of metabolism has to conform to this "principle of continuity". All of the molecules and energy sources we propose must lie close to the basic metabolism of modern cells. The only way to test these chemical theories is through experiments and there are so many experiments to try. Every single chemical step in a metabolic cycle is a potential experiment. Indeed, it is critical to be able to show that each step does work.

In the remainder of this lecture, I want to tell you about two of the most influential models of primitive metabolism, one involving baked amino acids and the second involving sulfur-carbon bonds. These models are called the "Protenoid World", as devised by the chemist Sidney Fox, and the "Thioester World" of Nobel Prize-winning chemist Christian de Duve. In the next lecture, I will continue with a group of closely related hypotheses based on what is known as the reverse citric acid cycle, hypotheses that include the elaborate "Iron-Sulfur World" of Günter Wächtershäuser, which builds on the general approach of Sidney Fox as well as some of the more specific ideas of Christian de Duve.

Sidney Fox entered the field of origins-of-life chemistry shortly after the Miller-Urey experiment of 1953, and he proposed the first comprehensive model of primitive metabolism. Fox's career was strongly influenced by Thomas Hunt Morgan, who was a world famous geneticist and served as a member of Fox's Caltech thesis committee. Morgan told Fox that the most important biological problems were tied to the behavior of proteins. Well, in the mid-1950s, shortly after becoming a protein chemist on the faculty at Florida State University, Fox began to study the role of proteins in life's origin.

Sydney Fox imagined a scenario in which amino acids from the primordial soup dried and baked on a cooling volcanic rock. He tried to duplicate these conditions in his laboratory by drying amino acids on a hot surface at 170°C. What he found was that his chemicals quickly cooked. They polymerized into a lumpy substance that he called "proteinoid" or "protenoid," thus implying a relationship to biological proteins. Fox announced this discovery in 1958 and for the next three decades, it would shape what many of his peers saw as a troubled career in origins research.

It is not surprising that Fox's protenoids became the centerpiece of his professional life because they did display fascinating behavior. Fox and his many collaborators at his Florida lab did many experiments on protenoids. They heated and cooled solutions to mimic prebiotic conditions, under which some protenoids acted like catalysts, to enhance reaction rates, though to much smaller degree than protein catalysts in modern cells. Protenoids self-organized, not unlike lipids; when placed in hot water they formed microbe-sized spheres, which he called "microspheres." Fox claimed that his spheres had a bilayer membrane of sorts. Sometimes these spheres appeared to divide. These intriguing cell-like spheres absorbed more protenoids and they could grow and divide to yield what was like a second generation of microspheres.

Now, the detailed structure of these microspheres and the mechanisms for self-replication were not obvious but these objects, which grew from nutrients in solution, represented a true metabolic model for life's origin.

So how does Fox's model score in the game of metabolism? Well, at first glance it seems to do rather well. The metabolic cycle is made with widely available amino acids. The volcanic environment near a tidal zone is certainly plausible for the early Earth. In addition, protenoids seem to be consistent with the principle of biological continuity because proteins are central to modern biochemistry. Also, Fox had lots of experiments with qualitative results that seemed to back up the model.

For more than a decade, Fox's career thrived. He got lots of funding from NASA, and he was able to establish his own Institute of Molecular Evolution at the University of Miami in 1964. He also was able to attract many graduate students who expanded his research program.

However, Sidney Fox fell into a dangerous trap for a scientist. He fell in love with his own theory. By the mid 1960s, his claims for the Protenoid World model became increasingly extreme. He stated for the press that the origin-of-life problem had been solved. He announced that protenoids are "alive in some primitive way". In addition, in his 1988 book, *The Emergence of Life,* he went so far as to claim that protenoid microspheres possess a kind of "rudimentary consciousness."

Many people in the origins-of-life community, particularly the influential Stanley Miller and his numerous students, were unconvinced. As early as 1959, Miller and Harold Urey had begun to criticize Fox's claims. They especially disliked the sensationalistic use of terms such as "protenoid" and "lifelike." They raised numerous objections regarding the rather subjective way that Fox and his students described behaviors, such as growth and reproduction. Also, they emphatically objected to the idea that the Protenoid World could have arisen before any sign of a genetic mechanism to pass information from one generation to the next. After all, how could the key molecules DNA and RNA have arisen in such a world?

Today, the majority opinion is that Sydney Fox went too far in his uncritical advocacy of the Protenoid World. Still, his metabolic model was extremely influential for at least two reasons. First, Fox developed the first comprehensive origin model in which metabolism came before genetics. Second, he introduced the important philosophical position that origin processes might be nonrandom and deterministic. That is a view that was later embraced by numerous other origins workers, and it is in sharp contrast to the picture of a random prebiotic soup. So, even if the model is wrong in crucial respects, Fox had a great influence on the field.

Let me tell you about a very different model of primitive metabolism, this one proposed by the great Belgian chemist Christian de Duve. His theory is outlined in his popular book, *Vital Dust*. De Duve made his scientific reputation, and won a Nobel Prize in the process, by studying the structural and functional organization of cells. He then shifted his work to the origins of cellular life. It is not really surprising that his background in cellular biology colors everything about his origins theory.

There are aspects of de Duve's theory that are somewhat ambiguous. For example, he takes a rather neutral stand in the question of whether metabolism preceded genetics. I think this gets at the question of semantics—when do you call a chemical system alive? For de Duve, that point is the simplest imaginable living entity with a cell with a full complement of genetic material to control cellular functions. However, he also proposes a stage in the emergence of life at which there is what he calls "protometabolism," as opposed to the more complex metabolism of modern cells but in that stage there is no genetic system. Therefore, metabolism, understood as proto-metabolism, comes first. It is really a question of where you want to place the divide between living cells and nonliving chemistry.

De Duve assumes that life began in some early environment, which he specifies only vaguely as a "volcanic setting". He assumes that this environment featured hydrogen sulfide gas and iron sulfide minerals along with the usual diverse suite of organic molecules like amino acids and lipids. Of special importance in de Duve's model is the production of a group of sulfur-containing molecules called "thioesters." These molecules feature a strong bond between one sulfur atom and one carbon atom. Now that is a bond that holds a significant amount of energy, energy that can be released if the bond is broken. De Duve's central hypothesis is that many energy-rich thioester molecules were available on the primordial Earth, hence his model has been called the "Thioester World."

So why would de Duve focus on thioesters? Well, there are good reasons that hark back to the principal of continuity. Thioesters are essential in the metabolism of all modern cells, forming peptides, fatty acids, sterols and other compounds. Thioesters transfer energy to promote metabolic reactions by breaking that strong carbon-sulfur bond. That is how thioesters promote the construction of larger molecules from smaller ones. So a steady supply of thioesters represents the chemical energy needed to drive protometabolism in the thioester world.

Thioesters are an attractive choice for another reason. They have a tendency to form bonds with the amino acids that were surely available in the prebiotic soup. When amino acids link to thioesters, the amino acids become more reactive. Indeed, amino-acid-thioester groups will assemble spontaneously into short protein-like chains, which de Duve has called "multimers," as opposed to longer polymers. He speculates that some of these molecules might have acquired important by pure chance catalytic properties, such as the ability to promote metabolism and self-replication. The high-energy bonds of thioesters might have played the role of ATP in a world without ATP and later served as an intermediate toward the formation of ATP. Eventually an autocatalytic cycle might have emerged.

De Duve's Thioester World model has been generally well received by the origins community. For one thing, people tend to listen to Nobel Prize winners. What's more, the Thioester World hypothesis is well aligned with Miller's primordial-soup paradigm. De Duve's scenario is dependent on a steady supply of amino acids, carbohydrates and so forth, as well as the energy-rich thioesters, themselves. Therefore, Miller and his allies were quite sympathetic.

So how does the Thioester World scenario score in our game of metabolism? Generally, it does quite well, since it relies on a reasonable prebiotic environment rich in plausible organic molecules. Also, the thioester angle is appealing because it looks a lot like the role of thioesters in modern cells. I think it is safe to say that there are lots of gaps in this model, as well. De Duve, like Fox before him, doesn't identify the specific molecules that make up his metabolic cycle. The whole protometabolism stage is vaguely defined. Thioesters promote the synthesis of some hypothetical catalytic multimers, which in turn trigger the synthesis of a genetic molecule like RNA. It is only at that point that we see the processes of modern biochemistry.

I think many origins-of-life researchers would agree that Christian de Duve's Thioester World scenario is quite attractive in its broad outlines but that it lacks many specific chemical details on which to commence an experimental program.

Therefore, because details matter, the next lecture will consider the highly detailed and remarkable model of Günter Wächtershäuser, a model known as the Iron-Sulfur World.

Lecture Twenty

Günter Wächtershäuser's Grand Hypothesis

Scope: One of the most contentious debates surrounding the origin of life involves which essential biological process came first—metabolism or genetics. The most elaborate and comprehensive theory of metabolism is contained in the Iron-Sulfur World of Günter Wächtershäuser.

Outline

I. Let's consider the two contending views in the debate over whether metabolism or genetics came first.

A. Those who favor genetics as the first step in life's origins are swayed by life's incredible complexity. Without a genetic mechanism to pass all the necessary information from one generation of cells to the next, there would be no way to reproduce that complexity.

B. Those persuaded by the metabolism-first scenario say that chemical simplicity must have preceded chemical complexity, if life began as a sequence of emergent events, each step more complex than the last. The chemistry of primitive metabolism is, at its core, much simpler than the chemistry of genetics. Metabolism requires relatively few small molecules working in a cycle to self-duplicate.

1. Many think that a core metabolic cycle lies at the heart of every modern cell. The so-called citric acid cycle survives as a chemical fossil from life's earliest stages.
2. Not only does this cycle make copies of itself, but it is also an engine of synthesis. The molecules of the citric acid cycle are the starting points for the synthesis of life's biochemistry, and that includes the molecular building blocks of DNA and RNA.

II. Without some experimental breakthroughs, the question will probably remain one of the hottest in the field. Meanwhile, the Iron-Sulfur World hypothesis, presented in a series of articles from 1988–1992, has changed the way we think about the origins of life and has added to the debate.

A. German chemist Günter Wächtershäuser has challenged many of the assumptions about life's beginnings. According to him, the first life was *autotrophic*, literally, "self-nourished." The cell makes every one of life's essential biomolecules from small building blocks: carbon dioxide (CO_2), water (H_2O), ammonia (NH_3), hydrogen sulfide (H_2S), and so forth. Chemical synthesis is accomplished in little steps, adding just a few atoms at a time. He soundly rejects the prebiotic soup of organic chemicals as having anything to do with life's origins.

B. In sharp contrast, most origins-of-life experts have assumed that the first cell was *heterotrophic*, which means "other nourishment." A heterotrophic cell obtains many of its essential biomolecules from its surroundings; it doesn't have to make them from scratch.

C. This difference is a fundamental point of conflict among researchers.

1. Supporters of a heterotrophic origin assert that it was easier for a primitive cell to eat the diverse molecular products that must have been available in the prebiotic soup, rather than to make them from small molecules. Modern heterotrophic cells are always much simpler than autotrophic cells; they don't need elaborate chemical machinery to make amino acids, sugars, and so forth.
2. True simplicity, advocates of autotrophic origins say, is found in building all biomolecules just a few atoms at a time. Autotrophisic cells do not depend on the chance resources of a local environment. Organisms make their organic molecules from scratch, the same way every time, from the simplest building blocks. The prebiotic soup is irrelevant. The chemistry of the autotroph is hardwired into the universe and emerges from geochemistry inevitably.

III. As with any metabolism-first model, the Iron-Sulfur World theory requires a source of energy, a supply of molecules, and a self-replicating cycle.

A. The model relies on the widely available chemical energy of minerals that are out of equilibrium with their environment. These minerals can react with their surroundings and release energy; many such minerals are formed in deep volcanic zones.

B. Wächtershäuser zeroes in on the common iron-sulfide mineral pyrrhotite with a 1:1 ratio of iron to sulfur. Pyrrhotite is often deposited at the mouths of hydrothermal vents, where it is unstable with respect to the surrounding seawater. The mineral gradually transforms to other more stable chemicals and releases energy.

1. In one reaction, pyrrhotite combines with a common volcanic gas, hydrogen sulfide (H_2S), to produce pyrite (which has a 1:2 ratio of iron to sulfur) plus hydrogen gas and energy.

$$FeS + H_2S \rightarrow FeS_2 + H_2 + \text{Energy}$$

2. The jolt of energy causes the hydrogen gas to react immediately with carbon dioxide in seawater. That reaction produces organic molecules, such as formic acid (HCOOH). Hydrogen and carbon dioxide plus energy react to make formic acid.

$$\text{Energy} + H_2 + CO_2 \rightarrow HCOOH$$

3. Energy produced in the first reaction doesn't just get lost to the environment but is used right away to make an organic molecule.
4. This kind of cooperative cascading of reactions is key to Wächtershäuser's model. Reaction after reaction builds up essential organic molecules from CO_2 and other simple gases.

IV. Many studies in the past 15 years have tested aspects of the Iron-Sulfur World model.

A. With a team of chemists at the University of Regensburg, Wächtershäuser studied the critical first reaction of iron sulfide and hydrogen sulfide. As he predicted, pyrrhotite plus hydrogen sulfide reacted to produce pyrite and hydrogen gas.

B. In 1996, Dutch scientists Wolfgang Heinen and Anne Marie Lauwers mimicked Wächtershäuser's scenario by mixing powdered iron sulfide and hydrogen sulfide in water with an atmosphere of carbon dioxide. As predicted, they produced a variety of synthetic organic compounds.

C. A number of other experiments have produced increasingly complex organic compounds under the influence of iron and nickel minerals.

1. Wächtershäuser's group made the 2-carbon molecule acetate and amino acids.

2. The Carnegie Institute produced the 3-carbon compound pyruvate.

D. At least part of Wächtershäuser's model has been confirmed, but a closed metabolic cycle that duplicates itself is still needed.

V. The centerpiece of the Iron-Sulfur World hypothesis is a metabolic pathway called the *reverse citric acid cycle.* (see diagram in Part One, Lecture Twelve, p. 65).

A. A simple self-replicating cycle of compounds closely related to the core metabolism of every modern cell, the citric cycle is an engine of biosynthesis that represents one of the most ancient biochemical aspects of cells.

B. This metabolic cycle is built from a sequence of 11 small molecules, each one made of carbon, hydrogen, and oxygen. The cycle works by gradually stripping off a few atoms at a time to make a succession of smaller molecules, say, from 6-carbon citric acid, to 5-carbon malate, to 4-carbon oxaloacetate, to 3-carbon pyruvate, to 2-carbon acetate, and so forth.

C. Each of these molecules in the cycle is the starting material for other essential biomolecules. Animals and plants employ this citric acid cycle to manufacture just about every kind of essential biochemical.

D. This cycle can run in the opposite way. Rather than starting with citric acid to make starting materials, small molecules build up to bigger ones step by step. Key for primitive metabolism is a cycle that doubles on every turn to make copies of itself. The same sequence of steps is present in life today; thus, the model obeys the principle of continuity.

E. In modern cells, however, every step in the citric acid cycle is promoted by a protein catalyst of incredible complexity. Without these specialized proteins to increase the chemical efficiency, it is hard to imagine the reverse citric acid cycle at all.

F. Someone has to demonstrate how that first metabolic cycle was able to go it alone, without the help of protein catalysts, and that demonstration must show that *every* step in the cycle works.

VI. Wächtershäuser relies on sulfide minerals to promote metabolic reactions.

- **A.** Most of the modern metabolic proteins that promote reactions in the citric acid cycle have at their core a small cluster of iron or nickel atoms, as well as sulfur atoms. Those clusters of atoms look exactly like tiny bits of sulfide minerals, and they trigger the desired chemical reactions. If modern, highly efficient proteins use sulfide clusters, perhaps ancient minerals played the same role, albeit less efficiently.
- **B.** Wächtershäuser invokes another chemical idea based on sulfur. In many chemical reactions, hydrogen sulfide can be substituted for water. Wächtershäuser proposes a metabolic cycle in which several crucial reactions use the sulfur-analogs of citric acid cycle compounds.
  1. His prediction is that hydrogen sulfide can promote faster reactions that release more energy.
  2. These sulfide reactions automatically couple to the formation of pyrite, the mineral whose formation releases energy essential to the Iron-Sulfur World.
  3. Together, all these reactions can form a self-consistent network.
  4. In Wächtershäuser's view, once a sulfide version of the reverse citric acid cycle is established, it is only a matter of time before the switch is made to water. Hydrogen sulfide is common only at volcanic vents, but water is everywhere.

VII. Impressed with Wächtershäuser's model, our Carnegie Institution group initiated a series of gold-tube experiments to discover how citric acid degrades in hot, pressurized water.

- **A.** For our first experiments, we avoided the hazardous sulfur system and studied the closely related water-based system. We saw evidence for two different reaction pathways. A significant fraction of the citric acid molecules broke down by what we called the *alpha pathway*. That begins with citric acid splitting into acetate plus oxaloacetate, which is exactly how the reverse citric acid cycle is supposed to work.
- **B.** Beyond that, any similarity ended. We found that oxaloacetate decomposes quite rapidly to pyruvate plus carbon dioxide, and

then, most of the pyruvate decomposes to acetate. Instead of building back up toward citric acid, we observed destruction of citric acid.

C. In a second round of experiments, we added sulfide minerals and carbon dioxide, both important ingredients in the Iron-Sulfur World. We couldn't find a combination of minerals and reactants that promoted the reaction from 3-carbon pyruvate to 4-carbon oxaloacetate, and we concluded that the reverse citric acid cycle doesn't work.

D. Some of the citric acid molecules followed a different pathway. This *beta pathway* operated much like the normal citric acid cycle—not the reverse cycle. Is it possible that the beta pathway might be part of a self-replicating citric acid cycle that has long since been replaced?

E. After more experiments, we found evidence that some steps of the beta pathway can be reversed in the presence of nickel sulfide. Perhaps there really is an abiotic route from carbon dioxide to citric acid. Someday, someone will try those risky experiments with hydrogen sulfide.

VIII. What did the Iron-Sulfur World look like?

A. Glasgow-based geoscientists Michael Russell and Allan Hall have proposed that *iron-sulphide bubbles* acted as the first membranes. In their scenario, the iron-sulfide bubbles form spontaneously at hydrothermal vents. The less acidic water of the vents comes into contact with the more acidic water of the ocean. The bubbles form spontaneously and act like primitive cell membranes that enclose metabolic chemicals. Energy comes from the contrast of acidity inside versus outside the bubble, promoting numerous metabolic reactions.

B. Wächtershäuser advocates "flat life." In this scenario, the first self-replicating entity is a thin layer of chemical reactants that grows on a sulfide mineral surface and expands laterally. It spreads out as a coating of organic molecules that's much too thin to see. Pieces of these layers might break off and reattach to other rocks. Given time and enough different minerals and environmental conditions, variations in the film would arise, as would competition among new "species" of flat life.

C. The concept of flat life is not yet a majority view, but a self-replicating chemical layer of molecules living on a mineral surface cannot be ruled out. Film-like molecular systems might even persist for billions of years beyond the predation of more efficient cellular life.

Essential Reading:

Hazen, *Genesis: The Scientific Quest for Life's Origin*, chap. 15.

Supplemental Reading:

Fry, *The Emergence of Life on Earth*, chap. 12.

Lahav, *Biogenesis: Theories of Life's Origin*, chap. 21.

Wills and Bada, *The Spark of Life: Darwin and the Primeval Soup*, chap. 4.

Questions to Consider:

1. In what ways does Wächtershäuser's Iron-Sulfur World hypothesis differ from other origin-of-life models? Do you agree with his statement that "something like this has never been done in the entire field"?
2. Why do some scientists suggest that flat life might exist on Earth but hasn't been discovered yet?

Lecture Twenty—Transcript
Günter Wächtershäuser's Grand Hypothesis

The study of life's origins is fascinating for many reasons. It is, of course, a grand question. How did life emerge on Earth? Everyone who is curious about the natural world would like to know the answer to that mystery. It is also fun because so many smart people have proposed so many creative ideas. Those untested speculations naturally lead to controversy and debate.

We have seen that the scientists who investigate life's origins divide themselves into several rival camps. You have some researchers who claim life is a cosmic imperative that emerges anywhere in the universe, if conditions are appropriate, while rival researchers see life on Earth as an improbable chance event and possibly even unique in the universe.

I have told you about experts who advocate that some sort of enclosing membrane must have preceded life; but other distinguished scientists, some of whom we will meet in this lecture, counter with their own models of "flat" surface life. Many origins-of-life researchers assume that life's origins must have depended on the Sun's energy but others are equally persuaded that Earth's internal heat was the triggering source of energy.

In each of these debates, we lack adequate observational or experimental evidence to make a choice. In addition, as is often the case in human discourse, the less the available evidence, the more polarized and inflexible people's positions become.

In the last lecture, I touched on what is, perhaps, the most fundamental and contentious scientific debate related to the origins of life, that is, the timing of the appearance of two essential biological processes, metabolism and genetics. In this lecture, I will continue with that theme. First, I want to consider briefly the two contending views in the debate over whether metabolism or genetics came first and then I want to examine the most elaborate and comprehensive theory of metabolism-first, the Iron-Sulfur World of Günter Wächtershäuser.

As we have seen, metabolism is the process that is used by every cell to manufacture biomolecular structures from a source of energy, such as chemical energy or sunlight, and from small molecular building blocks, such as carbon dioxide and water. It is axiomatic that a cell cannot survive and grow unless it has a reliable source of matter and energy. It is hard to imagine life without metabolism.

Genetics is different but equally important. Genetics relates to the transfer of biological information from parent cells to daughter cells. In modern cells, this transfer is accomplished using the two essential, information-rich polymers, DNA and RNA. An organism is so incredibly complex that it must have some sort of reliable means to pass on its genetic information. It is hard to imagine life without a genetic mechanism.

Here is the central problem. Metabolism and genetics constitute two separate, chemically distinct systems in modern cells. It is like the digestive system and nervous system in your body. They are physically distinct networks but, just as your stomach and your brain can't exist independently of each other, so too are metabolism and genetics linked in today's life.

So, metabolism uses energy and basic molecular building blocks to manufacture DNA and all the other genetic materials, but it's DNA that holds all the genetic instructions to make the hundreds of molecules that are essential to metabolism. This situation is similar to the old puzzle of the chicken and the egg. Which came first? It is almost impossible to imagine back to a time when metabolism and genetics were not intertwined. Therefore, one has to ask, did these two essential attributes of life arise simultaneously or did they arise independently? If they did arise separately, which one came first?

It is important to note that one possibility is the simultaneous emergence of metabolism and genetics. At first blush, this might seem like a great compromise position, but the chemical processes are rather different. Metabolism and genetics depend on completely different sets of molecules. So it's probably easier to think of life arising one small step at a time, with either metabolism or genetics preceding the other, but what is the most logical sequence of emergent steps?

Let's look at the debating points on both sides of this issue. On the one hand, those who favor genetics as the first step in life's origins are swayed by life's incredible complexity. Even the simplest living cell is astonishingly intricate, with hundreds of interconnected chemical pathways. It's easy to see why some researchers say that without a genetic mechanism to pass all the necessary information from one generation of cells to the next, there would be no way to reproduce all of that complexity. According to this viewpoint, metabolism without genetics is really just a bunch of reacting chemicals without direction or control. Remember that modern biologists have to deal with extremely complex cells all the time and their academic curriculum has been dominated for decades by the unifying concept of the genetic code. So it's no wonder that most people with biological training, especially those in molecular biology, are likely to adopt this genetics-first point of view.

Many other scientists, and I include myself in this group, are more persuaded by the metabolism-first scenario. If life began as a sequence of emergent events, each step of greater complexity than the last, then chemical simplicity must have preceded chemical complexity. I also think there's no question that the chemistry of primitive metabolism is, at its core, much simpler than the chemistry of genetics. Metabolism requires relatively few, small molecules working in a cycle to duplicate themselves. That is why Harold Morowitz and Günter Wächtershäuser, among many others, think that a core metabolic cycle lies at the heart of every modern cell. The so-called citric acid cycle, also known for historical reasons as the Krebs cycle or TCA cycle, survives as a chemical fossil from life's earliest stages. Not only does this cycle make copies of itself, but it is also an engine of synthesis. The molecules of the citric acid cycle are the starting points for the synthesis of life's biochemistry and that includes the molecular building blocks of DNA and RNA.

So, if you think that life arose through a sequence of ever more complex emergent steps, then basic metabolism seems more likely to have preceded genetics. I get the sense that many chemists, physicists, and geologists, who don't generally deal with DNA on a regular basis, are more likely to adopt this metabolism-first point of view.

As I am sure you have realized by now, origins-of-life scientists are not shy about stating their views on controversies like the metabolism-first versus genetics-first problem. Also, without some real experimental breakthroughs, this issue will probably remain one of the hottest controversies in the field. Meanwhile, many scientists focus on developing and elaborating models of life's origins. This brings us to the main character of this lecture, the German chemist Günter Wächtershäuser, who we met back in Lecture Eleven talking about the possible catalyzing role of minerals in deep, hot hydrothermal zones.

Günter Wächtershäuser's Iron-Sulfur World hypothesis for the origin of life stands alone in its striking originality, sweeping concept and unprecedented mass of detailed chemical predictions. Since it was first presented in a series of articles between 1988 and 1992, this theory has changed the way we think about the origins of life. Wächtershäuser has challenged many of the most deeply rooted assumptions about life's beginnings. He also knows he's doing it, "You don't mind if I brag a little," he said, "but something like this has never been done in the entire field."

Okay, what is so special about Wächtershäuser's model? First, he flies in the face of many biologists by arguing for a metabolism-first model of life's origins, but this isn't just any metabolism-first scenario. According to Wächtershäuser, the first life was "autotrophic," which means, literally, "self-nourished". An autotrophic cell makes all of its own essential biomolecules. Autotrophic metabolism must begin with chemical simplicity. The cell makes every one of life's essential biomolecules from small building blocks: carbon dioxide (CO_2), water (H_2O), ammonia (NH_3), hydrogen sulfide (H_2S), and so forth. So that means chemical synthesis is accomplished in little steps, adding just a few atoms at a time.

Why should that idea be upsetting to other researchers? The main reason is that it soundly rejects the prebiotic soup of organic chemicals as having anything at all to do with life's origins. If you are Stanley Miller and made your reputation on filling the soup with amino acids, sugars, and the like, that's a hard concept to swallow. For this reason, most origins-of-life experts have assumed that the first cell was heterotrophic, which means "other nourishment". A heterotrophic cell obtains many of its essential biomolecules from its surroundings. It eats amino acids, sugars, and so forth, so it does not have to make them from scratch.

This sharp contrast between autotrophic and heterotrophic life is quite profound and it represents a fundamental point of conflict among origins-of-life researchers. The arguments are compelling on both sides. On the one hand, supporters of a heterotrophic origin say that it's a lot easier for a primitive cell to eat the diverse molecular products that must have been available in the prebiotic soup, rather than to make them from small molecules. Why would you go to all the trouble of synthesizing lots of amino acids, for example, if they were readily available in the environment? Modern heterotrophic cells are always much simpler than autotrophic cells because they don't need as much elaborate chemical machinery to make amino acids, sugars and so forth. So it makes sense to assume that the simpler heterotrophic cells came first.

But advocates of autotrophic origins are equally insistent and their arguments are also compelling. True simplicity, they say, is found in building all biomolecules just a few atoms at a time. In that way, you only need a few simple kinds of chemical reactions. There is also a philosophical reason to prefer autotrophism because it's deterministic. Autotrophism does not depend on the chance resources of a local environment. Autotrophic organisms make their organic molecules from scratch, the same way every time, from the simplest building blocks. Such simple synthesis could occur on any viable planet or moon in a predictable, deterministic chemical path. For Günter Wächtershäuser and other committed advocates of autotrophy, the random, dilute prebiotic soup is irrelevant to the origin of life. That is because the chemistry of the autotroph is hardwired into the universe. Autotrophic life emerges from geochemistry, inevitably as basalt or granite.

Let's look at the details of Wächtershäuser's Iron-Sulfur World model. As with any metabolism-first model, it requires a source of energy, a supply of molecules, and a self-replicating cycle. So here is how he does it.

First, his model relies on the widely available chemical energy of minerals that are out of equilibrium with their environment. In other words, these are minerals that can react with their surroundings and release energy. Many of these minerals are formed in deep volcanic zones. Wächtershäuser zeroes in on the common iron-sulfide mineral called pyrrhotite; it has a one-to-one ratio of iron to sulfur. Pyrrhotite is often deposited at the mouths of hydrothermal vents, where it is unstable with respect to the surrounding seawater. So the mineral gradually transforms to other more stable chemicals and releases energy. In one such reaction, pyrrhotite combines with a common volcanic gas, hydrogen sulfide—that's H_2S—to produce the shiny mineral pyrite,—that has a one-to-two ratio of iron to sulfur—plus hydrogen gas and lots of energy. This is what that reaction looks like. Pyrrhotite and hydrogen sulfide on the left side react to pyrite plus hydrogen plus energy.

So you get this jolt of energy, and that boost causes the hydrogen gas to react immediately with carbon dioxide in seawater. That reaction produces organic molecules, such as formic acid (HCOOH), like this. Hydrogen and carbon dioxide plus energy react to make formic acid.

The real trick here is that the energy produced in the first reaction doesn't just get lost to the environment. You know, the way the heat of a fire just radiates away into the surroundings. Instead, the energy from Wächtershäuser's reactions is used right away to do something useful, to make an organic molecule. It is this kind of cooperative cascading of reactions that's key to Wächtershäuser's model. Reaction after reaction builds up essential organic molecules from CO_2 and other simple gases.

When he first proposed these speculative ideas, there was not much evidence to support them. But good ideas often trigger experiments and there have been many studies in the past fifteen years or so to test aspects of the Iron-Sulfur World model. Wächtershäuser, himself, initiated the first experiments by collaborating with a team of chemists at the University of Regensburg. They studied the critical first reaction of iron sulfide and hydrogen sulfide. Just as predicted, pyrrhotite plus hydrogen sulfide reacted to produce pyrite and hydrogen gas. That discovery appeared in a 1990 article in *Nature.*

Another clear prediction of the Iron-Sulfur World model is that iron-sulfide minerals can promote a variety of interesting organic reactions. So, it was in 1996, the Dutch scientists Wolfgang Heinen and Anne Marie Lauwers mimicked Wächtershäuser's scenario by mixing powdered iron sulfide and hydrogen sulfide in water, with an atmosphere of carbon dioxide. Just as Wächtershäuser had predicted, they produced a variety of synthetic organic compounds.

There have actually been quite a number of other experiments, both by Wächtershäuser and his colleagues in Germany and by our research group in Washington. The hallmark of this work is the production of increasingly complex organic compounds under the influence of iron and nickel sulfide minerals. Wächtershäuser's group made the two-carbon molecule acetate; also, they made amino acids. In one set of our gold-capsule experiments at high temperature and pressure, we produced a variety of organic molecules, including the important three-carbon compound pyruvate. A lot of our focus has been to see if other common hydrothermal minerals can promote the same reactions. So we did more than 300 gold-capsule experiments with lots of different minerals and found that interesting organic molecules are produced in almost every run.

Therefore, at least, part of Wächtershäuser's model has been confirmed. Minerals in a deep-ocean volcanic environment can and do trigger lots of interesting organic synthesis reactions. But, as we've seen in other metabolism-first models, you need much more than just a random assortment of chemical reactions. You must have a closed metabolic cycle that duplicates itself.

The centerpiece of the Iron-Sulfur World hypothesis is a metabolic pathway called the "reverse citric-acid cycle". The reverse citric-acid cycle is a simple, self-replicating cycle of compounds that is closely related to the core metabolism of every modern cell. It is an engine of biosynthesis that represents one of the most ancient biochemical aspects of cells. That is why many scientists have come to the same conclusions as Wächtershäuser. One of the most outspoken proponents of this view is Harold Morowitz, my colleague at George Mason University, whom we met in Lecture Twelve.

Let us take a closer look at this citric-acid cycle. This metabolic cycle is built from a sequence of 11 small molecules, each one made of carbon, hydrogen and oxygen. You can start the cycle anywhere but let's begin with the largest molecule, the six-carbon citric acid that gives the cycle its name. In our bodies, in our cells, the cycle works by gradually stripping off a few atoms at a time to make a succession of smaller molecules. So you go from six-carbon citric acid, to five-carbon malate, to four-carbon oxaloacetate, to three-carbon pyruvate, to two-carbon acetate, and so forth. Each of these various molecules in the cycle is the starting material for other essential biomolecules. So if you add ammonia to pyruvate you get the amino acid alanine; or if you combine carbon dioxide and acetate you get the building blocks of lipid molecules. Animals and plants employ this citric-acid cycle to manufacture just about every kind of essential biochemical.

It turns out that this cycle can also run in the opposite way—the reverse citric-acid cycle. Here we see a different synthesis strategy. Rather than starting with citric acid to make starting materials, you start with small molecules and build up to bigger ones. So, we take the two-carbon molecule acetate and add carbon dioxide to make three-carbon pyruvate. Then add another carbon dioxide molecule to pyruvate to make four-carbon oxaloacetate. By adding more small molecules, you work your way around to the six-carbon citric acid. Then the citric acid does something really neat. Citric acid breaks apart into two smaller molecules, an acetate molecule and an oxaloacetate molecule. The six carbons in citric acid split into two plus four and, all of a sudden, we have the starting materials for two more cycles. That is the key for primitive metabolism. You need a cycle that doubles on every turn to make copies of itself. Here, we have the essential ingredients of biochemistry. A microbe using the reverse citric acid cycle can manufacture all of its essential biomolecules from water and gas.

The reverse citric-acid cycle sounds almost too good to be true. It is a simple cycle that starts from molecules like acetate and pyruvate. Remember, experiments have shown those molecules can be synthesized in hydrothermal zones. Then you just build up larger molecules step by step and the same sequence of steps is present in life today, so the model obeys the principle of continuity. Well, why doesn't everybody agree that this is the solution and get on with it?

It turns out that it is not that simple. You see, in modern cells, every step in the citric-acid cycle—every one of those dozen or so chemical reactions—is promoted by a protein catalyst of incredible complexity. There is one of these proteins that speeds up the splitting of citric acid into the two carbons of acetate and the four carbons of oxaloacetate. Otherwise, citric acid would not break down very fast and when it did, it would split apart in various ways, most of them not useful to the cycle. Another protein helps by promoting the addition of carbon dioxide to pyruvate to make oxaloacetate. As you may remember from our high-pressure experiments in Lecture Twelve, pyruvate to oxaloacetate is a reaction that we've never been able to produce. So, without these very specialized proteins to increase the chemical efficiency, it's hard to see how the reverse citric-acid cycle could have gotten going at all.

So, before everyone in the origins game rushes to support the idea that metabolism preceded genetics, someone has to demonstrate how that first metabolic cycle was able to go it alone, without the help of protein catalysts. Also, you have to show that every step in the cycle works. If even one step fails, then the whole cycle fails.

Wächtershäuser has an answer. Perhaps the most original and clever aspect of Wächtershäuser's model is that he relies on sulfide minerals to promote his metabolic reactions. As it happens, most of the modern metabolic proteins that promote reactions in the citric-acid cycle have, at their very core, a small cluster of iron or nickel atoms, as well as sulfur atoms. Now, those clusters of atoms look exactly like tiny bits of sulfide minerals and it is those clusters of atoms that trigger the desired chemical reactions. So, if modern highly efficient proteins use sulfide clusters, perhaps ancient minerals played the same role, albeit less efficiently. Remember, you do not have to be as efficient if there's no competition for food.

So sulfide minerals may be a key. Wächtershäuser also invokes another chemical idea based on sulfur. As you might guess from the position of sulfur, which is just below oxygen in the periodic table, it turns out that in many chemical reactions, you can substitute hydrogen sulfide for water, which is just hydrogen oxide. That is, H_2S can replace H_2O in certain chemical reactions. The advantage is that the reactions with hydrogen sulfide may proceed more easily because the sulfur compounds, though much scarcer than water, are more reactive. So Wächtershäuser proposes a metabolic cycle in which several crucial reactions use the sulfur-analogs of citric-acid cycle compounds. His prediction is that you get faster reactions that release more energy with hydrogen sulfide. What's more, these sulfide reactions automatically couple to the formation of pyrite, the iron–sulfur mineral whose formation releases energy essential to the Iron-Sulfur World. Together, all these reactions can form a self-consistent network. In Wächtershäuser's view, once you establish a sulfide version of the reverse citric acid cycle, it is only a matter of time before you switch to water. Hydrogen sulfide is common only at volcanic vents, but water is everywhere.

So, are any of these ideas supported by experiments? Our Carnegie Institution group was very impressed with Wächtershäuser's model, so in the late 1990s we initiated a series of gold-tube experiments to discover how citric acid degrades in hot, pressurized water. Would it fragment into acetate plus oxaloacetate? Now we really would have liked to study the system with hydrogen sulfide gas, to conform as closely as possible to Wächtershäuser's model. The trouble is that hydrogen sulfide is extremely toxic and very difficult to load into capsules. It is potentially deadly if a gas cylinder leaks and you also can make even more toxic compounds at high temperature and pressure. So, for our first round of experiments we avoided the hazardous sulfur system and studied the closely related water-based system.

By this time, our experimental protocols were well established, so I loaded the capsules with a citric-acid solution, Hat Yoder did the runs, usually at 2,000 atmospheres and 200°C and George Cody analyzed the runs. After a lot of runs and analyses, George began to see evidence for two different reaction pathways. A significant fraction of the citric-acid molecules broke down by what we called the alpha pathway. That begins with citric acid splitting into acetate plus oxaloacetate, which is exactly how the reverse citric-acid cycle is supposed to work. But beyond that, any similarity ended. As in our previous experiments, we found that oxaloacetate decomposes quite rapidly to pyruvate plus carbon dioxide and then most of the pyruvate decomposes to acetate. So, instead of building back up toward citric acid, we observed rather efficient destruction of citric acid. The alpha pathway was literally a dead end.

In a second round of experiments, we took the logical step of adding sulfide minerals and carbon dioxide, which are both important ingredients in the Iron-Sulfur World. But that didn't work either. We could not find a combination of minerals and reactants that promoted the essential reaction from three-carbon pyruvate to four-carbon oxaloacetate. Without that crucial step, we had no choice but to conclude that the reverse citric-acid cycle doesn't work.

So it turns out that oxaloacetate is the real stopper. We have not been able to find a way to make oxaloacetate from pyruvate plus carbon dioxide. We tried increasing the pressure and that did not seem to help. We also tried all sorts of realistic mineral catalysts and none of them made any difference. Oxaloacetate always breaks down rapidly in water.

George Cody didn't give up. It turns out that some of the citric-acid molecules in our runs followed a different pathway, what George called the beta pathway. That pathway begins the same way as the normal citric acid cycle, with citric acid giving up a single water molecule to make aconitate, which is another six-carbon compound. Then you get a series of chemical reactions, each of which releases carbon dioxide to make compounds with five, then four, then three carbon atoms in succession. Again, that is a lot like the normal citric-acid cycle—not the reverse cycle.

Well, is this series of reactions another dead end for citric acid? Is it just another way of breaking citric acid down into lots of useless small molecules? Or is it possible that the beta pathway might be part of a different self-replicating citric-acid cycle, one that has long since been replaced? Well, we have been doing many more experiments and George Cody has found evidence that some steps of the beta pathway can actually be reversed in the presence of nickel sulfide. So, perhaps there really is an abiotic route from carbon dioxide to citric acid. We are still trying. Maybe, someday, someone will try those risky experiments with hydrogen sulfide. That will be the true test of Wächtershäuser's model.

Throughout this course, I have referred to the first life and the first cell almost interchangeably. But is that necessarily so? If Wächtershäuser is right, and life emerged as a cycle of chemical reactions on a mineral surface, what did the Iron-Sulfur World look like?

It is clear that cellular life requires some kind of enclosing membrane. It needs a chemical barrier that separates the controlled chemistry of the insides from the random soup of the outsides. That is why most origins-of-life researchers assume that a cell-like lipid vesicle surrounded whatever the first self-replicating metabolic cycle might have been. But that's not the only possibility. The Iron-Sulfur World scenario has spawned at least two alternative views.

On the one hand, Glasgow-based geoscientists Michael Russell and Allan Hall, who are both experts in sulfide ore deposits, have proposed that "iron-sulphide bubbles" act as the first membranes. Russell and Hall do not rely on reactive iron sulfides to drive metabolism, so this is different from Wächtershäuser's idea. In their scenario, the iron-sulfide bubbles form spontaneously at hydrothermal vents. What happens is that the less acidic water of the hydrothermal vents meets the more acidic water of the ocean. The bubbles form spontaneously and they act like primitive cell membranes that enclose metabolic chemicals. Then energy comes from the strong contrast of acidity inside versus outside the bubble. That is an energetic difference that can promote all sorts of metabolic reactions.

Wächtershäuser has a very different view. He advocates what has been called "flat life". In this scenario, the very first self-replicating entity is a thin layer of chemical reactants that grows on a sulfide mineral surface. This flat life expands laterally. It spreads out from mineral grain to mineral grain as a coating of organic molecules that is much too thin to see. You can imagine how pieces of these layers might break off and then reattach to other rocks. They would behave just like cloned colonies. Given enough time and enough different minerals and environmental conditions, it is easy to imagine that variations in the film would arise. So you would have competition among new "species" of flat life.

The concept of flat life is not yet a majority view. Indeed, it is still somewhat heretical. Scientists like Dave Deamer, who have spent a good part of their careers studying lipid membranes, are not going to accept flat life without more proof. But a self-replicating chemical layer of molecules living on a mineral surface can't be ruled out, and it raises a fascinating possibility. It is very possible that a simple growing layer of molecules will be more resistant to high temperatures than life based on nucleic acids, which tend to decompose at about 100°C. If that is so, this means that it is very possible that extensive colonies of flat life will survive, even today, in deeper parts of Earth's crust. It is entirely possible that film-like molecular systems that represent a completely different domain of life will persist for billions of years because they survive at high temperatures beyond the predation of more efficient cellular life.

Now some biologists reject that idea out of hand. But how would they know? A self-replicating layer of molecules would be invisible under any light microscope and, at best, it would look like a nondescript film in more powerful atomic microscopes. You would not be able to detect flat life in any standard biological assay procedure, either, because those assays rely entirely on the presence of DNA and proteins, so one has to wonder if it is possible that flat life is plentiful in Earth's crust, yet remains overlooked.

Well, as you can see, we have a tremendous amount still to learn about the emergence of self-replicating metabolic systems. Furthermore, as we will consider in the next lecture, many scientists still argue passionately that genetics came first, in what is commonly called The RNA World.

Lecture Twenty-One
The RNA World

Scope: Most scientists studying the origins of life believe that the first living entity was characterized by some kind of genetic molecule. That is, a molecule carried the information needed to make more copies of itself. The prevailing genetics-first scenario goes by the name *RNA World.*

Outline

I. To understand what it means to say that life began with genetics, let's take a review the way that modern cells use their genetic molecules to store information, copy that information, and manufacture useful biomolecules.

A. DNA is the fundamental molecule that stores and copies all the biological information in every cell. James Watson and Francis Crick discovered DNA's famous double-helix structure in 1952.

1. The vertical sides of the DNA "ladder" are formed by alternating molecules of the sugar, deoxyribose, and phosphate groups, composed of a phosphorus atom surrounded by four oxygen atoms. That alternating sequence—sugar, phosphate, sugar, phosphate—provides a sturdy support.

2. The rungs of the ladder are made of pairs of four different bases: adenine, cytosine, guanine, and thymine, usually designated by the letters A, C, G, and T. The shapes of the bases exactly match up, so A always links to T, and C always links to G. Every rung of the ladder is, thus, either C-G or A-T.

3. Once the ladder is formed from the vertical supports and the pairs of bases are matched up to make rungs, the assembly is twisted into the double helix. Each of the basic units is called a *nucleotide*, which consists of three smaller molecules: one phosphate group, one deoxyribose sugar, and one base. There is no limit to the length of a DNA strand.

4. DNA can store information and encode any conceivable message in its strings of A, C, G, and T. DNA also can copy that information.

B. One of the most amazing discoveries in the history of science is that every living organism shares the exact same chemical process for turning the information stored in DNA into proteins. That process, called the *genetic code*, is pretty clear.

1. Segments of DNA called *genes* carry the code to make proteins. Groups of three DNA letters designate specific amino acids. [For example, CCC designates praline, and CGC designates arginine.]

THE GENETIC CODE

	C	G	A	T	
C	Proline	Arginine	Histidine	Leucine	C
	Proline	Arginine	Glutamine	Leucine	G
	Proline	Arginine	Glutamine	Leucine	A
	Proline	Arginine	Histidine	Leucine	T
G	Alanine	Glycine	Aspartic Acid	Valine	C
	Alanine	Glycine	Glutamic Acid	Valine	G
	Alanine	Glycine	Glutamic Acid	Valine	A
	Alanine	Glycine	Aspartic Acid	Valine	T
A	Threonine	Serine	Asparagine	Isoleucine	C
	Threonine	Arginine	Lysine	Methionine	G
	Threonine	Arginine	Lysine	Isoleucine	A
	Threonine	Serine	Asparagine	Isoleucine	T
T	Serine	Cysteine	Tyrosine	Phenylalanine	C
	Serine	Tryptophan	STOP	Leucine	G
	Serine	STOP	STOP	Leucine	A
	Serine	Cysteine	Tyrosine	Phenylalanine	T

2. The cellular machinery to build proteins, one amino acid at a time, based on the information carried by DNA, requires another kind of molecule, called *ribonucleic acid*, or *RNA*.

C. Three different kinds of RNA participate in protein synthesis: Messenger RNA, transfer RNA, and ribosomal RNA.

RNA

An mRNA Molecule

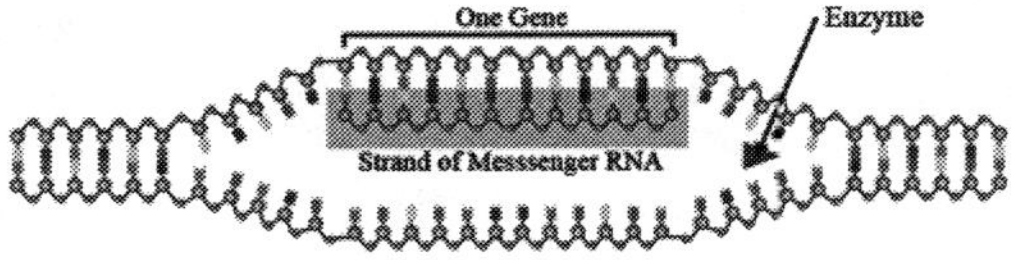

A tRNA Molecule

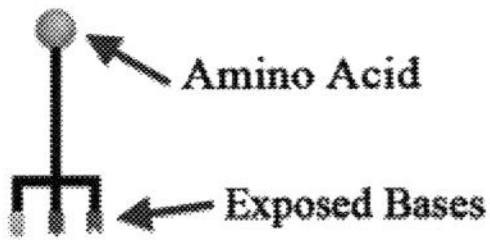

A Ribosome

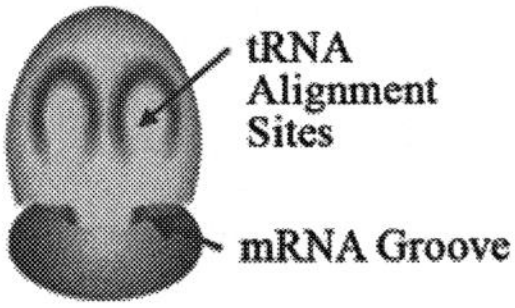

1. Messenger RNA copies the base sequence of a DNA segment (gene) letter by letter. RNA nucleotides quickly make a copy of the information from only one side of the double helix. A long, single-stranded messenger RNA molecule carries all the information about one gene. The genetic code defines a one-to-one correspondence between three bases and an amino acid.
2. Transfer RNA and ribosomal RNA play the central roles in protein synthesis.
3. The transfer RNA molecule is the physical manifestation of the genetic code—at one end you have three bases; they're physically linked to the corresponding amino acid at the other end.
4. The ribosome provides the chemical machinery needed to link amino acids together in the exact order dictated by the messenger RNA. Eventually, a complete protein is assembled.

D. The story of the genetic code, and how genetic information is converted into proteins, reveals that proteins cannot be made without DNA and RNA, and cells cannot operate without proteins. That's why most origins experts dismiss the idea of a metabolism-first scenario. Even the simplest known cell must pass volumes of coded information from one generation to the next, and the only way we know to store and copy that much information is with a genetic molecule, such as DNA or RNA.

II. The central problem in understanding a genetics-first origin of life is the identification of a stable genetic molecule that can make copies of itself. Some kind of polymer is required that carries all the information needed to do so and catalyzes the self-replication reaction.

A. In 1968, Leslie Orgel, from the Salk Institute for Biological Studies in San Diego, published a now-classic paper entitled "Evolution of the genetic apparatus." Orgel summarized four broad approaches to the problem of jumpstarting such a genetic organism.

1. The first possibility is the emergence of a self-replicating peptide. This idea, that proteins came first and then somehow "invented" DNA, is appealing because amino acids, the building blocks of proteins, were widely available in the prebiotic environment. The problem is that the random prebiotic assembly of amino acids would have been a messy business. Without the careful control of ribosomes to assemble amino acids in just the right way, they link together in irregular, undisciplined clusters.
2. The second possibility is the simultaneous emergence of both proteins and DNA. DNA and proteins are inextricably linked in modern cells; perhaps they started out that way. This scenario is seen as extremely unlikely, because it requires the simultaneous emergence of two improbable macromolecules.
3. Graham Cairns-Smith offered the idea that clay minerals with genetic-like sequences of elements reproduce, evolve, and eventually act as templates for organic assembly (see Lecture Seventeen). This idea is unsupported by experimental evidence.
4. The fourth genetics-origin model, favored by Orgel, postulates that some nucleic-acid molecule like RNA acts both as a carrier of genetic information and as a catalyst that promotes self-replication.

B. The citric acid cycle lies at the heart of all modern metabolism and studies of protometabolism. In a similar way, top-down studies of molecular genetics have zeroed in on RNA as the essential core molecule of ancient genetics.

III. The early 1980s saw the discovery of RNA *ribozymes*, which are strands of RNA that not only carry genetic information but also act as catalysts.

A. Sidney Altman of Yale and Thomas Cech of the University of Colorado independently demonstrated that a particular segment of RNA has the ability to promote key biochemical reactions. Most known enzymes are proteins, but they discovered that RNA could act as an enzyme, too. This fundamental finding has inspired a new vision of life's origins.

1. Modern cells rely on two complexly interrelated molecules: DNA carries the information to make proteins; proteins perform key chemical functions, such as making DNA. This interdependence leads to a classic chicken-and-egg dilemma: Which came first, proteins or DNA? RNA, it turns out, has the ability to do both jobs.

2. The theory that RNA served the dual roles as the first genetic molecule and the first self-replicating molecule is called *RNA World.* This theory quickly emerged following the discovery of ribozymes. It is preeminently a genetics-first model.

B. The RNA World champions the central role of genetic material in the dual tasks of catalyst and information transfer. Three assumptions are common to all versions of the theory.

1. First, at one time, RNA, not DNA, stored genetic information.

2. Second, ancient RNA replication followed the same rules as modern DNA replication by matching pairs of bases. Thus, A matched up with U, and C matched up with G.

3. Ancient RNA played the same catalytic roles as modern proteins.

C. In the RNA World scenario, the first life-form was a self-replicating strand of RNA, most likely enclosed in a protective lipid membrane. Modern metabolism and the citric acid cycle are thought to have emerged later as a way to make RNA replication more efficient.

IV. The more we learn about RNA, the more versatile it appears to be.

A. One of the biggest surprises came from the study of ribosomes. Recall that they're made from several RNA strands and about 50 proteins, all carefully assembled into one molecular machine.

1. Many biologists had assumed that the proteins play the key role as the enzymes that do the actual assembly work, whereas the strands of RNA merely hold the ribosomes together.
2. Recent studies, however, have proved just the opposite. RNA controls the critical step of linking amino acids to make protein. In essence, RNA does the key tasks in protein assembly. That discovery reinforces the presumed ancient role of RNA in biochemistry.

B. RNA has other key biochemical roles. For example, RNA nucleotides play structural roles in a variety of essential biological catalysts called *coenzymes*. These catalysts promote reactions in the citric acid cycle, including the problematic synthesis of citrate from oxaloacetate. Coenzymes also help to make lipids and other essential biomolecules. Recently, scientists at Yale discovered *riboswitches*. These RNA segments change shape when they bind to specific molecules in the cell, acting as chemical sensors that turn genes on and off.

C. Given these discoveries, many scientists have concluded that RNA is an ancient molecule that contributes to virtually every aspect of cellular chemistry. However, problems remain. The emergence of RNA is a crucial step in life's origins, yet no one knows where the RNA came from.

1. Decades of chemical experiments demonstrate that RNA is a difficult molecule to make. The RNA World could not have emerged fully formed from random processes in the primordial soup. Yet decades of study show no plausible mechanism to make all three components of RNA—the sugar ribose, a phosphate group, and a base—much less link them all together in just the right way and string them end-to-end to make an RNA strand.
2. Some critical transition stage must have bridged the prebiotic milieu and the RNA World.

V. RNA is also an incredibly complex molecule.

A. Some scientists argue that a simple, self-replicating metabolic system must have emerged first, followed by some form of genetic molecule both structurally simpler and chemically more stable than RNA. Much later, the mechanisms of RNA genetics and ribozymes come into play to make the system more efficient. Here are three reasons one might reject the RNA World:

1. The first reason is the system's chemical complexity.
   a. The primitive citric acid cycle uses simple molecules in the carbon-oxygen-hydrogen system. Even if a fourth element, sulfur, must be added, it seems vastly easier to jumpstart this cycle than genetics.
   b. By contrast, the RNA World scenario depends on exact sequences of chemically complex nucleotides in the carbon-oxygen-hydrogen-nitrogen-phosphorus system.
   c. Metabolism seems to be much closer to the heart of biochemistry. Modern cells use metabolism to synthesize nucleic acids, but this RNA synthesis is several steps removed from the core metabolic cycle, the citric acid cycle. This onion-like arrangement of a simple core metabolism surrounded by more complex layers of synthesis suggests that metabolism came first and other chemical pathways were added later.
2. A second reason is that many of the presumed protometabolic molecules can be synthesized in experiments that mimic prebiotic environments. RNA nucleotides, by contrast, have never been synthesized from scratch, despite decades of focused effort.
3. Even if we could identify a plausible prebiotic pathway to RNA nucleotides, no one has found a mechanism to link those individual nucleotides end to end in an RNA strand. There is no obvious way to make RNA sequences in any prebiotic environment.

B. RNA is too deeply rooted in all of life's chemical complexity to be anything but ancient. Thus, the RNA World was certainly a critical stage in life's emergence, but it must have happened relatively late. It was a transitional stage that occurred when life had already been well established on Earth. It is likely that RNA came well after the emergence of a stable, evolving metabolic world. However, RNA also appeared well before the modern genetic world of DNA and proteins.

VI. The greater mystery lies in the gap between primitive metabolism and RNA. Before we can fully understand the origin of the RNA World, we must focus on what came before—the pre-RNA World. By what chemical process did the first self-replicating, information-bearing system emerge? And if it was not RNA, then what was it? The next lecture will focus on this mystery.

Essential Reading:

De Duve, *Vital Dust: Life as a Cosmic Imperative*, part II.

Fry, *The Emergence of Life on Earth*, chap. 11.

Hazen, *Genesis: The Scientific Quest for Life's Origin*, chap. 16.

Supplemental Reading:

Davies, *The Fifth Miracle: The Search for the Origin and Meaning of Life*, chap. 5.

Lahav, *Biogenesis: Theories of Life's Origin*, chap. 17.

Wills and Bada, *The Spark of Life: Darwin and the Primeval Soup*, chap. 6.

Questions to Consider:

1. Why do many scientists consider RNA to be a likely cornerstone for the first living entity?
2. What are some of the arguments, pro and con, for a genetics-first origin of life?

Lecture Twenty-One—Transcript
The RNA World

The last two lectures have focused on the hypothesis that metabolism came first. In the view of many origins researchers, a self-replicating metabolic cycle was the first entity to fulfill all of the requirements for life—that is, the ability to grow, the ability to reproduce and the ability to evolve. So we have the Protenoid World model of Sidney Fox, the Thioester World model of Christian de Duve and the Iron-Sulfur World model of Günter Wächtershäuser. In all those models, some kind of self-replicating chemical system came first.

Not everyone believes that metabolism came first. In fact, I guess that the majority of scientists studying the origins of life are in the opposite camp. They are more persuaded that the first living entity was characterized by some kind of genetic molecule. That is, a molecule carried the information needed to make more copies of itself. In this lecture, the twenty-first of our course, I want to introduce the idea that life began with genetics. In particular, I want to examine the prevailing genetics-first scenario, which goes by the name RNA World.

What exactly does it mean to say that life began with genetics? To answer that, we need to take a step back and review the way that modern cells use their genetic molecules to store information, to copy that information and to manufacture all sorts of useful biomolecules. I also cover this material in more detail in Lectures Forty-Nine and Fifty of my Teaching Company course called *The Joy of Science.*

To understand modern genetics, we have to start with DNA. That is the fundamental molecule that stores and copies all of the biological information in every cell. DNA is an amazing molecule. James Watson and Francis Crick discovered its famous double helix structure in 1952. You can read about this discovery in Jim Watson's best-selling account, *The Double Helix*.

Let's look in some detail at this structure. The vertical sides of the DNA ladder are formed by alternating molecules of the sugar, deoxyribose and phosphate groups, which as we saw earlier, are composed of a phosphorus atom surrounded by four oxygen atoms. That alternating sequence—sugar, phosphate, sugar, phosphate—provides a sturdy support. Then the rungs of the ladder are made of pairs of bases. There are four different bases, adenine, cytosine, guanine and thymine; which are usually designated by the letters A, C, G, and T. The shapes of the bases exactly match up, so A always links to T and C always links to G. Every rung of the ladder is thus either C-G or A-T.

Once you form the ladder, from the vertical supports, and match up the pairs of bases to make the rungs, the whole assembly is twisted into the double helix. By the way, each of the basic units of DNA is called a nucleotide. Each nucleotide consists of three smaller molecules—one phosphate group, one deoxyribose sugar and one base. Now, there's no limit to the length of a DNA strand. It can extend up to billions of nucleotides long.

When Watson and Crick figured out this elegant structure, they immediately recognized its power and beauty. DNA is able to store information in its long string of letters, A, C, G, and T. You can encode any conceivable message in the letters of DNA but DNA is also able to copy all that information.

Here is how it does it. Every cell has a toolkit of molecules that manipulates DNA. One of those tools effectively unzips the double helix, giving you two single strands of DNA, but each side of the chain holds all the information of the entire molecule because the letters are complementary. The base A is complementary to T, and C is complementary to G. Each double helix strand, in effect, carries two copies of the message, so when you unzip the DNA, you can assemble two complete double helix strands just by inserting new complementary nucleotides.

That is exactly what happens when a cell divides. The DNA strands split. Then a T nucleotide bonds to an exposed A, while a G nucleotide bonds to an exposed C, and so forth. So, where before there was one A-T or C-G rung, afterwards there are two. Ultimately, this process ends with two identical sets of DNA, one for each daughter cell.

Okay, so DNA can store and copy genetic information at the molecular level. That is pretty amazing but storing information is only part of the genetic challenge. That information also has to be translated into some kind of useful molecular structures. After all, cells are made of molecules—proteins, lipids, carbohydrates and so forth. To make life you somehow have to manufacture those molecules.

To do this you need a second type of nucleic acid, called RNA. Three different types of RNA, called messenger RNA, transfer RNA and ribosomal RNA work together to make proteins. Proteins act as the molecular workhorses of cells. They make all the other molecules and they serve in lots of other cellular functions, as well, so if you can make proteins, you can make everything.

To me, one of the most amazing discoveries in the history of science is that every living organism—from slime mold, to sequoias, to us—shares the exact same chemical process for turning the information stored in DNA into proteins. That process is called the "genetic code". Whatever the nature of the genetic code, it somehow had to convert the four-letter alphabet of DNA—the molecules designated A, C, G, and T—into proteins. Somehow, those letters have to match up to a sequence of the 20 amino acids that make protein. How does that happen?

The first thing you need to know is that the long strands of DNA consist of many relatively small segments called genes. Each gene is a segment of DNA that carries the information to make one protein, but how could that be? There are 20 different amino acids but only four different bases. Well, you have to conclude that more than one genetic letter is required to specify which of the 20 amino acids to use. Now, two genetic letters isn't enough either, because that only gives you four times four, or 16, different combinations of two bases. So that brings us to groups of three bases: there are four times four times four, or 64, different combinations of three bases. That is plenty to indicate which one of the 20 amino acids to use.

So the concept of the genetic code is pretty clear. Segments of DNA code for proteins. Groups of three DNA letters designate specific amino acids. But you still have to wonder, how are the proteins actually made? You need some kind of cellular machinery to build proteins, one amino acid at a time, based on the information carried by DNA. How do you do it?

The key is another kind of molecule, called ribonucleic acid or RNA. RNA is closely related in structure to DNA but there are three significant differences. For one thing, each RNA molecule is a single-stranded polymer, as opposed to the double-stranded DNA. Also, in RNA the sugar molecule in the backbone is ribose, instead of deoxyribose in DNA. One more difference is that the base uracil, which is designated by the letter U in RNA, substitutes for thymine, which is the letter T in DNA.

Now, the RNA story is complicated by the fact that there are three different kinds of RNA that participate in protein synthesis. First, you have messenger RNA, which copies the base sequence of a DNA segment—that's a gene—letter by letter. In this process a protein opens a specific segment of DNA in the middle, just like a broken zipper.

So then, RNA nucleotides quickly come in to make a copy of the information from one side of the double helix. What you ultimately get is a messenger RNA molecule that is a long, single-stranded molecule that carries all the information about one gene. In a sense, this process is like making a Xerox copy of one recipe in a large cookbook.

So you've copied the instructions but somehow the cell has to take the messenger RNA and turn those instructions into a protein. That means the genetic code has to relate specific groups of three nucleotides on the messenger RNA to specific amino acids. The complete genetic vocabulary consists of 64 different, three-letter words, each of which is a three-letter combination of A, C, G, and U. It turns out that 61 of those words define specific amino acids, while the other three words indicate the end of a gene on the DNA strand. By the way, one of the genetic words, AUG, codes for the amino acid methionine, which marks the beginning of every protein.

Also, as you've probably figured out, there's a lot of redundancy in the genetic code. After all, there are 64 genetic words and only 20 amino acids, so most amino acids can be specified by more than one combination of three letters. For example, phenylalanine is coded by UUU or by UUC. Also, six different words code for leucine; that includes all four words that begin with CU, plus UUA and UUG. By 1964 the complete genetic code had been determined; that's only about a decade after the discovery of the DNA structure. It is amazing to think that this chemical language is common to every living thing on Earth.

So, we've seen how the genetic code defines a one-to-one correspondence between three bases, on the one hand, and an amino acid, on the other. The next step is to understand the actual mechanism that uses this information to assemble a protein. It is here we need to introduce the other two kinds of RNA, called transfer RNA and ribosomal RNA. These forms of RNA play the central roles in protein synthesis.

Transfer RNA molecules are really the key. Each of these small molecules has an amino acid attached to one end and three exposed bases on the other.

There are 61 different kinds of transfer RNA molecules, one for each of the 61 genetic words that specify amino acids. So here is how it works. The three exposed bases at one end of transfer RNA are perfectly designed to attach to the complementary three bases on a messenger RNA molecule. The transfer RNA molecule is the physical manifestation of the genetic code. At one end, you have three bases, and they are physically linked to the corresponding amino acid at the other end. Think about it. If you line up a bunch of transfer RNA molecules, then you are also lining up a sequence of amino acids. If you want to assemble a protein with 30 amino acids, for example, it requires you to line up 30 transfer RNA molecules. Those 30 transfer RNA molecules match up to a sequence of 90 bases on a messenger RNA.

Well, to complete this operation there must be some molecular structure that does the assembly. It has to be some kind of template that matches up messenger RNA, on the one hand, with lots of transfer RNA molecules, on the other. It turns out that that structure is called the ribosome, which includes the third kind of RNA, called ribosomal RNA. The ribosome is complicated. It is made up of several different RNA strands as well as about 50 proteins. When all these pieces are assembled together, the ribosome looks something like two balls. There is a smaller one perched on top of a larger one with a prominent groove in between. There are also alignment sites on the larger ball that perfectly fit transfer RNA molecules.

The ribosome provides the chemical machinery needed to link amino acids together in the exact order dictated by the messenger RNA. Here is how the process works. First, a messenger RNA molecule fits nicely into the groove between the two balls of the ribosome. As the long messenger RNA molecule feeds into the ribosome, appropriate transfer RNA molecules fit into grooves on the sides of the larger ball.

Then the ribosome snaps the amino acids together. The long messenger RNA is fed through, sort of like the tape in an old stock ticker. One by one the amino acids are assembled into a long protein. One amino acid is added, then another, then another and so forth. Eventually, this process assembles a complete protein. Note that once this process is completed, the ribosome is ready to find another messenger RNA and make another protein.

What is perhaps most remarkable about this story is that every living thing uses the exact same genetic mechanism. Every living cell makes proteins with the same kind of messenger RNA, the same kinds of transfer RNA and the same ribosomes. The genetic code itself, the matching of three genetic letters to amino acids, is also exactly the same in every living cell. That means that human genes can be inserted into yeast or bacteria and the one-celled organisms will produce the appropriate human protein.

Okay, that is the story of the genetic code and how genetic information is converted into proteins. You cannot make proteins without DNA and RNA, and you cannot operate cells without proteins. That is why most origins experts dismiss the idea of a purely metabolic life form in favor of a genetics-first scenario. Even the simplest known cell has to pass volumes of coded information from one generation to the next and the only way we know to store and copy that much information is with a genetic molecule like DNA or RNA.

Perhaps the most influential thinker in the field of genetics and the origins of life is Leslie Orgel, who is based at the Salk Institute for Biological Studies in San Diego. In 1968, he contributed a now-classic paper entitled, "Evolution of the Genetic Apparatus." The ideas in that article have guided generations of researchers and they continue to exert a lot of influence in the field. Orgel recognizes that the central problem in understanding a genetics-first origin of life is the identification of a stable genetic molecule that can make copies of itself. You need some kind of polymer that carries all the information needed to make copies of its own self and that also catalyzes the self-replication reaction. Accordingly, he has summarized four broad approaches to the problem of jump-starting such a genetic organism.

The first possibility is the emergence of a self-replicating peptide of the kind proposed by Sidney Fox and actually synthesized by Reza Ghadiri's group at the Scripps Research Institute. This idea, that proteins came first and then somehow "invented" DNA, is appealing because amino acids, the building blocks of proteins, were widely available in the prebiotic environment. The problem is the same one faced by Fox. The random prebiotic assembly of amino acids would have been a messy business. Modern cells have learned how to form neat, chain-like polymers of amino acids to form the proteins essential to life. But without the careful control of ribosomes to assemble those amino acids in just the right way, they link together in irregular, undisciplined clusters.

The second of Orgel's possibilities is the simultaneous emergence of both proteins and DNA. DNA and proteins are inextricably linked in modern cells, so maybe they started out that way. But most origin workers see this scenario as extremely unlikely because it requires the simultaneous emergence of two improbable macromolecules.

Then there is Graham Cairns-Smith's clever Clay World model, which provides an intriguing third option. Recall, from Lecture Seventeen, that Cairns-Smith invokes clay minerals with genetic-like sequences of elements. He says, "These clay particles reproduce and evolve, and eventually act as templates for organic chemistry." It is a fascinating idea but so far the Clay World scenario is totally unsupported by experimental evidence.

Then there is the fourth genetics-origin model, the one favored by Orgel and many followers. That model postulates that some nucleic-acid molecule like RNA acts both as a carrier of genetic information and as a catalyst that promotes self-replication. Orgel proposed this model way back in 1968, long before any experimental evidence supported such a notion. Also, it is not too surprising that he came to this conclusion. "I must confess to a strong, longstanding bias in favor of [this] explanation," he said. "It is, at the very least, the model that can be studied most easily in the laboratory." Without laboratory studies, none of these models could gain much support.

So, how is one to choose among the various models of Orgel? Proteins first? DNA first? Some combination of both? Or perhaps neither, as in the Clay World? Remember that when evaluating various origin-of-life models, scientists aren't restricted just to chemical experiments. Many of the models we have talked about, including the metabolism-first models of Wächtershäuser, de Duve, and others, are also influenced by top-down studies of biochemistry. What are the most deeply embedded, primordial biochemical pathways? The principle of continuity demands an unbroken path from Earth's primordial geochemistry to modern biochemistry. So, we find that the citric-acid cycle that lies at the heart of all modern metabolisms is a prime target for studies of protometabolism. In a similar way, top-down studies of molecular genetics have now zeroed in on RNA as the essential core molecule of ancient genetics.

A few discoveries in the history of origins research have absolutely electrified the community—Stanley Miller's experiment, for example, and David Deamer's self-assembling lipids. But no advance has been more influential than the early 1980s discovery of RNA "ribozymes," which are strands of RNA that not only carry genetic information but also act as catalysts. Sidney Altman of Yale and Thomas Cech of the University of

Colorado, who shared the Nobel Prize in 1989, made the discovery independently. They independently demonstrated that a particular segment of RNA has the ability to promote key biochemical reactions. Most known enzymes are proteins but they discovered that RNA could act as an enzyme, too, hence the name, ribozyme. This fundamental finding has inspired a new vision of life's origins.

We have seen that modern cells rely on two complexly interrelated molecules. On the one hand, DNA carries the information to make proteins. On the other hand, proteins perform key chemical functions like making DNA. This interdependence leads to a classic chicken-and-egg dilemma. Which came first, proteins or DNA? RNA, it now turns out, has the ability to do both jobs.

The theory that RNA served the dual roles as the first genetic molecule and the first self-replicating molecule is called the RNA World. This theory quickly emerged following the discovery of ribozymes. It is preeminently a genetics-first model. The RNA World champions the central role of genetic material in the dual tasks of catalyst and information transfer. Most origins-of-life experts would say the RNA World model is the theory to beat.

Over the years, the expression "RNA World" has come to mean rather different things to different people. Nevertheless, three assumptions are common to all versions of the theory. First, all versions of the RNA World assume that once upon a time, RNA, rather than DNA, stored genetic information. Second, it is assumed that ancient RNA replication followed the same rules as modern DNA replication by matching pairs of bases. So A matches up with U, and C matches up with G. The third assumption is that ancient RNA played the same kinds of catalytic roles as modern proteins do.

In the RNA World scenario, the first life-form was a self-replicating strand of RNA, most likely enclosed in a protective lipid membrane. According to most versions of this hypothesis, modern metabolism and the citric-acid cycle emerged later as a way to make RNA replication more efficient.

The RNA World idea caught on very fast and I think there were at least two reasons for this acceptance. For one thing, the discovery of ribozymes came almost a generation after the Miller-Urey experiment, and there were still few good leads on how to make the transition from the random prebiotic soup to organized cellular life. I imagine that the origins-of-life community was poised to try something new, and RNA certainly provided an original angle. Another reason for the RNA World's support was the evidence that RNA plays a dual role in modern life. RNA is both a catalyst and a carrier of information today, so why not at the dawn of life? This genetics approach proved especially seductive to the new generation of biologists, who were born and raised in the age of molecular genetics. To many biologists, life and genetics are almost synonymous, so to them the idea of an RNA World makes a lot of sense.

The more we learn about RNA, the more remarkably versatile it appears to be. One of the biggest surprises came from the study of ribosomes. Recall that they are made from several RNA strands and about 50 proteins, all carefully assembled into one molecular machine. Many biologists had assumed that the proteins play the key role as the enzymes that do the actual assembly work, while the strands of RNA merely hold the ribosomes together, but recent studies have proved just the opposite. It turns out that RNA controls the critical step of linking amino acids to make a protein. In essence, RNA does the key tasks in protein assembly. That discovery strongly reinforces the presumed ancient role of RNA in biochemistry.

Then it turns out that RNA has other key biochemical roles. For example, RNA nucleotides play essential structural roles in a variety of essential biological catalysts called "coenzymes". It turns out that these catalysts promote reactions in the citric-acid cycle, including the problematic synthesis of citrate from oxaloacetate. Coenzymes also help to make lipids and other essential biomolecules. Recently, scientists at Yale discovered yet another RNA trick, called "riboswitches." These amazing RNA segments change shape when they bind to specific molecules in the cell. In this way they act as chemical sensors that turn genes on and off.

Given this flurry of discoveries, many scientists have come to the conclusion that RNA is a very ancient molecule that contributes to virtually every aspect of cellular chemistry. Little wonder, then, that a thousand articles amplify the idea and hundreds of researchers have pursued variations on the theme. Every origins-of-life conference features sessions dedicated to RNA World studies. The RNA World model seems to be the paradigm to beat.

But problems remain. The emergence of RNA represents a crucial step in life's origins, yet no one knows where the RNA came from. Decades of frustrating chemical experiments have demonstrated repeatedly that RNA is a frustratingly difficult molecule to make. The RNA World could not possibly have emerged fully formed from random processes in the primordial soup. In spite of decades of focused study, no plausible mechanism has been found to make all three components of RNA—the sugar ribose, a phosphate group and a base—much less link them all together in just the right way and then string them end-to-end to make an RNA strand. There must have been some critical transition stage that bridged the prebiotic milieu and the RNA World.

RNA is an incredibly versatile molecule, to be sure, but it is also an incredibly complex molecule. It is easy to see why some scientists argue that a simple self-replicating metabolic system must have emerged first, followed by some form of genetic molecule that was both structurally simpler and chemically more stable than RNA. Then, much later, the mechanisms of RNA genetics and ribozymes come into play as a way to make the whole system more efficient.

Let's review three reasons why one might decide against the RNA World. The first reason is the system's chemical complexity. The primitive citric-acid cycle uses rather simple molecules in the carbon-oxygen-hydrogen system. Even if you have to add a fourth element, sulfur, it seems vastly easier to jump-start this cycle than genetics. By contrast, the RNA World scenario depends on exact sequences of chemically complex nucleotides in the carbon-oxygen-hydrogen-nitrogen-phosphorus system. Metabolism seems to be much closer to the heart of biochemistry. Modern cells use metabolism to synthesize nucleic acids but this RNA synthesis is several steps removed from the core metabolic cycle, the citric-acid cycle. This onion-like arrangement of a simple core surrounded by successively more complex layers of synthesis suggests that metabolism came first and other chemical pathways were added later.

A second argument against the RNA World is that many of the presumed protometabolic molecules, including acetate and pyruvate, can be synthesized in experiments that mimic prebiotic environments. But RNA nucleotides, by contrast, have never been synthesized from scratch, in spite of decades of focused effort. In fact, RNA bases tend to break down under conditions that favor sugars, like ribose, and ribose tends to break down under conditions that favor the bases. It is very difficult to understand how sugars, bases and phosphate groups combined in just the right way to make RNA nucleotides.

A third problem in RNA synthesis has to be overcome. Even if we could identify a plausible prebiotic pathway to RNA nucleotides, no one has found a plausible mechanism to link those individual nucleotides end-to-end into an RNA strand. Now, it is true that James Ferris has managed to link up RNA nucleotides on clays; I introduced that work back in Lecture Seventeen. But Ferris relied on so-called "activated" nucleotides that have an extra reactive chemical group attached. It is just not clear how such activated nucleotides could have been made, even if you could solve the problem of making nucleotides in the first place. So the RNA World hits a roadblock. There is just no obvious way to make RNA sequences in any prebiotic environment.

Yet, the RNA World must have happened! RNA is too deeply rooted in all of life's chemical complexity to be anything but ancient. So, the RNA World was most certainly a critical stage in life's emergence but it must have happened relatively late. It was a transitional stage that occurred when life had already been well established on Earth. I think it is likely that RNA came well after the emergence of a stable, evolving metabolic world but it also appeared well before the modern genetic world of DNA and proteins. It is a matter of stages of progressive complexity.

The greater mystery lies in the seemingly intractable gap between primitive metabolism and RNA. Before we can fully understand the origin of the RNA World, therefore, we must focus on what came before, the pre-RNA World. By what chemical process did the first self-replicating, information-bearing system emerge? If it wasn't RNA, then what was it?

So the next lecture will focus on the mystery of the pre-RNA world.

Lecture Twenty-Two

The Pre-RNA World

Scope: A huge gap exists between the diverse molecules of the primordial soup and the RNA World. No one has figured out a way to synthesize RNA from the molecules in the soup. No known mineral acts as a template to assemble sugar, phosphate, and base into a nucleotide, much less link nucleotides together into a strand of RNA. This lecture focuses on the hypothetical first genetic molecule that must have come before RNA in the *Pre-RNA World.*

Outline

I. The challenge of Pre-RNA World research is to come up with a genetic molecule that carries information in the same way as RNA but is much more stable and more likely to have been synthesized from the prebiotic soup.

A. Experts seem to agree on three basic features. First, the molecule is a long polymer like RNA, with information carried as a sequence of similar molecules. Second, the information is carried by the same nucleotide bases used today by DNA and RNA—A, C, G, and T (or U). Third, this pre-RNA molecule is replicated by the same sort of base pairing used in cells today.

B. Given these assumptions, the most logical strategy is to start with RNA and try different backbones, because RNA consists of a sequence of bases, attached to a long sugar-phosphate backbone.

1. Swiss chemist Albert Eschenmoser has attempted to synthesize more than a dozen variants of RNA with different sugar-phosphate backbones. He has identified at least seven new kinds of stable polymers, each of which represents a possible alternative to RNA.
2. Only one of Eschenmoser's new molecules is potentially significant: a nucleic acid that uses the 4-carbon sugar threose, called *threose nucleic acid* (TNA).
3. Threose can be made in a single reaction by joining a pair of 2-carbon molecules. As a result, TNA appears to be much more likely than RNA to arise spontaneously from the prebiotic soup. No one has synthesized threose nucleotides,

much less a TNA strand, in any prebiotic process, but it *is* a simpler molecule than RNA.

4. In 1991, Danish chemist Peter Nielsen and colleagues synthesized a novel genetic molecule with a backbone of amino-acid molecules. This variant is called *peptide nucleic acid* (PNA). PNA features RNA-like bases, but they're bound to the amino acid backbone. Discovery of PNA has highlighted the chemical richness of plausible genetic molecules. We don't have to rely entirely on a sugar-phosphate backbone.

C. The range of plausible molecules for the Pre-RNA World has been expanded, but no one has yet managed a prebiotic synthesis of any of these alternative nucleotides, much less a viable genetic polymer.

D. There is good reason to believe that neither the threose of TNA nor the peptides of PNA provide a bridge between the soup and RNA. There is no obvious way for these molecules to emerge by self-organization, and what's needed is a system that forms quickly and easily from readily available building blocks. Neither sugars and phosphates nor amino acids self-organize. However, there may be another way.

II. The central idea for a new scenario for the Pre-RNA World—known as the *PAH World*—offers another model for an alternative backbone for a genetic polymer in the tradition of TNA and PNA.

A. In May 2004, Australian chemist Simon Nicholas Platts formulated an idea based on polycyclic aromatic hydrocarbons in deep space. Recall from Lecture Five on the Martian meteorite that these cyclic molecules are referred to as *PAHs*.

1. PAHs would have been abundant among prebiotic molecules. Each of these flat molecules consists of two or more fused 6-member rings of carbon atoms with hydrogen atoms around the edge.
2. PAHs are most often relatively insoluble in ocean water, but their solubility can be increased. If sunlight shines on PAHs, they can be chemically modified. Hydrogen atoms can be stripped off the edges, and new molecular fragments can take their place. A typical scenario is for the PAH edges to become

bonded to OH. In that case, the solubility of PAHs increases dramatically.

3. The prebiotic world would have had many PAHs that were insoluble, plus modified PAHs that were somewhat soluble. PAHs must have represented a good bit of carbon on the primitive Earth. Platts wondered if PAHs played a key role in the ancient ocean.
4. A PAH with OH around the edges is also an amphiphile. Recall from Lecture Sixteen that amphiphiles are molecules that both love and hate water, similar to Dave Deamer's self-organizing lipids. In this case, the flat surfaces of the carbon rings repel water (they're hydrophobic), but the OH-bearing edges of the PAHs are attracted to water (they're hydrophilic).
5. When amphiphilic PAH molecules are placed in seawater, the hydrophobic parts are kept away from water and the hydrophilic parts are left in contact with water, because the molecules tend to assemble into a stack. The flat hydrophobic parts shield each other from water, and the hydrophilic edges are on the outside that touches water. In this way, the PAHs self-organize.
6. Platts predicted that once the PAHs were stacked, the system would bind to small, flat molecules at the edges. For example, the RNA bases, A, C, G, and U, could attach to the PAH edges and, in that way, become concentrated from the prebiotic soup.
7. Each PAH in a stack is free to move around sideways, in a motion similar to rubbing one's hands together. The molecules attached to the edges would attach and break free, sort of like a game of molecular musical chairs.
8. Because of this sideways motion, only molecules that are very flat can stay attached. More bulky ones would tend to get stripped off. There is a natural selection for bases to decorate the sides.
9. Because the RNA bases are amphiphiles, they also have a tendency to stack more or less on top of each other. A mini-stack of RNA bases forms at the sides of the much bigger PAH stack.
10. Platts noticed that the spacing between the layers in a stack of PAHs is 0.34 billionths of a meter—exactly the same spacing

between the bases of DNA and RNA. The arrangement looks like the information-rich genetic sequence of DNA or RNA. That's the key to the pre-RNA World.

B. Platts fleshed out this model by proposing that other small molecules could start to form a backbone linking all the bases together into a true information-rich molecule.

THE "PAH WORLD" HYPOTHESIS

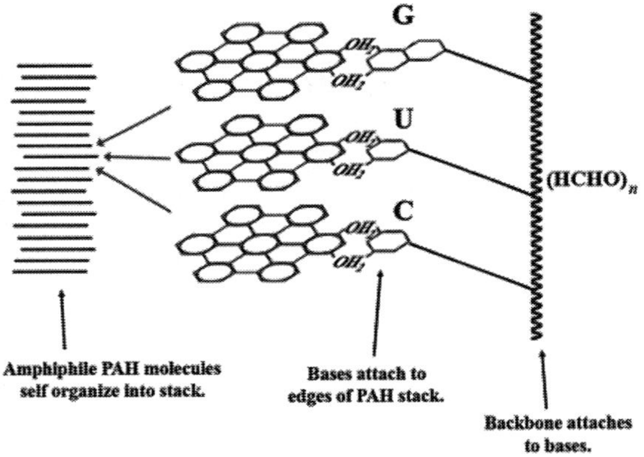

1. He imagined a backbone of formaldehyde or, maybe, amino acids.
2. If the stack of PAHs moved to a place in the ocean with different temperature or acidity, as might be experienced by moving from a deep hydrothermal environment to a shallower zone, then the whole RNA-type arrangement of bases and backbone could break away and become a free-floating, pre-RNA genetic molecule.
3. That molecule could play the same roles as RNA. It could fold back on itself to match up base pairs. It could act as a catalyst. Eventually, strands of this genetic molecule might self-replicate and grow.
4. Though lacking any experimental support at the time, the PAH World hypothesis provided a geochemically plausible scenario. It is conceptually simple, and it does provide a chemical path from the dilute primordial soup to an RNA-like genetic sequence.

C. A scientific hypothesis is useless unless it makes unambiguous predictions. Nick Platts's hypothesis made many testable predictions.

1. Amphiphilic PAHs must self-assemble into stacks.
2. The stacked PAHs must be similar in size and shape.
3. The PAH edges must attract a variety of molecules.
4. There must be a preferential selection for flat molecules, such as the RNA bases.

D. Already, significant publications show that PAHs can self-organize into stacks of disks—a process known as *discotic self-organization*. This confirmation of the first prediction adds credibility to the model.

E. Platts imagined a sequence of experiments that could confirm the essential points of his theory.

1. He would get a sample of some relatively small PAH. The best choice was hexabenzocoronene (HBC), a PAH with a snowflake arrangement of 13 rings.
2. He would mix this HBC molecule with water in a flask, then shine a strong ultraviolet lamp on it to make the edges chemically reactive—to decorate the edges with OH groups. That process should turn the HBC into amphiphiles, which could be examined for discotic self-organization.
3. Platts was confident those procedures would work, because they were similar to previous experiments in the discotics literature. Then he proposed to add RNA base molecules to see if the PAHs bonded to them. Finally, he wanted to add amino acids or formaldehyde, either of which might serve as a backbone.
4. If the bases linked together, he would actually have formed a long sequence reminiscent of a genetic polymer.

F. There were problems from the start. For example, pure PAHs, such as HBC, are almost impossible to find commercially and are incredibly expensive. PAHs can be made, but the mixture is uncontrolled. Since he constructed his original hypothesis, Platts has been awarded a Ph.D. in chemistry and became a postdoctoral fellow in David Deamer's laboratory. It may be decades before his PAH World model is sufficiently tested to make it into the textbooks or be proved invalid.

III. Whatever the outcome of Nick Platt's effort, some sort of genetic polymer eventually began to make copies of itself. Self-replication must be a key to understanding the transition to life. At some point, self-replicating molecules must have started to compete with each other. Competition triggers evolution. In the next lecture, we'll examine competitive systems of molecules that actually evolve.

Essential Reading:

Hazen, *Genesis: The Scientific Quest for Life's Origin*, chap. 17.

Supplemental Reading:

Lahav, *Biogenesis: Theories of Life's Origin*, chap. 17.

Wills and Bada, *The Spark of Life: Darwin and the Primeval Soup*, chap. 6.

Questions to Consider:

1. Why do many scientists argue that there must have been a Pre-RNA World?
2. What are the most attractive features of Nick Platts's PAH World scenario? What are its biggest drawbacks?

Lecture Twenty-Two—Transcript
The Pre-RNA World

In the last lecture, we saw that the RNA World represents a crucial step in the origins of life. Working from the top down, we find abundant evidence that RNA is a truly ancient molecule that can fulfill both essential prebiotic chemical roles. It can make copies of itself, it can act as a catalyst to promote metabolic reactions, and it can pass genetic information from one generation of cells to the next. In the next lecture we will see how RNA actually evolves from one generation to the next. So, the top down approach now takes us all the way back to a plausible primitive system that predates our modern DNA-protein world.

We have also learned a great deal about the probable characteristics of the earliest stages of life's emergence. We certainly have a good idea about how to make the prebiotic soup, which was stocked with all sorts of interesting molecules from many sources, from deep space to the deep oceans. We have some good ideas about how those molecules might have been assembled into a variety of macromolecules and membranes.

There's still a huge, mysterious gap between the diverse molecules of the primordial soup and the RNA World. No one has figured out a way to synthesize RNA from the molecules in the soup. No known mineral acts as a template to assemble sugar, phosphate and base into a nucleotide, much less link nucleotides together into a strand of RNA. Stanley Miller has put the problem concisely, "Identifying the first genetic material will provide the key to understanding the origin of life." He says, "RNA is an unlikely candidate."

The focus of this lecture, Lecture Twenty-Two, is that hypothetical first genetic molecule, the one that must have come before RNA. Researchers refer to this stage in life's emergence as the "Pre-RNA World." I have to warn you, this is a subject with a great deal of speculation but few certainties, but that's actually a good thing. As much as any aspect of origins research, efforts to understand the Pre-RNA World illustrate the excitement and creativity of the field.

The challenge of Pre-RNA World research is to come up with a genetic molecule that carries information just like RNA but is much more stable and more likely to have been synthesized from the prebiotic soup. Many models have been proposed, as we will see, but all experts seem to agree on three basic features. First, the molecule must be a long polymer like RNA, with information carried as a sequence of similar molecules. Second, the information is carried by the same nucleotide bases that are used today by DNA and RNA—A, C, G, and T, or U. Third, the same sort of base pairing that's used in cells today replicates this pre-RNA molecule.

Given these assumptions, the most logical strategy is to start with RNA and try substituting different backbones. After all, RNA consists of a sequence of bases, which are attached to a long sugar-phosphate backbone. In a remarkably original research program, the Swiss chemist Albert Eschenmoser has done exactly that. He has attempted to synthesis more than a dozen variants of RNA with different sugar-phosphate backbones. One by one, he has systematically replaced the five-carbon sugar, ribose, with many other four-, five- and six-carbon sugars. In the process, he has identified at least seven new kinds of stable polymers. Each one of these polymers represents a possible alternative to RNA.

Most of these new nucleic acids are unlikely candidates for the pre-RNA world since they're no more stable or more likely to form in a prebiotic environment, but one of Eschenmoser's new molecules is potentially significant. That is a nucleic acid that uses the four-carbon sugar "threose," which is why the molecule was dubbed TNA, for "threose nucleic acid". In Lecture Nine we saw that ribose has to be synthesized through a rather cumbersome sequence of chemical steps but threose can be made in a single reaction by joining a pair of two-carbon molecules. As a result, TNA appears to be a much more likely molecule than RNA to arise spontaneously from the prebiotic soup. Now I should emphasize that no one has synthesized threose nucleotides, much less a TNA strand, in any simple prebiotic process. But it is a simpler molecule than RNA.

Well, there are other solutions to the backbone problem as well. For one thing, you do not have to use a sugar-phosphate backbone. In 1991, the Danish chemist Peter Nielsen and colleagues synthesized a novel genetic molecule with a backbone of amino-acid molecules. This variant has been given the name "peptide nucleic acid,", or PNA. PNA features RNA-like bases but they are bound to the amino acid backbone. I think you can see why the reliance on amino acids, which were readily available in the soup, appealed to many members of the origins community. It is also safe to say that the discovery of PNA has highlighted the chemical richness of plausible genetic molecules. You do not have to rely entirely on a sugar-phosphate backbone.

A variety of work, including that of Eschenmoser, on alternative sugars, and Nielsen, on peptide backbones instead of sugar, has expanded the range of plausible molecules for pre-RNA world. You might be interested to know that molecules such as TNA and PNA are now being used to provide new kinds of synthetic genetic molecules that can interact with modern cells, yet not interfere with cellular function. This ability is a potential boon to medical research and it shows, once again, how fundamental work on life's origins can lead to useful technologies. Even so, no one has yet managed a prebiotic synthesis of any of these alternative nucleotides, much less a viable genetic polymer. That means the opportunity for new pre-RNA world ideas is wide open.

I want to tell you in more detail about a third proposal for a solution to the pre-RNA World. In spite of the clever ideas that are out there, there is good reason to believe that neither the threose of TNA nor the peptides of PNA provide a bridge between the soup and RNA. There is just no obvious way for these molecules to emerge by self-organization. That's really what you want. You want a system that forms quickly and easily from readily available building blocks. Sugars and phosphates don't self organize. Amino acids do not self-organize, either, but there may be another way.

So, I want to introduce another scientific personality, the Australian chemist Simon Nicholas Platts. Everybody calls him Nick. Unlike many other scientists in this lecture series, Nick is a relative unknown, at least so far. The story begins in late 2004, at a time when Nick was in a bit of trouble. He had been a graduate student for almost five years, and he was almost 40 years old, but he hadn't finished his Ph.D. thesis, and he was about to be deported to Australia with no degree and no job.

Most scientists take a pretty straight-arrow path from high school to college to grad school to post-doc to tenure-track job. If you make it through, you are pretty much guaranteed a satisfying life in science. But Nick Platts's life had taken a different path. He loved chemistry research but he also loved teaching, so he went from a master's degree in chemistry to teaching at Melbourne Grammar School. It was not until his 35th birthday that he resolved to go back to school to get a doctorate degree and do some research. Nick was especially fascinated by life's chemical origins, so he applied to work with Professor James Ferris at Rensselaer Polytechnic Institute in Troy, New York. Recall that we met Ferris in Lecture Seventeen because of his work assembling RNA on clay minerals.

Nick's new teachers and fellow students at RPI found him to be an outgoing and supportive colleague. He was always ready to help others with their research, and he was especially eager to teach undergraduate chemistry courses; but all those commitments thwarted progress on his thesis. So, after three years working in Troy, he was given a chance to move to our Carnegie Institution lab in Washington, DC, as a NASA-sponsored pre-doctoral fellow. That coveted position would allow Nick to acquire his doctorate from RPI but to do the research in a new setting with a lot fewer distractions. At least, that is what we all hoped.

That was in early 2003, and Nick had two years to complete a project. He was full of ideas, and initially he was eager to study the origins of life's homochirality—that is the problem of life's selection of left-handed amino acids and right-handed sugars. Well, Nick certainly started with a flurry of activity, but he seemed gradually to lose interest and progress slowed and stopped. As his advisors, George Cody, Marilyn Fogel and I felt that we had failed.

Well, by his own recollection, everything changed dramatically for Nick Platts on Tuesday, May 25, 2004. That is the date when he formulated the central idea for a new scenario for the pre-RNA World—what has come to be known as the PAH World. As we will see, this is yet another model for an alternate backbone for a genetic polymer in the tradition of TNA and PNA but this model relies on a logical process of self-assembly.

The seed for the PAH World concept was planted in November 2001, when Dr. Max Bernstein presented a seminar at RPI. Bernstein is a colleague of Lou Allamandola at NASA Ames Research Center, where he worked on the deep-space production of organic molecules. Bernstein's talk focused on the widespread occurrence of polycyclic aromatic hydrocarbons in deep space. Recall from Lecture Five on the Martian meteorite that these cyclic molecules are referred to as PAHs. Well, Bernstein's talk got Nick thinking about PAH chemistry. But the information sort of simmered for a while.

A couple of years later, in September of 2003, Nick was attending an origin-of-life conference in Trieste, Italy, celebrating the 50th anniversary of Stanley Miller's experiment, and he began thinking anew about the pre-RNA world on the return flight to the States. It took a few more months to iron out the details but the actual drawings that tied the whole hypothesis together were completed in a flash, on the evening of May 25, 2004. I am telling you this story, not because this was the big breakthrough everyone was waiting for, nor even because Nick Platts's ideas are correct. On the contrary, so far there is no strong evidence that what I'm about to tell you is true. But this is a great example of how ideas are formulated and how they're tested.

Now Nick's new idea is not very radical. In many ways, it follows in the tradition of other models of pre-RNA genetic polymers by proposing an alternate backbone, but he does offer a truly original chemical twist based on polycyclic molecules. We know that PAHs would have been abundant in the mix of prebiotic molecules. Recall that each of these flat molecules consists of two or more fused six-member rings of carbon atoms with hydrogen atoms around the edge. PAHs are, most often, relatively insoluble in ocean water but there's a way to increase their solubility. If sunlight shines on PAHs they can be chemically modified. Hydrogen atoms can be stripped off the edges and new molecular fragments can take their place. A typical scenario is for the PAH edges to become bonded to OH. In that case, the solubility of PAHs increases dramatically.

Let us think about PAHs. The prebiotic world would have had lots of PAHs that were essentially insoluble, plus modified PAHs that were somewhat soluble. In fact, by any calculation, PAHs must have represented a lot of carbon on the primitive Earth and there really doesn't seem to be much you can do with them. So I suspect that a lot of origins researchers considered it; too bad that so many potentially useful carbon atoms were tied up in the relatively inert, biologically useless PAHs.

Nick Platts wondered if everyone else had missed something. What if PAHs did play a key role in the ancient ocean? What Nick realized was that a PAH with OH around the edges is not only somewhat soluble in water but it is also an amphiphile. Recall from Lecture Sixteen that amphiphiles are molecules that both love and hate water. They are like Dave Deamer's self-organizing lipids. In this case, the flat surfaces of the carbon rings repel water, so that part is hydrophobic but the OH-bearing edges of the PAHs are attracted to water; they're hydrophilic.

So think about what happens when lots of these amphiphilic PAH molecules are placed in seawater. They want to do what any amphiphile wants to do—keep the hydrophobic parts away from water, while the hydrophilic parts are in contact with water. The simple solution is for the molecules to assemble into a pile of PAHs, like a stack of dishes in the cupboard. That way, the flat hydrophobic parts shield each other from water and the hydrophilic edges form the outside part that touches water. In this simple way the PAHs self-organize.

Well, so what? How do stacks of PAHs contribute anything to the pre-RNA world? The answer lies in the chemically reactive edges around the PAHs. Nick Platts predicted that, once the PAHs were stacked, then the system would tend to bind to small flat molecules at these edges. So, for example, you could attach the RNA bases, A, C, G, and U, to the PAH edges and, in that way, concentrate them from the prebiotic soup.

Here is the trick. Each PAH in a stack is free to move around sideways, kind of shuffling back and forth. It is something like rubbing your hands together. Then the molecules attached to the edges would attach and break free, sort of like a game of molecular musical chairs. But, because of this sideways motion, only molecules that are themselves very flat can stay attached. Anything that is more bulky would tend to be stripped off. So there is a natural selection for lots of bases to decorate the sides. There is also another twist. The RNA bases are, themselves, amphiphiles, so they also have a tendency to stack more or less on top of each other. So you can imagine a mini-stack of RNA bases forming at the sides of the much bigger PAH stack.

Here is something else Nick Platts noticed. The spacing between the layers in a stack of PAHs is 0.34 billionths of a meter. This is exactly the same spacing that's found between the bases of DNA and RNA. So we see the PAHs self-organizing into stacks, and then bases attaching and self-organizing on the sides of PAH, and that arrangement of bases is exactly what you find in RNA. So we automatically get an arrangement that looks, for all the world, like the information-rich genetic sequence of DNA or RNA. That is the key to the pre-RNA World.

Nick fleshed out this model by proposing that other small molecules could start to form a backbone linking all the bases together into a true information-rich molecule. He imagined a backbone of formaldehyde, or maybe of amino acids—that part of his model is still rather fuzzy. Then, if the whole stack of PAHs moved to a place in the ocean with different temperature or acidity, as might be experienced by moving from a deep hydrothermal environment to a more shallow zone, then the whole RNA-type arrangement of bases and backbone could break away and become a free-floating, pre-RNA genetic molecule. That molecule could play the same sort of role as RNA. It could fold back on itself to match up base pairs. It could act as a catalyst. Eventually, strands of this genetic molecule might be able to self-replicate and grow.

Now, this is all very speculative stuff, and the later steps are much less obvious than the earlier ones. Also, at the time he proposed this model, the proposal lacked any experimental support. Still, the PAH World hypothesis does seem to provide a geochemically plausible scenario. It is conceptually simple and it does provide a chemical path, all the way from the dilute primordial soup to an RNA-like genetic sequence.

So, imagine you are there in May of 2004, and Nick Platts has just figured out the basic outlines of his model. What would you have done next? It is interesting to see how scientific ideas evolve, so I want to tell you about what happened over the next year or so.

Nick's first instinct was to talk to other people and to get as much feedback as possible. Now this strategy was necessary but it also carried some risks. After all, it was just an idea and the more people knew about it, the more chance there was that someone might steal part or the entire model. But Nick was in no position to develop and publicize it on his own. So he started making phone calls and visiting people's offices at the Carnegie Institution.

I certainly remember his visiting me; it was on the afternoon of Thursday, May 27. He was very excited and agitated as he described his hypothesis. He went over the basics—the self-assembly of PAHs, the way flat molecules like bases were selected along the edges, the exact spacing of the bases. He told me to keep everything secret but he also rattled off a list of the many people he already had called to talk over the idea. I have to tell you, he looked like he hadn't slept in days.

My immediate reaction was cautious, so I asked him if he could think of any simple experimental tests. He was much more concerned about staking a claim to the basic idea and he wanted to write up a short speculative paper right away. This was a long shot but it's happened before. The prime choices would be to provide just an outline of seven hundred words for the "Brief Communications" section in *Nature* or a "Brevia" in *Science*. There are many periodicals in which he could submit a longer piece but he had no data. So Nick decided to go with *Nature* first.

Twenty-four hours later Nick had a seven-hundred-word draft that he had titled "Edge-derivatised and stacked polycyclic aromatic hydrocarbons as essential scaffolding at the origin of life." He handed it to several of us at the lab and we went to work editing. I suggested that he change the title to "The PAH World," in the tradition of origins research and to make the title a little more appealing.

A scientific hypothesis is useless unless it makes unambiguous predictions. It turns out that Nick Platts's hypothesis made many testable predictions. The very first one was that amphiphilic PAHs have to self-assemble into stacks. What's more, the stacked PAHs have to be similar in size and shape, the PAH edges have to attract a variety of molecules and there has to be a preferential selection for flat molecules like the RNA bases. The model would be a lot stronger if one of these predictions could be confirmed.

George Cody came up with the first piece of supporting evidence. George's own thesis had investigated coal and coal is loaded with PAHs. As you can imagine, many people have studied the chemistry of coal and it turns out that there is already a huge literature on the way that PAHs can self-organize into stacks of disks. That process has been known for almost a quarter century as "discotic self-organization", so there are many publications that have already confirmed Nick's first prediction. You might think that Nick's prediction of stacked PAHs would not count, since it had already been discovered, but it actually adds a lot of credibility to his model.

Nick had gotten written and telephone comments from at least a dozen experts and he submitted the PAH World paper to *Nature* on Thursday, June 3rd, just nine days after his original idea had crystallized. Now, a submission to *Nature* requires a list of suggested reviewers who are experts. He also had the option of suggesting people who *should not* review the paper, for whatever reasons.

Nick did what a lot of us do. At the same time he submitted the manuscript to *Nature*, he also sent it out to all the proposed reviewers so they would have a better chance to form their reactions. Nick also wanted to stake his claim to the idea, whether or not *Nature* accepted the paper, so he began e-mailing the manuscript, designated "in review," to a dozen astrobiologists and origins experts to request their comments. He was especially interested in the reactions of the leading RNA World proponents, including Jerry Joyce and Leslie Orgel, who we met in the last lecture, and Jack Szostak, who we will meet in the next lecture. After all, they would have the most to gain from a solution to the pre-RNA World problem.

It is interesting to me how similar the responses were. Many scientists began by congratulating Nick on a clever and original idea but most replies were cautiously skeptical, and almost everyone cited the need for experimental evidence. Let me read you some of these replies because they reveal how scientists think.

Jack Szostak at Harvard responded in less than an hour. He said, "I think it's worth pursuing experimentally—it would certainly be cool if an effect could be demonstrated."

Then the next day, Leslie Orgel wrote to Nick, "An experimental demonstration of your scheme might be interesting, but I wouldn't advise publishing without showing that it works well."

Within a few hours Jerry Joyce wrote, "I thought it was interesting and certainly appealing. However, it would help tremendously if there were some experimental support." By the way, Jerry Joyce and Leslie Orgel play tennis together on a regular basis, so they may well have talked with each other about the manuscript.

A day later, Andy Knoll wrote back with essentially the same message, "For now it is a fascinating speculation, but the ideas seem amenable to what you guys do best—careful experimentation."

Well, you can see the trend here. Everyone wanted Nick to run some experiments and wait for solid evidence before publishing the idea. Evidently, the editors at *Nature* agreed. It took only a week for them to decide to reject the manuscript. Actually, they told Nick that there had been a flood of "Brief Communications" and they were rejecting everything at that point. It softened the blow, I suppose, but we all knew that *Nature* always makes room for the most important advances. The bottom line was that the paper had not yet made the grade.

So it is obvious that experimental evidence was essential to gain support for the PAH World hypothesis. In an ideal world, Nick would have taken a year or two to pursue these experiments. But science, like every other human endeavor, is done under real constraints. In Nick's case, he had to complete his Ph.D. thesis by July, and he needed to have a job shortly thereafter or he would have to return to Australia.

Indeed Nick had imagined a single sequence of experiments that could confirm the essential points of his theory. He wanted to get a sample of some relatively small PAH, with maybe about a dozen interlocking rings. The best choice he thought was "hexabenzocoronene," or "HBC." That's a PAH with a snowflake arrangement of thirteen rings. Nick proposed to put a lot of this HBC molecule into a flask with some water and then shine a strong ultraviolet lamp on it to make the edges chemically reactive; that is to decorate the edges with OH groups, and that should turn the HBC into amphiphiles. Then look for discotic self-organization of the amphiphiles.

Now, Nick was confident those procedures would work because they were like previous experiments in the discotics literature. Then he proposed to add RNA base molecules to see if the PAHs bond to them. Then, finally, he wanted to try the long-shot experiment and add amino acids or formaldehyde, either of which might serve as a backbone. If the bases linked together then he would actually have formed a long sequence, reminiscent of a genetic polymer.

Given the short time frame, Nick Platts got to work immediately but there were problems right away from the start. One of the biggest difficulties is that pure PAHs like HBC turn out to be almost impossible to find commercially and, when you do find them, they are incredibly expensive. Nick located one chemical supplier in Europe who was willing to sell tiny amounts of HBC for a thousand dollars. It was the equivalent of several million dollars per ounce, so that wasn't an option.

He did find research labs in Europe and Japan that might give us the chemical, but that was going to take months. As an alternative, Nick tried to manufacture his own PAHs by burning acetylene in air and catching the soot. You can make many PAHs that way but it is an uncontrolled mixture. In any event, there was no way he was going to complete the experiments before his thesis was due. So it was back to writing.

He still wanted to publish his idea, so he prepared a revised version of the manuscript that had been rejected by *Nature* and submitted it to *Science* on Thursday, July 15. By the way, that manuscript was also rejected, but this time it took several weeks and the reviews really were not all that bad.

Also, as word of the idea spread, a lot of people wanted to hear more, so Nick presented an informal talk to the Carnegie campus on Friday afternoon, July 2nd and once again for a NASA Astrobiology Institute video seminar on Monday, July 12th. These talks were important because they gave Nick a chance to refine his presentation and hear various questions and objections. These talks also helped to establish Nick's intellectual property rights for his model.

Then there was the huge hurdle of a thesis. Most people take a year or more to write a dissertation, but Nick had about a month. He scrambled to write up an expanded description of his hypothesis, and he worked hard to assemble whatever support he could from the published literature. He headed back to Troy, New York, for a July 20 thesis defense. Well, his thesis committee was actually quite impressed, and he was awarded a Ph.D. in chemistry based largely on the PAH World model.

Having gotten his Ph.D. Nick also needed a job, preferably one that would allow him to pursue his idea. By a stroke of good fortune, Nick became a post-doctoral fellow in David Deamer's laboratory, where he began running experiments to test the PAH World hypothesis. It may be decades before the model is sufficiently tested for it to make it into the textbooks and, of course, it may prove to be invalid.

So what do we learn from this episode? For one thing, it is obvious that science is a very human endeavor. It does not just march along, inevitably revealing "The Truth". Scientists face time constraints, their personal lives interfere and chemicals can cost a lot of money. For another thing, you can see clearly the need for both inspiration—the original, creative idea—and perspiration. Experiments are hard work. This episode also reveals a lot about how scientists think. Everyone enjoys a good idea but ideas in science mean little without experimental backing. I think it is remarkable that the first question asked by virtually every scientist who heard Nick's idea was, "Have you done any experiments?"

It is much too soon to tell if the PAH World model of Nick Platts will work, but whatever the outcome of that effort, we will certainly wind up knowing more than we did before. We also can be sure that some sort of genetic polymer, be it TNA, or PNA, or RNA, eventually began to make copies of itself.

Self-replication must be a key to understanding the transition to life. At some point, self-replicating molecules must have started to compete with each other. As examples from many arenas demonstrate repeatedly, whether in the history of ideas, the design of automobiles, or the development of life itself, competition triggers evolution. So, in the next lecture we will examine competitive systems of molecules that actually evolve.

Lecture Twenty-Three
Natural Selection and Competition

Scope: A critical step in the transition from nonlife to life is evolution. All living systems change over time. What's special about living chemical systems and helps to drive their evolution is competition. In this lecture, we explore the consequences of competition and the ways in which the struggle for resources can lead to rapid evolution by natural selection.

Outline

I. *Evolution* means change over time; it is not the same as natural selection. The British naturalist Charles Darwin proposed that life evolved by the *process* of natural selection, and that's the prevailing view today.

A. On the British naval vessel *HMS Beagle,* Darwin undertook an expedition to South America and the Pacific Islands from 1831–1836.

B. On the voyage, Darwin noted the variations among living species from different regions. He began to formulate a new theory of evolution as a fact of the natural world and published his revolutionary book, *On the Origin of Species by Means of Natural Selection*, in 1859.

II. Darwin's theory of evolution by natural selection is based on four commonsense observations:

A. All species produce more offspring than can survive to maturity.

B. Individuals of a species display variations in physical characteristics, differing slightly in their coloration, speed, and so on.

C. An individual's traits are likely to resemble those of its parents.

D. Some traits are more advantageous than others. Advantageous traits provide an individual with a better chance of surviving to produce more offspring and, thus, a better chance of passing those traits from one generation to the next.

E. Where do these four observations lead? Darwin concluded that after many generations, gradual small changes—variations that occur because of natural selection of individuals—can add up to large changes in species.

F. Natural selection was part of the critical sorting-out process that led to the origin of life. It is a powerful mechanism for the emergence of complexity.

III. Before life emerged, the oceans and their shorelines held a great variety of organic molecules derived from many sources. All of that chemical complexity had to be sorted out by two closely connected processes—molecular selection and molecular evolution.

A. Molecular selection occurred partly because some molecules are inherently unstable or unusually reactive. Many were extremely soluble in the oceans. Others bonded permanently to the surfaces of chemically unhelpful minerals or clumped into tar-like masses of no use at all in the emergence of life.

B. Earth's many cyclical processes probably amplified these processes of molecular selection. Many cycles were at work—day and night, hot and cold, sunshine and rain, high tide and low. Over time, the chemical mix was refined.

C. In every different geochemical environment, each kind of organic molecule had sources that added to its inventory and antagonists that destroyed that kind of molecule.

D. Perhaps for hundreds of millions of years, there was kind of stable state. New supplies of each molecular species balanced the losses. The concentration of amino acids, sugars, bases, and so on may have been fairly constant. A molecular equilibrium of this type features nonstop reactions among molecules, but the system doesn't necessarily change with time.

IV. At some point, a different process of molecular selection came into play. A self-replicating cycle of molecules and chemical reactions emerged and forever changed the character of molecular selection. Even relatively unstable molecules not only persist but also *increase* in concentration if they can make copies of themselves. Molecules that self-replicate would have thrived at the expense of other, less competitive molecular species.

A. Two or more self-replicating cycles of molecules, each of which competed for atoms and energy, was inevitable, because any reasonably complex cycle would have variations. New molecular species would offer alternative chemical pathways. Changes in environment would have caused slight variations in the original cycle. Competing molecular networks would have vied for the same limited supply of resources.
   1. That kind of molecular competition would have mimicked modern life's constant struggle for survival—the struggle that Darwin described in his theory of natural selection.
   2. Molecules behaved in a way analogous to living species. More molecules are born than can survive; some cycles of molecules are more stable than others.
   3. In the long run, the more competitive molecules win at the expense of the less competitive molecules. Thus, ever more efficient cycles of molecules emerged and flourished at the expense of less efficient variants.
   4. Cycles of molecules evolved in a process that slowly shifted the molecular balance. Molecular evolution by natural selection had begun on Earth.

B. Experts in the origin of life draw the arbitrary line between living and nonliving systems at the point in history when self-replicating chemical cycles emerged and began to evolve by this process of natural selection.

C. Replication of a molecular cycle is not sufficient for the origin of life. That cycle must have a considerable degree of variability for competition to take place and for more efficient variants of the cycle to survive. Molecular natural selection kicks in. The system gradually changes and becomes better at collecting atoms and energy.

V. How do we test molecular evolution in a laboratory? How do we get one kind of molecule to change gradually into another kind of molecule?

A. This has been done since the mid-1960s, when Professor Sol Spiegelman and his colleagues at the University of Illinois investigated a virus called Qβ that attacks the common bacterium *E. coli*.

1. Like other viruses, Qβ is a shell of proteins that holds a tiny loop of genetic material—in this case, RNA. Qβ causes *E. coli* cells to make copies of itself.

THE Qß VIRUS

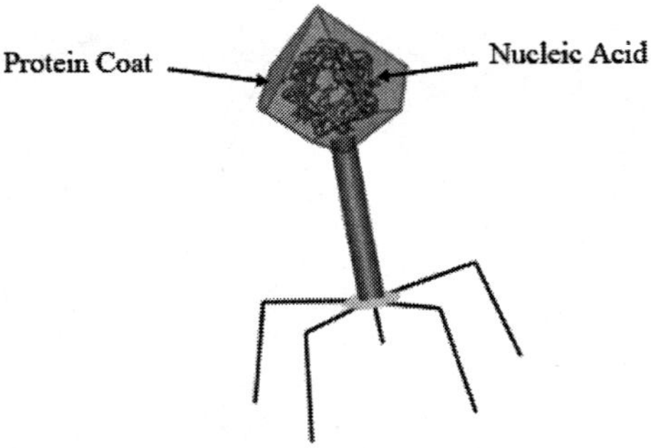

2. To make copies of the Qβ virus, a protein called Qβ replicase must make copies of the Qβ RNA. The behavior of the two suggested to Spiegelman an experiment.
3. First, Qβ replicase and a strand of Qβ RNA are placed into a test tube with RNA nucleotides.
4. Eventually, vast numbers of Qβ RNA strands appear, but the Qβ replicase is sloppy. Each new Qβ RNA strand includes random mistakes, resulting in a new assortment of RNA strands.

B. Spiegelman forced the new collection of RNA molecules to evolve, allowing the chemical mix of Qβ replicase, RNA strands, and RNA nucleotides to make copies for 20 minutes.

1. During that short time, some of the new RNA sequences bonded strongly to Qβ replicase, and they tended to duplicate rapidly.
2. Other RNA strands did not interact well with Qβ replicase and weren't copied very efficiently.
3. The cycle was repeated: After 20 minutes, the solution became enriched in RNA strands that easily replicate, and that collection of RNA was transferred into a third beaker with fresh Qβ replicase and RNA building blocks.
4. Spiegelman and coworkers repeated the process 74 times in all. They also gradually reduced the time allowed for the

reaction to proceed. In every cycle, Qβ replicase preferentially selected and copied RNA strands that were better at being copied.

C. By the experiment's end, the most efficiently copied RNA strands had been reduced to a sixth of their original length and were being copied 15 times faster than before.

D. In similar experiments, where collections of RNA were subjected to heat, acidity, or other stresses, Spiegelman's group could cause the Qβ RNA molecules to evolve. Soon, the origins-of-life community was referring to these evolving RNA strands as *Spiegelman monsters*.

E. Spiegelman's group had shown something profound. A cyclical process of competitive pressure can induce molecular systems to evolve, making it easier to imagine a primitive Earth on which cycles of heating or wetting induced a similar kind of evolution.

VI. An even more ambitious project is headed by Harvard Medical School biologist Jack Szostak, who wants to engineer an evolving life-form in his laboratory.

A. Since the early 1980s, Szostak and his students have studied RNA, inspired by the discovery of ribozymes. More recently, they have concentrated on self-replicating RNA.

1. In 2001, they made an impressive RNA sequence that grabs onto a shorter piece of RNA up to 14 nucleotides long and copies it.

2. Eventually, they expect to make self-replicating RNA strands more than 100 letters long—shorter than the genetic material of modern cells but still a significant advance. When Szostak and his colleagues do this, synthetic life will be just around the corner.

B. Self-replicating RNA is an important step to understanding life's origins, but no known geochemical environment could have supported such an isolated molecule. There is no supply of nucleotides, and the RNA strand wouldn't have survived long under most natural chemical conditions.

C. The key to survival in modern cells is protection by a lipid membrane. Szostak's group has tested ways that membranes might have encapsulated RNA strands. Lipid vesicles squeezed through tiny pores can stretch out, divide, and grow larger, a process similar to cell division.

D. Szostak's latest scenario for the first life-form to evolve by natural selection involves a lipid vesicle that contains self-replicating RNA. Experts used to think that RNA played all the roles that DNA and proteins play today. Szostak realized that RNA can promote cell growth by a much simpler process—just building up internal pressure.

1. As RNA pushes out on the membrane, pressure is applied to neighboring protocells to promote transfer of lipids from cells with less internal pressure.
2. Competition for space drives natural selection. Protocells with more RNA out-compete those with less.

E. Szostak's students prepared two sets of vesicles, one filled with a solution of the sugar sucrose and another filled with pure water. When they mixed the two together and confined the solution, the sucrose-laden vesicles got bigger by incorporating lipids from their water-filled neighbors. The same thing happened when they repeated the experiment with strands of RNA. The vesicles swollen with RNA grew at the expense of adjacent empty vesicles.

F. Before Szostak's work, most researchers assumed that RNA would have to learn how to control cellular functions, such as making membranes and guiding metabolism, before a protocell could evolve by natural selection. Szostak's results point to a much simpler scenario.

1. The only vital task for RNA in the protocell is self-replication.
2. The more RNA a vesicle can hold, the more likely it will be to replicate and grow.
3. Efficient RNA replicators have an advantage; thus, molecular natural selection and Darwinian evolution could take over.

VII. Researchers move ever closer to synthesizing a self-replicating, encapsulated, evolving chemical system in the lab. Will this kind of synthetic life tell us how life began on the primitive Earth?

A. A synthetic RNA organism would give credibility to the RNA World scenario. It would show that a strand of RNA, or perhaps some kind of precursor genetic molecule, could have formed an evolving, self-replicating chemical system. However, laboratory conditions may not be a plausible replica of the prebiotic environment. Researchers still influence the results by providing a supply of RNA nucleotides and vesicle-forming lipids.

B. In the final lecture, we'll see that much of what we've learned can be organized into three plausible scenarios for the origins of life.

Essential Reading:

Fry, *The Emergence of Life on Earth*, chaps. 8 and 9.

Hazen, *Genesis: The Scientific Quest for Life's Origin*, chap. 18.

Supplemental Reading:

Lahav, *Biogenesis: Theories of Life's Origin*, part IV.

Wills and Bada, *The Spark of Life: Darwin and the Primeval Soup*, chap. 6.

Questions to Consider:

1. Are Spiegelman's monsters alive? Why or why not?
2. Would a self-replicating, evolving strand of RNA be alive? Why or why not?

Lecture Twenty-Three—Transcript
Natural Selection and Competition

We have reached the penultimate lecture, number twenty-three in this series on life's ancient origins. We have come a very long way. From the top-down approach we have seen how evidence from fossils, isotopes and biomolecules has pushed the origin of life back to within the first billion years of Earth history.

From the bottom-up perspective, we have seen how simple chemical reactions led to the emergence of chemical systems of increasing complexity. First, we had mixtures of simple organic molecules. Then those molecules were selected and organized into macromolecules and membranes. Eventually, some select group of molecules, either a primitive metabolic cycle like the reverse citric acid cycle or a primitive genetic polymer like PNA or TNA, began to make copies of itself. Even if we do not know all the details, that much of life's history seems clear.

However, there is another critical step in the transition from nonlife to life, one that we haven't yet addressed. That is evolution. All living systems evolve. All living systems change over time. Now, we have seen what appears to be one inevitable kind of change on the early Earth. That is emergent complexity. I have argued right from Lecture Eight on that there was an inexorable increase in chemical complexity on the ancient Earth, from simple organic compounds to large interacting systems of molecules. This increase in complexity occurs because lots of energy from the Sun and the Earth's inner heat flowed through the oceans, the atmosphere, and rocks near the surface.

But this kind of increasing organizational complexity only takes you so far. What is special about living chemical systems, and what really helps to drive evolution in these systems, is competition. In this lecture, I want to explore the consequences of competition and to show how the struggle for resources can lead to rapid evolution by natural selection.

What is natural selection? There is a common misunderstanding that evolution means the same thing as natural selection, and that is just not true. "Evolution" just means change over time. There is overwhelming evidence that life on Earth has evolved. I talk about that evidence in Lecture Fifty-Six of my Teaching Company course, *The Joy of Science*.

Just to review, the most compelling evidence comes from fossils. The sedimentary rocks of our planet contain huge numbers of fossilized organisms that are extinct. Dinosaurs are abundant in the fossil record in rocks of ages between about 250 and 65 million years old. Dinosaurs never appear in rocks older or younger than that. Ammonites are another huge and common group of animals. They have beautifully coiled shells and they also occur in rocks as recent as 65 million years old but there are no ammonites in younger rocks.

Then there are trilobites; they are my favorite fossils, with their amazing eyes and armored bodies. Trilobites dominated the oceans for much of the period between about 550 and 250 million years ago, but you never find a trilobite in rocks younger than about 250 million years old. These are just a few of the countless examples of ancient creatures that were different from life today.

Furthermore, fossils are not just different from life today. They also display a clear record of increasing complexity over the past three-plus billion years. For most of that period, the only life forms were single cells. Multi-celled organisms start to appear about a billion years ago. Then about 550 million years ago the first animals with hard parts, including mollusks and arthropods are found. Four hundred million years ago, the only vertebrates were primitive bony fish but gradually, over time, we see the origins of amphibians, then reptiles, then birds and mammals. Therefore, there is an obvious increase in the complexity of life.

There is other evidence for change, for evolution, as well. Molecular phylogeny, for example, points to common ancestors in the past. So do vestigial organs like the wings of penguins and the tailbones of you and me. So, if you believe the observational evidence of our planet, then life has changed over time. That means that evolution is an observational fact.

But what was the process by which life evolved? That is a different question altogether. The famous British naturalist, Charles Darwin, who lived from 1809 to 1882, proposed that life evolved by natural selection. This is the prevailing view today.

Let me review a bit about Darwin and his theory. Though he was a rather mediocre student at Cambridge University, Darwin came from an influential family and he was recommended to be the naturalist aboard the British naval vessel *HMS Beagle*. The *Beagle* commenced a five-year exploring expedition to South America and Pacific Islands, from 1831 to 1836. So the young Charles Darwin had several years to read and think about the natural world. On the voyage, he collected numerous animal, plant and rock specimens and he made innumerable observations in his notebooks.

One of the influential books that Darwin read on the voyage was Charles Lyell's *Principles of Geology*, which emphasized that large, geological effects can be produced by gradual change over many millions of years. Darwin saw evidence for this principle firsthand in the formation of coral islands and in the effects of earthquakes on raising coastal regions of western South America. So Darwin was primed to think about the effects of vast stretches of time.

On the voyage of the *Beagle,* he also noted the variations among living species from different regions. Most dramatically, he observed differences among species of birds and tortoises from different islands in the Galapagos group.

Now, at the time of Darwin's travels, most naturalists accepted the view that species are unchangeable. Indeed, many scholars accepted the idea that life on Earth is the result of a single creation event, as described in the Bible. But Darwin's observations on his voyage caused him to question that idea. Why would God have created similar but different species on each of the Galapagos Islands, he wondered.

Charles Darwin began to formulate a new theory of evolution. He started a notebook of observations and recorded a litany of evidence from fossils, from comparative anatomy, from vestigial organs and from the geographical distribution of closely related species. He quickly convinced himself that evolution is a fact of the natural world. Then Darwin set about trying to understand the mechanism of evolution. As early as 1838, his theory of evolution was framed and by 1842, he wrote a brief outline of a new theory of evolution, called natural selection. Nevertheless, Darwin hesitated to publish the work. He realized that there was much more data that could be gathered together and he knew how much resistance his ideas would meet.

In fact, it was not until 1858 that he finally decided to publish, and that was only after he received a letter from another naturalist, Alfred Russel Wallace, who had independently arrived at a similar theory. So it was that

Darwin wrote the revolutionary book, *On the Origin of Species by Means of Natural Selection*, which was published in 1859.

A striking aspect of Darwin's *Origin of Species* is that it was published in a large popular edition that was accessible to any educated Englishman. Like a lawyer arguing a case before a jury, Darwin set down the case for the process of evolution by natural selection.

Darwin's theory of evolution by natural selection is based on four common sense observations. The first observation is that all species produce more offspring than can possibly survive to maturity. All resources, including food, water and space, are limited. Some species like spiders and fish and oak trees produce thousands of offspring for every individual that reaches adulthood, while others, like large mammals, tend to be much more careful of their relatively few children, but every species produces more offspring than can possibly survive.

The second observation is that individuals of a species display variations in physical characteristics. Individuals differ slightly in their coloration, their speed, their eyesight and every other physical characteristic.

The third observation is that an individual's traits are likely to resemble those of its parents.

Finally, the fourth observation is that some traits are more advantageous than others. Advantageous traits provide an individual with a better chance of surviving to produce more offspring and thus a better chance of passing those traits from one generation to the next.

Where do these four observations lead us? Darwin's conclusion was that after many generations, gradual small changes—variations that occur because of natural selection of individuals—can add up to large changes in species. That, in a nutshell, is the process of evolution by natural selection.

Darwin presented his theory of "natural selection" by first reminding his readers of the power of "artificial selection". In this way, he appealed to the British passion for breeding animals such as racing pigeons. He described the process of selective breeding, which was so well known to British farmers and sportsmen. Then he drew parallels between artificial selection and the natural process of selection. Darwin's audience knew that humans could produce a new breed of dog or cattle in a few dozen generations by selecting for desirable traits. Darwin's theory is that nature does the exact same thing, though in a less directed process.

I want to tell you about a dramatic example of natural selection at work. In England, there is a common small moth, called the peppered moth. That moth exists in both dark and light varieties. Light varieties of the moth were predominant during the decades of the pre-industrial revolution; that is, when most tree trunks were relatively light in color. But during the industrial revolution, Britain's factories spewed out smoke from coal-burning power plants and the tree trunks got darker and darker. That meant that predators were better able to see the light-colored individuals. Just as Darwin would have predicted, natural selection took place and within a few decades, the moth population had become predominantly dark in color. Today, factories in Britain have cleaned up their emissions, trees have light-colored trunks again, and the moth population is once again predominantly light colored. It all goes back to the competition for limited resources. The individuals most able to compete for resources are most likely to pass their genes on to the next generation.

Even today, 150 years after the publication of Darwin's theory, many people view the theory of evolution by natural selection as a dangerous and subversive doctrine. There is a sense that natural selection is too random and undirected. It seems to substitute a kind of materialism for faith, but natural selection isn't anything of the sort. On the contrary, natural selection was part of the critical sorting-out process that led to the origin of life. Natural selection is a powerful mechanism for the emergence of complexity.

Yet, it is not so easy to understand exactly what happened 4 billion years ago. The late, great Carl Sagan put it this way, "It is evident that once a self-replicating, mutating molecular aggregate arose, Darwinian natural selection became possible, and the origin of life can be dated from this event. Unfortunately, it is this event about which we know the least."

Let's try to imagine what it was like on the primitive Earth more than 4 billion years ago, at a time before life had emerged. We can be confident that the oceans and their shorelines held a great variety of organic molecules that were derived from many different sources. Somehow, all of that chemical complexity had to be sorted out. That sorting out process occurred by two closely connected processes, molecular selection and molecular evolution.

We have seen in Lectures Fifteen to Eighteen that molecular selection, by which a few molecules were concentrated and organized, proceeded on many fronts. We have lipids that self-organized and other molecules that were concentrated on mineral surfaces, but let's think about the larger issue of molecular selection.

Part of that selection occurred because some molecules are inherently unstable or unusually reactive. Those compounds quickly disappeared. Many other molecules were extremely soluble in the oceans and so they were diluted and removed. Still, other molecular species were removed from contention because they bonded permanently to the surfaces of chemically unhelpful minerals or they clumped together into tar-like masses that were of no use at all in the emergence of life.

Earth's many cyclical processes probably amplified these processes of molecular selection. We had tidal pool cycles of wetting and evaporation that concentrated molecules, and then the Sun's ultraviolet radiation fragmented the least stable molecular species. Deep in the ocean, periodic pulses of hydrothermal seawater would have delivered new supplies of chemicals to deep-ocean volcanic vents. Then, different molecules were adsorbed and detached on reactive mineral surfaces and that tended to concentrate a select subset of molecular species. Lots of cycles were at work—day and night, hot and cold, sun and rain, high tide and low tide. Over time, the chemical mix was refined.

It was all a matter of sources and sinks for each kind of molecule. In every different geochemical environment, each kind of organic molecule had sources that added to its inventory, and sinks that destroyed that kind of molecule. We can imagine that for a long period, perhaps for hundreds of millions of years, there was a kind of stable state. New supplies of each molecular species balanced its loss. Therefore, the concentration of amino acids, sugars, bases and so forth may have been pretty constant. A molecular equilibrium of this type features nonstop reactions among molecules, to be sure, but the system of molecules does not necessarily change with time.

But at some point, a different process of molecular selection came into play. We certainly do not know all the details but a self-replicating cycle of molecules and chemical reactions emerged. That self-replication forever changed the character of molecular selection. That is because even relatively unstable molecules cannot only persist but also increase in concentration if they can make copies of themselves. Self-replication changes the entire ball game. Molecules that self-replicate would have thrived at the expense of other less competitive molecular species.

As soon as you have one self-replicating cycle, then you can imagine an even more interesting possibility. What if you had two or more self-replicating cycles of molecules, each of which competed for atoms and energy? I think that such competition was inevitable because any reasonably complex cycle would have variations. New molecular species would offer alternative chemical pathways. Changes in environment would have caused slight variations in the original cycle. So, you'd have competing molecular networks that would have vied for the same limited supply of resources. That kind of molecular competition would have mimicked modern life's constant struggle for survival, the same kind of struggle that Charles Darwin described in his theory of natural selection. So, molecules behaved in a way analogous to living species. More molecules are born than can survive. Some cycles of molecules are more stable than others, so in the long run, the more competitive molecules win at the expense of the less competitive molecules. Thus, it was that ever more efficient cycles of molecules emerged and flourished at the expense of less efficient variants. Cycles of molecules evolved in a process that slowly shifted the molecular balance. Molecular evolution by natural selection had begun on Earth.

This dynamic phase of molecular evolution differed fundamentally from the earlier, more passive process of molecular selection. Can you see how competition among self-replicating cycles served to drive evolutionary change? You get new, more efficient cycles. That is exactly why many experts in the origin of life draw that very arbitrary line between living from non-living systems at the period in history when self-replicating chemical cycles emerged and began to evolve by this oh-so-powerful process of natural selection.

A key point here is that replication of a molecular cycle is not sufficient for the origin of life. In addition, that cycle has to have a considerable degree of variability. That way, a competition takes place and more efficient variants of the cycle survive. Molecular natural selection kicks in. The system gradually changes and becomes better at collecting atoms and energy. Okay, that is the theory but it doesn't matter how convincing a theory sounds, you need some kind of experimental test. But how do you test molecular evolution in a laboratory? How do you get one kind of molecule to change gradually into another kind of molecule?

It has actually been done since the mid 1960s. That was when Professor Sol Spiegelman and his colleagues at the University of Illinois investigated a virus called Qβ. You know how viruses can give you the cold or flu? Well Qβ is a virus that attacks the common bacterium *E. coli*. Like other viruses, Qβ is just a shell of proteins and that shell holds a tiny loop of genetic material; in this case, it is RNA. The only thing Qβ does is infect *E. coli* cells and get those cells to make lots of copies of itself.

Now, in order to make copies of the Qβ virus, there has to be a protein that makes copies of the Qβ RNA. That protein is called "Qβ replicase." The behavior of Qβ RNA and Qβ replicase suggested to Spiegelman a neat experiment.

First, put some Qβ replicase and a Qβ RNA strand into a test tube along with RNA nucleotides; those are the individual building blocks of RNA. If you wait awhile, the Qβ replicase does its thing and you get vast numbers of new Qβ RNA strands. But it turns out the Qβ replicase is pretty sloppy and it makes tons of mistakes. That means that each new Qβ RNA strand will have random mistakes. So you wind up with a new assortment of RNA strands.

So the next thing that Spiegelman and co-workers did was to force their new collection of RNA molecules to evolve. They allowed their chemical mix of Qβ replicase, RNA strands and RNA nucleotides to make copies for just twenty minutes. During that short period, some of the new RNA sequences were able to bind strongly to Qβ replicase and they tended to duplicate rapidly. Meanwhile, other RNA strands do not interact well with Qβ replicase and they are not copied very efficiently.

Then you repeat the cycle. After twenty minutes the solution becomes enriched in RNA strands that easily replicate and that collection of RNA is transferred into a third beaker with fresh Qβ replicase and RNA building blocks. Spiegelman and co-workers repeated the process, 74 times in all. They also gradually reduced the time allowed for the reaction to proceed. In every cycle, Qβ replicase preferentially selected and copied RNA strands that were better at being copied.

So the system evolved. By the experiment's end, the most efficiently copied RNA strands had been reduced to a sixth of their original length and they were being copied fifteen times faster than before. There were also other similar experiments. By subjecting collections of RNA to heat, or acidity, or other stresses, Spiegelman's group could cause their Qβ RNA molecules to evolve. Soon the origin-of-life community was referring to these evolving RNA strands as "Spiegelman monsters."

Let me be clear. These fascinating systems are nowhere close to being alive. It not only takes overt manipulation of beakers and solutions but you also have to add all sorts of specialized chemicals like Qβ replicase and nucleotides. But Spiegelman's group had shown something profound. A cyclical process of competitive pressure can induce molecular systems to evolve. It was much easier to imagine a primitive Earth on which cycles of heating or wetting induced a similar kind of evolution. These results have also pointed the way to an amazing new world of artificially evolved molecules that are designed to perform specific chemical tasks.

I want to tell you about another effort in molecular evolution that is even more ambitious. This project is headed by Harvard Medical School biologist Jack Szostak, who actually wants to engineer an evolving life form in his laboratory. Since the early 1980s Szostak and his many gifted students have been studying RNA. He was inspired by the discovery of ribozymes. Recall that ribozymes are RNA enzymes, that is, RNA acting as a catalyst. In his first series of ribozyme experiments, he focused on designing a chain of RNA that can copy other RNA molecules. That is an important step in origins research because replicating RNA using such a so-called "RNA replicase" is a first step in designing a self-replicating chemical system.

More recently, Jack Szostak and his colleagues have concentrated on self-replicating RNA. In 2001, they made an impressive RNA sequence that grabs onto a shorter piece of RNA up to 14 nucleotides long and copies it. Also, they are sure that it's just a matter of time before they can make self-replicating RNA strands more than 100 letters long. Now, that is a lot shorter than the genetic material of modern cells but, still, it would be a huge advance. When Szostak and his colleagues accomplish that feat, then synthetic life will be just around the corner.

Self-replicating RNA is a complex and important step to understanding life's origins. It is not alive but it sure is getting closer. The problem is that no known geochemical environment could have supported such an isolated molecule. There is no supply of nucleotides and the RNA strand itself would not have survived long under most natural chemical conditions.

The key to survival in modern cells is protection by a lipid membrane. Szostak's group has taken a page from David Deamer's work on lipids and has been experimenting with ways that membranes might have encapsulated RNA strands. They find that lipid vesicles squeezed through tiny pores can stretch out, divide and then grow larger. That process looks a lot like cell division.

What is interesting is that this process is much faster in the presence of fine-grained clay minerals, and some of the clay particles wind up inside the vesicles. Now, recall from Lecture Seventeen that James Ferris showed how clays could attract and help to assemble RNA strands. Imagine the possibility that the cell-like structures spontaneously incorporate RNA-bound clay particles.

So here is Jack Szostak's latest scenario for the first life form to evolve by natural selection. First, you have a lipid vesicle that contains self-replicating RNA. Now, you have to understand that previous experts thought that RNA has to play lots of different roles in such a protocell—indeed, all the roles of catalysis, regulation, information storage and more that DNA and proteins play today.

But Szostak realized that RNA could promote cell growth by a much simpler process. Just build up internal pressure. As RNA pushes out on the membrane, pressure is applied to neighboring protocells. This contact pressure would have promoted transfer of lipids from cells with less internal pressure. It is the competition for space that drives a process of natural selection. Protocells with more RNA would out-compete those with less.

You can actually test these ideas. Szostak's students prepared two sets of vesicles, one filled with a solution of the sugar sucrose and another filled with pure water. When they mixed the two types of vesicles together and confined the mix, the sucrose-laden vesicles got bigger by incorporating lipids from their water-filled neighbors. The same thing happened when they repeated the experiment with strands of RNA. The vesicles swollen with RNA grew at the expense of adjacent empty vesicles.

This discovery is highly significant. Before Szostak's work, most researchers assumed that RNA would have to learn how to control all sorts of cellular functions, such as making membranes and guiding metabolism, before a protocell could evolve by natural selection. But Szostak's results point to a much simpler scenario. The only vital task for RNA in the protocell is self-replication. The more RNA a vesicle can hold, then the more likely it will be to replicate and grow. It will out-compete its neighbors in this molecular version of the-rich-get-richer scenario. A tremendous advantage is conferred on efficient RNA replicators, so molecular natural selection and Darwinian evolution could take over.

Researchers move ever closer to synthesizing a self-replicating, encapsulated, evolving chemical system—a living system in the lab—as each year goes by. Have no doubts; that will be an epic achievement. Such synthetic life will also raise many important ethical questions about the hidden dangers of scientific research and, perhaps, about the meaning of life as well. But you have to wonder. Will this kind of synthetic life tell us how life began on the primitive Earth?

I have to admit that a synthetic RNA organism would give a lot of credibility to the RNA World scenario. It would show that a strand of RNA, or perhaps some kind of precursor genetic molecule, could have formed an evolving, self-replicating chemical system. But remember, this laboratory-created life will not have emerged spontaneously from a sequence of chemical reactions among lots of simple molecular building blocks. The laboratory conditions may not be a plausible replica of the prebiotic environment either. After all, researchers still influence the result by providing a supply of RNA nucleotides and vesicle-forming lipids.

As we come to the end of this penultimate lecture, there is no question that molecular natural selection was a critical stage in the emergence of modern cellular life. Even before molecules began to reproduce themselves, cycles of light, moisture, temperature, and so on were at work selecting some molecules over others for hundreds of millions of years. The appearance of self-replication accelerated the process of molecular selection. Self-replication with variability opened the way for more direct and intense competition among related molecules, again accelerating the process of molecular selection even more.

Under idealized conditions, researchers can already design evolving molecular systems in the lab. Researchers have used a virus to demonstrate that a cyclical process of competitive pressure can induce molecular systems to evolve. Other researchers studying RNA have suggested that competitive pressures for RNA might have been something as simple as increasing internal pressure for space within each protocell.

Huge gaps remain in our understanding, still. But, in the final lecture, we'll see that much of what we've learned so far can be organized into three plausible scenarios for the origins of life.

Lecture Twenty-Four

Three Scenarios for the Origin of Life

Scope: In this last lecture, we'll see how the various debates and "Worlds" that we've considered in previous lectures contribute to three broad scenarios for the origins of life. We'll also gain a sense of the vibrant state of research on how life continued to develop *after* the emergence of the first living entity. Finally, we'll touch on how NASA and other astrobiology groups conduct the search for life on other worlds.

Outline

I. The central mystery of life's origins lies in the unknown transition from a geochemical world to the evolving biochemical world. We can choose from among three possible scenarios for this transition.

- **A.** The first possibility is that life began with metabolism, and genetic molecules were incorporated later. This metabolism-first scenario states that life began autotrophically.
    1. Life's original building blocks were simple molecules, such as water and carbon dioxide, whereas minerals, perhaps in an active volcanic zone or hydrothermal vent, provided chemical energy.
    2. Life began with a self-replicating chemical cycle, such as the reverse citric acid cycle, perhaps on a mineral surface. Variants of this cycle competed for resources, and the system became more efficient and more complex. All subsequent chemical complexities, including genetic mechanisms and encapsulation into a cell-like structure, emerged later by the process of natural selection. Life emerged as an evolving chemical coating on rocks.
    3. The true predictive tests of this metabolism-first origin scenario rest on chemical synthesis experiments. Researchers must discover a way to start a self-replicating cycle of molecules that mimics the core citric acid metabolism.
    4. Several of the essential steps have been demonstrated, but the key missing experiment remains the synthesis of 4-carbon oxaloacetate from 3-carbon pyruvate.

5. Perhaps it can be done using sulfur analogs in an environment rich in hydrogen sulfide. Or maybe a 3-carbon molecule, such as propene, can make a useful 4-carbon molecule.
6. If that step can be demonstrated and if a self-sustaining cycle of reactions can be maintained, metabolism-first will be the model to beat.

B. The second scenario is that life began with a self-replicating strand of some genetic molecule—the RNA World.

1. Life relied on an abundance of biomolecules in the environment. Metabolism was incorporated later. A template, such as a clay mineral or a stack of PAHs, helped to assemble information-rich polymers from organic molecules in the prebiotic soup.
2. Ultimately, one of these polymers acquired the ability to self-replicate. Then, as variants of the genetic polymer became more efficient at self-replication, new chemical complexities arose through natural selection. In this genetics-first version, life began as an evolving polymer with a functional genetic sequence.
3. This scenario has great appeal because of its reliance on the multiple ancient roles of RNA—a molecule that can carry information and act as a catalyst.
4. The model lacks the crucial support of one experiment that demonstrates the prebiotic synthesis of a genetic polymer like RNA. If such an experiment succeeds, most experts will agree that the problem of life's chemical origins has been more or less solved.

C. In the third scenario, life began as a cooperative chemical phenomenon, arising between metabolism and genetics.

1. It is possible that neither a primitive metabolic cycle nor a primitive genetic molecule could have progressed far by themselves. Metabolic molecules without enzymes tend to react in uncontrolled ways and may be too unconstrained to sustain advanced cellular life. Genetic molecules are much too unstable, and their individual building blocks—the nucleotides—don't seem to emerge spontaneously from the soup.
2. Yet what if crudely self-replicating genetic molecules became attached to a crudely functioning surface-bound metabolic

coating? A kind of cooperative chemistry might have occurred.

3. The genetic molecules might have made use of chemical energy produced by surface metabolism to make copies of themselves, and a surface metabolite might protect the delicate genetic molecules.
4. This system could also evolve. Variations that by chance improved the chemical efficiency of the system would have been preserved preferentially. Gradually, both the genetic and metabolic components would have become more efficient and interdependent.
5. Such a *dual-origins* model might seem to introduce a needless complication by requiring the emergence of two chemical systems. Still, it may be that this kind of cooperative coupling of metabolism and genetics occurred early in the history of cellular life.
6. If experiments establish easy synthetic pathways to *both* a simple metabolic cycle *and* an RNA-like genetic polymer, such a symbiosis may be the most likely origins scenario of all.

II. Whatever the first life-form was like, an unbroken chemical pathway must have run from that early, primitive state to the simplest of modern cells.

A. The top-down approach provides the most important clues here. Chemical studies of cells reveal a few shared biochemical characteristics in *all* cells. Every cell has the deeply embedded citric acid cycle—the metabolic heart of life. Every cell has RNA that acts as a catalyst, as well as RNA that carries genetic information.

B. Every known cell also features a complex interplay between DNA, which carries genetic information, and proteins, which do most of the chemical work. In modern cells, the RNA converts the information of DNA into the structure of proteins. The universal conclusion is that the RNA World was followed by the much more complex and modern, but much more efficient, DNA-Protein World.

C. To understand the emergence of the DNA-Protein World, researchers resort to the second type of top-down evidence, molecular phylogeny, which involves comparing sequences of DNA or RNA for similarities and differences (see Lecture Fifteen). Any sections of the genetic message that are shared by all cells presumably represent the most ancient instructions essential to all life.

1. The genetic code is the set of 64 different genetic words, each one consisting of a combination of three letters selected from A, C, G, or T. Recall that each letter also has a matching base: A always matches T, and C always matches G. Various combinations of three letters correspond to the 20 different amino acids that make proteins.
2. If we look closely at this genetic code (see diagram in Lecture Twenty-One, p. 54), an intriguing pattern emerges. For combinations where the first two letters are either C or G (CC, CG, GG, and GC), those first two letters apparently suffice to code some of the amino acids.
3. The combination GG for the first two letters *always* corresponds to glycine; the third letter makes no difference. Glycine is the simplest amino acid; its side group is just a hydrogen atom. It is also the amino acid produced in greatest abundance by the Miller-Urey experiment. That's why many researchers think glycine is the most primitive, essential amino acid and had to have been a major part of the earliest proteins.
4. If the first two letters are GC, they always correspond to alanine, which is the second most abundant amino acid produced in experiments. Its side group is CH3; thus, alanine is also assumed to be one of the key initial amino acids.
5. The other two combinations, CC and CG, always indicate proline and arginine, respectively. They're among the simpler amino acids.
6. Thus, the genetic code contains a kind of mini-code of two bases resulting in four of the simpler amino acids.

Mini-Genetic Code Using Two Bases

	C	**G**
C	**proline**	**arginine**
G	**alanine**	**glycine**

7. Those four amino acids, in turn, can combine to form a variety of functional proteins.

D. In short, an earlier, intermediate state may have occurred between the RNA World and the modern DNA-Protein World. The structure of the modern genetic code, with a simple core and layers of increasing chemical complexity, gives us good reason to believe in such a transition. The progression is from a simpler to a more complex organization, and it obeys the principle of continuity. The modern genetic code with four letters is an elaboration of the earlier code with only two letters. Much more sophisticated and efficient proteins can be made with 20 amino acids, which is why the more complex code replaced the simpler version long ago.

E. Other hints about the history of life are buried in the DNA and RNA of today's cells. Many scientists are looking closely at the RNA that makes ribosomes and the transfer RNA that matches specific groups of three genetic letters to specific amino acids. This idea of layered complexity, with modern biochemistry improving on much simpler beginnings, provides a strategy for understanding how we got to modern life.

III. It is possible that the ultimate solution to the origins question will come not from labs on Earth but from probes of space.

A. Mars was once a warm, wet planet, and it was probably habitable long before Earth (see Lecture Five). It is possible that life in our Solar System first arose there, and Earth may have been seeded with that life by meteorites. If so, then any microbial life we find on Mars may be almost indistinguishable from microbes on Earth. It is also possible that microbes from Earth were transferred to Mars on chunks of rocks blasted into space by impacts here.

B. Finding Earth-like life on Mars would certainly be interesting, but the real Holy Grail would be the discovery of a completely independent origin of life—a "second genesis."

C. Second genesis is an independent chemical emergence of a self-replicating, evolving system. In our lifetimes, that discovery might happen in one of two ways.

1. Astronomers might find observational evidence for a living planet around a distant star using telescopes with giant mirrors. With spectroscopy, we'll be able to see each planet's color and measure the distinctive absorption lines from its

atmosphere. If we find a planet with an oxygen-rich atmosphere like ours, we'll have discovered compelling evidence for a second genesis.

2. Or, we may recover samples of a completely alien form of life from Mars, Europa, or some other body in our Solar System. How would we know if a Martian microbe represented a second genesis? Biochemistry must be key. A second genesis would use a suite of amino acids that is different from the idiosyncratic set of 20 used by cells on Earth. The details of the genetic code—which genetic letters specify which amino acids—would also be different.

D. Hard evidence of a second genesis would give us a gold mine of biological data that might reveal countless details about life's inevitable origin and its shared chemical characteristics. Is the citric acid cycle an essential metabolic first step? Is RNA the universal genetic molecule? Which amino acids are optional? What other polycyclic molecules besides sterols might play a role in alien life?

IV. A central theme of this lecture series is the remarkable power of emergence to drive increases in the complexity of natural systems. The theory of emergence argues for an inexorable evolution of the cosmos, from atoms to stars to planets to life. Each emergent step arises from interactions among countless agents. Each emergent event produces an outcome that's much greater than the sum of its parts. Each emergent process is reasonable and sequential, and each step increases the degree of order and complexity.

A. We recognize this progression only in hindsight. Emergent phenomena are all but impossible to predict from observations of earlier stages.

B. Is human intelligence the highest possible emergent state? Perhaps the universe holds levels of emergence beyond individual consciousness and beyond even the collective accomplishments of human societies. If that's true, then the story of life's origins and evolution is far from over.

Essential Reading:

Hazen, *Genesis: The Scientific Quest for Life's Origin*, chap. 19.

Supplemental Reading:

Dick and Strick, *The Living Universe: NASA and the Development of Astrobiology*, chaps. 5 and 6.

Fry, *The Emergence of Life on Earth*, chap. 14.

Lahav, *Biogenesis: Theories of Life's Origin*, part IV.

Wills and Bada, *The Spark of Life: Darwin and the Primeval Soup*, chap. 10.

Questions to Consider:

1. How would discovery of a second genesis change our view of life in the universe?
2. Which of the three scenarios for life's origins is most appealing to you? What evidence would be convincing?

Lecture Twenty-Four—Transcript
Three Scenarios for the Origin of Life

We have come to the twenty-fourth, and final, lecture of this series on the origins of life. I think you'll agree that we have covered a lot of territory, but it's also obvious that the scientific story is not even close to being over. As in so many arenas of scientific research, what we know today is dwarfed by what remains to be discovered. We're at a stage where the problem is clearly defined. We know what we have to do, and literally hundreds of scientists are tackling the question full time. Many advances are being made and there is no doubt that we will see real progress in our lifetimes. Just during the time I have been delivering this course, new work has appeared from NASA's Cassini-Huygens and Deep Impact missions that reveal amazingly rich sources of organic molecules and water in our own solar system.

I have three main objectives in this concluding lecture. First, I want to show you how the various debates and "Worlds" that we have considered in previous lectures contribute to one or another of three broad scenarios for the origins of life. Second, I want to give you a sense of the vibrant state of research on what came just after the origins of life. An army of scientists around the world is trying to understand how life continued to develop after the emergence of the first living entity. There were many important chemical developments, to be sure—for example the introduction of DNA and proteins and the emergence of photosynthesis. Then there were billions of years of biological evolution, from relatively simple single cells to all the complexity of multi-cellular plants and animals.

Third, I also want to touch briefly on how NASA and other astrobiology groups are conducting the search for life on other worlds. One of the central objectives of astrobiology, beyond an understanding of origins on Earth, is the search for life on Mars; on Jupiter's moon, Europa, on Saturn's moon, Titan, and other nearby planets and moons. This effort may reveal to us whether life is indeed a cosmic imperative, or if it is relatively rare in the universe. These stories are not necessarily part of the story of life's emergence on Earth but they certainly play a major role in the broader field of astrobiology.

The time has finally come to summarize this lecture series. As I mentioned at the beginning of this course, the Santa Fe Institute theorist Stuart Kaufman wrote one of my favorite quotations regarding the origins of life in 1995. He said, "Anyone who tells you that he or she knows how life started on … Earth is either a fool or a knave." Leslie Orgel, who played such a key role in the RNA World model, put an interesting twist on this idea. He said, "Anyone who thinks they know the solution to this problem is deluded, but anyone who thinks it is insoluble is also deluded."

Well, I certainly have not given you the answer to life's origins, nor could anyone else at this point in history, but I agree that the chemical and physical steps that led to the origins of life are knowable and, someday, the story will be known. We have learned a tremendous amount and the basic outlines of the origins process are coming into focus.

In Lecture Three, we explored the tricky question, "What is life?" While there's no correct answer, we did identify a few essential defining characteristics. For one thing, life is chemical in nature. That is why we've focused so much attention on the kinds of chemical reactions that might have transformed simple, small carbon-based molecules into larger, more complex molecular structures.

Then, life requires self-replication. So we need a molecule, or more likely a cycle of many molecules, that makes copies of itself. What's more, that self-replicating cycle has to display some degree of chemical variability. In that way, different versions of the system will compete for resources and the entire chemical cycle will have the ability to evolve. Gradually, the entire chemical system will become more efficient at processing matter and energy. In addition, new variations of the system will become adapted to a variety of different environments, including a range of different temperatures, pressure, salinity, acidity and other chemical factors.

That much is certain, but there's a lot we don't know about that first living system. Was it autotrophic, making its own molecules from water and volcanic gases? On the other hand, was it heterotrophic, scavenging amino acids, sugars, lipids, and other molecules from the surroundings? We also do not know if self-replication began as a metabolic cycle, like the reverse citric acid cycle, or as a genetic molecule similar to RNA. We are not even sure if the first life form was encapsulated in a primitive membrane like modern cells or if it was a flat layer growing sideways on mineral surfaces.

So, what do we know for sure? Let's go back to the very first lectures, when I introduced the top-down approach. We worked our way backward from modern cells to fossil cells and their components—their atoms, isotopes and biomolecules. What we find is that even the oldest rocks on Earth hold evidence for life that is surprisingly modern in character. The rare traces of cell-like fossils look like modern cells. The carbon isotopes and biomolecules, likewise, are reminiscent of cells today. All traces of life's emergent chemistry seem to have been eradicated. In fact, the major contribution from the top-down approach seems, to me, to be that life must have emerged in a rather narrow window, perhaps between about four and 3.9 billion years ago.

The bottom-up approach provides a complementary and chemically more satisfying, approach to studying life's origins. In Lecture Nine, we saw how Stanley Miller and Harold Urey catalyzed this field with their brilliant experiment using water and simple gases in a glass, tabletop experiment. They made biomolecules by mimicking a presumed prebiotic environment. In the process, they brought scientific studies of life's origins into the modern experimental age.

We have learned, subsequently, that organic molecules must have been synthesized in abundance and they must have accumulated in a host of prebiotic environments. Origins researchers have also demonstrated how biomolecular systems—including lipid membranes and genetic polymers—might have self-organized, or perhaps formed on mineral surfaces. Then, as molecular complexity increased, it seems plausible that primitive metabolic cycles of self-replicating molecules emerged, as did self-replicating genetic molecules. The details remain fuzzy but much of the basic outline seems to be in place.

What is certain is that the central mystery of life's origins lies in the unknown transition from a geochemical world with oceans, air, rocks, and many interesting organic molecules, to the evolving biochemical world in which systems of molecules self-replicate, compete, and evolve. Even if we don't know the details, it seems to me that we have to choose from among three possible scenarios.

The first possibility is that life began with metabolism and genetic molecules were incorporated later. This metabolism-first scenario, which is championed by Günter Wächtershäuser, Harold Morowitz and others, states that life began autotrophically. Life's original building blocks were simple molecules such as water and carbon dioxide, while minerals, perhaps in an active volcanic zone or hydrothermal vent, provided chemical energy. Life began with a self-replicating chemical cycle like the reverse citric acid cycle, perhaps on a mineral surface. Variants of this cycle competed for resources and the system became more efficient and more complex. All subsequent chemical complexities, including genetic mechanisms and encapsulation into a cell-like structure, emerged later by the process of natural selection. In this version of life's origin, life first emerged as an evolving chemical coating on rocks.

The true predictive tests of this metabolism-first origin scenario rest on chemical synthesis experiments. Researchers have to discover a way to start a self-replicating cycle of molecules that mimics the core citric-acid metabolism. We may be getting close. Several of the essential steps have been demonstrated but there is still that key missing experiment—that is, the synthesis of four-carbon oxaloacetate from three-carbon pyruvate. Perhaps it can be done using sulfur analogs in an environment rich in hydrogen sulfide. Or maybe, as George Cody suggests, some other three-carbon molecule like propene was used to make a useful four-carbon molecule. If that step can be demonstrated and if a self-sustaining cycle of reactions can be maintained, then metabolism-first will be the model to beat.

The second scenario is that life began with a self-replicating strand of some genetic molecule, the RNA World. According to this hypothesis, life relied on an abundance of biomolecules already present in the environment. Metabolism was incorporated later. A template like a clay mineral or a stack of PAHs helped to assemble information-rich polymers from organic molecules in the prebiotic soup. Ultimately, one of these polymers acquired the ability to self-replicate. Then, as variants of the genetic polymer became more efficient at self-replication, new chemical complexities arose through natural selection. In this genetics-first version, life began as an evolving polymer with a functional genetic sequence.

This scenario has great appeal because of its reliance on the multiple ancient roles of RNA, a molecule that can carry information and act as a catalyst. Indeed, the genetics-first model lacks only the crucial support of one experiment, an experiment that demonstrates the plausible prebiotic synthesis of a genetic polymer like RNA. If such an experiment succeeds, then the biggest gap in our understanding of life's origins will have been filled. In which case, most experts in the field will agree that the problem of life's chemical origins has been more or less solved.

Finally, there's a third scenario in which life began as a cooperative chemical phenomenon, arising between metabolism and genetics. It is possible that neither a primitive metabolic cycle nor a primitive genetic molecule could have progressed far by themselves. On the one hand, metabolic molecules by themselves, without enzymes, tend to react in a variety of uncontrolled ways. So they may be much too unconstrained to sustain advanced cellular life. On the other hand, genetic molecules are much too unstable and their individual building blocks, the nucleotides, do not seem to emerge spontaneously from the soup.

However, what if crudely self-replicating genetic molecules became attached to a crudely functioning surface-bound metabolic coating? Then a kind of cooperative chemistry might have occurred. In that event, the genetic molecules might have made use of chemical energy produced by surface metabolism to make copies of themselves. A surface metabolite might also offer a degree of protection for the delicate genetic molecules.

This system could evolve. Variations that, by chance, improved the chemical efficiency of the system would have been preserved preferentially. Gradually, both the genetic and metabolic components would have become more efficient and more interdependent.

Such a "dual origins" model might at first seem too wishy-washy. It might seem to introduce a needless complication by requiring the emergence of not one, but two chemical systems. But it may be that exactly this kind of cooperative coupling of metabolism and genetics occurred early in the history of cellular life. So, if experiments establish easy synthetic pathways to both a simple metabolic cycle and an RNA-like genetic polymer, then my guess is that such a symbiosis may represent the most likely origins scenario of all. In this case, calling the process the origins of life—plural, with an "s"—would make even more sense.

Well, whatever the first life-form was like, there must have been an unbroken chemical pathway from that early, primitive state to the simplest of modern cells. The law of continuity requires that. So let's think about what must have come after the emergence of Earth's first living object.

We have seen that the top-down approach provides the most important clues in this effort. We can examine two lines of evidence. First, chemical studies of cells reveal a few shared biochemical characteristics in all cells. We have seen that every cell has the deeply embedded citric acid cycle; that's the metabolic heart of life. In addition, every cell has RNA that acts as a catalyst, as well as RNA that carries genetic information. These are surely among the most ancient biochemical tricks.

Every known cell also features a complex interplay between DNA, which carries genetic information and proteins, which do most of the chemical work. In modern cells, the RNA plays what seems to be a much less prominent role. It acts sort of like a broker that converts the information of DNA into the structure of proteins. The universal and obvious conclusion, therefore, is that the RNA World was followed by the much more complex and modern, but much more efficient, DNA-Protein World.

Now, to understand the emergence of the DNA-Protein World, researchers have to resort to the second type of top-down evidence that is molecular phylogeny. Recall, from Lecture Fifteen, that molecular phylogeny involves comparing sequences of DNA or RNA and looking for similarities and differences. Any sections of the genetic message that are shared by all cells presumably represent the most ancient instructions, the information that is essential to all life.

I think you can imagine how complex and time-consuming this task might be. First, you need to compare the genetic information of hundreds of different species. Then you need to tease out similarities that may be hidden in strings of millions or billions of genetic letters. That is a huge job.

But let me give you a sense of the kind of conclusions that are now coming out. Consider the genetic code, itself. The genetic code is the set of 64 different genetic words, each one consisting of a combination of three letters—A, C, G, or T. Recall that there are two different sets of matching bases; A always matches T and C always matches G. Also, various combinations of three letters correspond to the 20 different amino acids that make proteins. So, A-A-A codes for the amino acid lysine, A-A-C codes for the amino acid asparagine and so forth.

So, here is the complete genetic code that is used by every known living organism and it is pretty complex. Well, if you look closely at this genetic code, an intriguing pattern emerges. For combinations where the first two letters are either C or G, those first two letters apparently suffice to code some of the amino acids. There are exactly four combinations of C and G. You have CC, CG, GG, and GC.

So what amino acids do they represent? It turns out that the combination GG for the first two letters always corresponds to glycine. The third letter makes no difference. Glycine is the simplest amino acid. Its side group is just a hydrogen atom. In addition, glycine is also the amino acid that is produced in greatest abundance by the Miller-Urey experiment. That is why many researchers think glycine is the most primitive, essential amino acid and it had to be a major part of the earliest proteins.

Then, if the first two letters are GC, then it always corresponds to alanine. And alanine is the second most abundant amino acid produced in experiments. Its side group is just CH_3, so alanine is also assumed to be one of the key initial amino acids.

The other two combinations, CC and CG, always indicate proline and arginine, respectively. Also, they are among the simpler amino acids. So, we have built into the genetic code a kind of mini-code of two bases and four of the simpler amino acids. Those four amino acids can combine, all by themselves, to form a variety of functional proteins.

This situation suggests that there may have been an earlier, intermediate state between the RNA World and the modern DNA-Protein World. Today we have four bases and 20 amino acids. But what if there was a primitive period with just two bases, and four amino acids—glycine, alanine, proline, and arginine?

This layered structure of the modern genetic code, with a simple core and layers of increasing chemical complexity, gives us good reason to believe in this transition. We go from a simpler to a more complex organization. We also obey the principle of continuity; the modern genetic code with four letters is an elaboration of the earlier code with only two letters. It is easy to see why the more complex code survives because you can make much more sophisticated and efficient proteins with 20 amino acids. The more complex code is better, so it incorporated and replaced the simpler version long ago.

Well, there are lots of other hints about the history of life that are buried in the DNA and RNA of today's cells. Many scientists are looking especially closely at the RNA that makes ribosomes and also the transfer RNA that matches specific groups of three genetic letters to specific amino acids. This idea of layered complexity, with modern biochemistry elaborating and improving on much simpler beginnings, provides a great strategy for trying to understand how we got to modern life. Now, the details are way beyond the scope of this lecture, but you could consult Noam Lahav's excellent book, *Biogenesis*, for more information. Christian de Duve's book, *Vital Dust*, also has a good introduction to this topic. Anyway, I think you get the idea. We really do have a good chance of filling in the gaps between the RNA World and modern life.

We do not yet know the answer to how life emerged but day by day we are finding out more. New experiments expose parts of the truth, while winnowing out the possibilities. Whatever the correct scenario may be, we can be sure that, ultimately, competition began to drive natural selection. Once life emerged, it would continue to evolve to this day.

For the last twenty-three lectures, I have been describing a systematic laboratory approach to understanding life's origins, both from the top down and from the bottom up. But it's possible that the ultimate solution to the origins question will come not from labs on Earth, but from probes of space.

Think about what it would mean to discover life on another planet. Of course, the most dramatic discovery would be to make contact with an alien intelligent civilization. I think you will all agree that such an event would radically transform our understanding of life's origins and distribution in the universe. It is always possible that we will meet a benign, advanced alien species that could explain to us all the intricacies of life's chemical origins. Indeed, such an encounter might put many scientists out of work!

What seems, to me, a lot more likely is that we will find convincing evidence for primitive life on another planet or moon. A lot of scientists at NASA would make a bet that we will find evidence of life on Mars if we look carefully enough but the prevailing view is that it will be Earth-like microbial life. As I mentioned in Lecture Five, Mars was once a warm, wet planet and it was probably habitable long before Earth. It is very possible that life in our Solar System first arose on Mars. Given the significant influx of meteorites from Mars, Earth may well have been seeded with that life. If so, then any microbial life we find on Mars may be almost indistinguishable from microbes on Earth.

In addition, we can't discount the possibility that microbes from Earth have been transferred to Mars on chunks of rocks blasted into space by impacts here. The probability of an Earth-to-Mars transfer is much lower than Mars to Earth for two reasons. For one thing, Earth is almost ten times more massive than Mars, so it is much harder to eject a rock from the surface. It is also vastly easier to transfer a rock towards the Sun, from Mars to Earth, than the other way around.

In spite of these difficulties, NASA scientists are confident that life has been transferred to the surface of Mars. This is because even with the most rigorous sterilization protocols, every Mars lander and rover is loaded with microbes from Earth. Every interior surface, every crack and pit has microbes. Unless those microbes are exposed directly to sunlight, they can survive for many years. So the *Viking* landers, and *Pathfinder*, and the *Spiri,* and *Opportunity* rovers all have brought life to Mars. Now, it is doubtful that those microbes will thrive and evolve to populate the entire planet but we cannot be sure. It is just a fact of space exploration that there's absolutely no way that we can visit other worlds without introducing this kind of contamination.

Finding Earth-like life on Mars would certainly be interesting, but I don't think such a discovery would transform our thinking. We already know that microbial life has emerged and evolved. A physical transfer of life from one planet to another would say more about the resilience of microbial life than it would about the origins of life.

The real Holy Grail in the search for alien life is what is known in the business as "second genesis." That is the discovery of a completely independent origin of life. Second genesis is an independent chemical emergence of a self-replicating, evolving system. In our lifetimes, that discovery might happen in one of two ways—actually one of three ways, if you count the possibility of meeting ET.

One possibility is that astronomers might find observational evidence for a living planet around a distant star. We are just now entering the age when the discovery of planets around other stars is possible. Now most of these discoveries are made by indirect means. We can observe the periodic wobble of a star as a planet orbits around; a planet's gravity perturbs its star ever so slightly. Or we can measure the slight dimming of a star as a planet passes in front of the bright disk. These techniques are now revealing the presence of distant planets around distant stars by the hundreds but they say nothing about whether or not the planets harbor life.

But the next generation of telescopes will be designed to do much more. Plans are now in the works for ground-based telescopes with mirrors approaching the dimensions of a football field. The next generation of space telescopes, if NASA can get the necessary funding, will also be much larger than the Hubble Space Telescope. The idea is that these giant instruments will allow us to actually image Earth-like planets orbiting nearby stars. Using spectroscopy, we will be able to see the planet's color and measure the distinctive absorption lines from the planet's atmosphere. If we find a planet with an oxygen-rich atmosphere like ours, then we'll have discovered compelling evidence for a second genesis.

But there's another way that we might discover a second genesis. It is possible that we will recover actual samples of a completely alien form of life from Mars or Europa or some other body in our own Solar System. The first Mars sample return mission might take place as early as 2013. Who knows what secrets a shovel full of Martian soil might hold?

So how would we know if a Martian microbe represents a second genesis? It is very likely that alien cells would be small, round or elongate objects and rather nondescript, so they'd look a lot like Earth life. Biochemistry would be the key. Any organism that emerged by an independent origin process will display a different biochemistry. To take the extreme view, it is possible that alien life will possess radically different chemistry, with a very different complement of metabolic and genetic molecules. In that case, we will know the life is alien pretty quickly.

But even if we find that the citric-acid cycle, lipid membranes, and RNA are inevitable aspects of life everywhere in the universe, some of the chemical details will certainly be different. For example, all life forms will likely use the commonest amino acids, such as glycine, alanine and proline. But a second genesis will surely employ a suite of amino acids that's different from the idiosyncratic set of 20 that cells on Earth use. Also, the details of the genetic code, exactly which genetic letters specify which amino acids, will also be different for a second genesis. Biochemists can sort out rather easily those details.

If we find hard evidence of a second genesis, then we will have a gold mine of biological data. On the one hand, a second, entirely independent origin of life will reveal countless details about life's inevitable origin and its shared chemical characteristics. Is the citric-acid cycle an essential metabolic first step? Is RNA the universal genetic molecule? On the other hand, we will also discover some of the chance chemical choices and quirks of our own biochemistry. Which amino acids are optional? What other polycyclic molecules, besides sterols, might play a role in alien life? If we find physical evidence of a second genesis in our own Solar System, then suddenly Earth biology will become just one example of what must be countless billions of independent origins of life throughout space and time.

A central theme of this lecture series has been the remarkable power of emergence to drive increases in the complexity of natural systems. The theory of emergence argues for an inexorable evolution of the cosmos, from atoms to stars to planets to life. Each emergent step arises from interactions among countless agents. Each emergent event produces an outcome that is much greater than the sum of its parts. Each emergent process is reasonable and sequential and each step increases the degree of order and complexity.

It all makes sense but the fact is that we can recognize this majestic progression only in hindsight. Emergent phenomena are all but impossible to predict from observations of earlier stages. Given hydrogen atoms, for example, it would require an extraordinary conceptual leap to predict the behavior of stars or the variety of planets. Then, given a planet, no theoretician could predict the emergence of cellular life. For the same reason, given a collection of cells, no one could extrapolate to the emergence of consciousness and self-awareness. In every case, the inherent novelty and layered complexity of emergent phenomena seem impossible to predict.

I want to leave you with one closing thought about the emergence and evolution of life. Are we it? Is human intelligence the highest possible emergent state? Or is it possible that there will be higher orders of emergence? If so, then it is also possible that our individual brains can no more comprehend such a state, than a single neuron can comprehend consciousness.

Perhaps the universe holds levels of emergence beyond individual consciousness and beyond even the collective accomplishments of human societies. If that is true, then the story of life's origins and evolution is far from over. Indeed, perhaps the story has only begun.

Chronology—Key Events in Research on Life's Origins

1828 .... German chemist Friedrich Wöhler demonstrates that the molecules of life are no different in principle from any other chemicals. This discovery shows that life is a chemical process.

1859 .... Publication of British naturalist Charles Darwin's On the Origin of Species by Means of Natural Selection, which introduced the concept of evolution by natural selection. This evolutionary principle applies to collections of molecules, as well as to living cells.

c. 1860 .... French biochemist Louis Pasteur refutes the theory of spontaneous generation, thus pushing the origin of life back to an unknowably remote time and place. Ironically, any chemical theory of life's origin must involve spontaneous generation.

1906 .... German chemist Walter Löb synthesized amino acids by using sparks in a gas mixture, thus anticipating the Miller-Urey experiment by almost half a century. However, his experiment was not designed to mimic a prebiotic environment.

1924 .... Publication of Russian chemist Aleksandr Oparin's The Origin of Life, which elaborated on the idea that life arose from a body of water that gradually became enriched in organic molecules.

1929 .... Publication of British biochemist J. B. S. Haldane's essay "The Origin of Life" in The Rationalist Annual.

1952 .... University of Chicago chemists Harold Urey and Stanley Miller perform the Miller-Urey experiment, the first experiment designed to mimic the production of organic molecules under plausible prebiotic conditions. The initial publication, with Stanley Miller as sole author, appeared in Science in 1953 under the title "A Production of Amino Acids Under Possible Primitive Earth Conditions."

1953 .... Cambridge University crystallographers James Watson and Francis Crick discover the double-helix structure of DNA.

1958 .... Protein chemist Sidney Fox presents the first version of his Protenoid World theory of life's origin, by which self-replicating collections of amino acids have lifelike qualities. The origins community largely rejected the theory, but it was the first model to treat life's origin as a deterministic process.

1960 .... Chemist John Oró of the University of Houston synthesizes adenine by heating a concentrated hydrogen cyanide solution.

1966 .... Scottish chemist Graham Cairns-Smith presents the first version of his Clay Life hypothesis, in which evolving clay minerals are the first life-form on Earth. This model was subsequently revised and expanded in a series of books and articles.

1967 .... Sol Spiegelman and colleagues at the University of Illinois demonstrate that a virus called Qβ can be induced to evolve in carefully controlled laboratory experiments.

1968 .... Leslie Orgel of the Salk Institute contributes his now-classic paper entitled "Evolution of the Genetic Apparatus." Orgel argues in favor of a genetics-first origin of life, based on an RNA-like polymer.

1970s... Development of the theory of emergent systems by Russian-born chemist Ilya Prigogine. Prigogine recognized that complex patterns arise when energy flows through a system of many interacting particles. The origin of life is a dramatic example of emergence.

1976 .... NASA's two Mars Viking landers provide ambiguous evidence regarding the presence of microbial life on Mars.

1977 .... Professor Carl Woese of the University of Illinois proposes a dramatic revision of the tree of life, with three domains—the Bacteria, the Archaea, and the Eukarya. The first two of these domains are exclusively single-celled organisms, many of which are extremophiles.

1977 .... Professor Thomas Gold of Cornell University proposes that petroleum arises from nonbiological processes in Earth's deep interior. He follows up this speculation in 1992 with descriptions of the "deep, hot biosphere."

1979 .... Publication of "An Hypothesis Concerning the Relationship Between

Submarine Hot Springs and the Origin of Life on Earth," by Oregon State University scientists Jack Corliss, John Baross, and Sarah Hoffman. This paper was a significant challenge to the prevailing view of a surface origin of life.

1981 .... The independent discovery by Sidney Altman of Yale University and Thomas Cech of the University of Colorado of ribozymes, which are strands of RNA that act as catalysts. They shared the Nobel Prize in 1989 for this finding.

1985 .... Publication of Belgian chemist Christian de Duve's Vital Dust: Life as a Cosmic Imperative, which elaborated on the Thioester World scenario for life's origin.

1986 .... Nobelist Walter Gilbert proposes the RNA World scenario for life's origin, by which a self-replicating strand of RNA represents the first living entity. Leslie Orgel, Gerald Joyce, and others quickly elaborated on the theory.

1988 .... Initial publication of the Iron-Sulfur World scenario for life's origin by German chemist Günter Wächtershäuser. Subsequent papers, notably in 1990 and 1992, greatly expanded on the chemical details of this model.

1988 .... James Ferris and coworkers at Rensselaer Polytechnic Institute discover the ability of clay minerals to induce RNA nucleotides to link together. Subsequent experiments document chains of more than 50 RNA units.

1989 .... David Deamer and coworkers at the University of California demonstrate the self-assembly of amphiphilic molecules that were extracted from the Murchison meteorite, thus demonstrating a possible prebiotic pathway to primitive cell membranes.

1991 .... Swiss chemist Albert Eschenmoser and colleagues synthesize more than a dozen variants of RNA with different sugar-phosphate backbones. Of special note was their synthesis of TNA, a polymer with the simple 4-carbon sugar threose instead of the 5-carbon ribose of RNA.

1991 .... Danish chemist Peter Nielsen and colleagues synthesize a novel genetic molecule with a backbone of amino-acid molecules. This variant has been given the name peptide nucleic acid, or PNA.

1993 .... Paleontologist William Schopf of UCLA publishes descriptions of 11 purported new species of fossil microbes from the 3.45-billion-year-old Apex Chert—supposedly the oldest fossils on Earth.

1996 .... The publication in Science of "Search for Past Life on Mars: Possible Relic Biogenic Activity in Martian Meteorite ALH84001" by a team of NASA scientists led by David McKay. The article immediately creates intense controversy over the authors' claims.

1997 .... Formation of NASA's Astrobiology Institute (NAI), which is headquartered at the NASA Ames Research Center at Moffett Field, California. The NAI becomes a primary funding organization for origins-of-life research.

Chronology—Key Events in Life's Origins

4.55 billion years ago .... Formation of the Solar System, including the Sun and Earth, from a giant, rotating, disk-shaped nebula. The Sun and planets gradually formed as gravity pulled matter together into clumps.

4.5 billion years ago .... Formation of the Moon by a giant impact of a Mars-sized object with Earth.

4.1 to 3.9 billion years ago .... The gradual end to the great bombardment of Earth's surface and the last of the globe-sterilizing giant impacts. End of the Hadean Eon, beginning of the Archean Eon.

3.85 billion years ago .... The oldest rocks on Earth, which occur in outcrops along the coast of Greenland. Some of these rocks contain small amounts of carbon that have the diagnostic signature for life. It is possible, therefore, that primitive life had established a foothold by this early date.

3.45 billion years ago .... The oldest purported fossils, which occur in western Australia. Some of these microscopic cell-like structures have been the subject of much recent debate.

2.7 billion years ago .... The oldest known molecular *biomarkers*, which are distinctive biological organic molecules that have been recovered from ancient rocks.

2.5 billion years ago .... Continent-sized landmasses first coalesced. End of the Archean Eon.

Glossary

abiosignatures: Evidence in the form of isotopes, molecules, or other chemical or physical structures that a rock never held life.

amino acids: Key molecular building blocks of proteins.

amino group: A molecular group of one nitrogen atom linked to two hydrogen atoms.

amphiphile: A molecule that has surface regions that attract water and other surface regions that repel water. Such molecules tend to self-organize in water.

astrobiology: The search for the origin, distribution, and future of life in the universe.

autocatalyst: A molecule that catalyzes its own synthesis.

autocatalytic network: A group of molecules, each of which catalyzes the formation of some other molecule in the group to form a self-replicating network.

autotrophic: Literally, "self-nourishment." Used to describe cells that synthesize essential biomolecules from simple building blocks, such as water and carbon dioxide.

base: A small, flat, cyclic molecule that serves as a letter in the DNA and RNA genetic molecules. Bases include adenine (A), cytosine (C), guanine (G), thymine (T), and uracil (U).

biosignatures: Any fossil structure, isotope, molecule, or other unambiguous sign of life.

calcite: A common mineral, calcium carbonate, that forms the hard portions of many shells.

carbohydrates: Comprised of carbon, oxygen, and hydrogen, carbohydrates are key biomolecules that provide a source of energy and support structures in cells. The commonest building blocks of carbohydrates are sugars.

carboxyl Group: A molecular group of a carbon atom linked to one oxygen atom and one OH group.

chirality: The property of handedness. Any molecule that is not superimposable on its mirror image is said to be chiral.

citric acid cycle: The core metabolic cycle in modern life. See **reverse citric acid cycle**.

clay minerals: Common layer-structure minerals that form as fine-grained weathering products in many geological environments.

complementary molecules: A pair of molecules that fits together like pieces of a jigsaw puzzle.

cross-catalytic system: A chemical system of two molecules, each of which catalyzes the synthesis of the other.

cyanobacteria: The primitive photosynthetic microbes that are thought to be responsible for generating Earth's oxygen-rich atmosphere.

DNA (Deoxyribose nucleic acid): The genetic polymer that stores and duplicates genetic information in every cell.

DNA-Protein World: A stage in life's development that followed the RNA World. We now live in the DNA-Protein World.

emergence: The process by which simple systems of many interacting particles spontaneously become more complex.

evolution: Any change of a system over time. Hence, the origin of life is often referred to as *chemical evolution*, while the subsequent development of life is *biological evolution*.

extremophiles: Life, especially single-celled organisms, that lives at extremes of temperature, acidity, or other physical variable.

flat life: A self-replicating, evolving layer of molecules that coats energetic mineral surfaces. A key idea of the Iron-Sulfur World model.

genetic code: The means by which genetic information stored in DNA is converted into the protein molecules that do the cell's chemical work.

heterotrophic: Literally, "other nourishment." Used to describe cells that obtain essential biomolecules from their surroundings.

homochirality: A collection of chiral molecules of a single handedness. A key characteristic of life on Earth is the homochirality of amino acids, sugars, and other essential molecules.

hopanes: A class of polycyclic molecules; these molecules sufficiently stable to be preserved for billions of years as biomarkers.

Iron-Sulfur World: The origin-of-life model, principally by Günter Wächtershäuser, in which the chemical instability of iron-sulfur minerals provides metabolic energy.

life: While no single definition is universally accepted, a NASA-sponsored panel concluded that "life is a self-sustained chemical system capable of undergoing Darwinian evolution."

lipids: Key biomolecules that include fats and oils. Lipids self-assemble to form cell membranes.

metabolism: The process by which cells obtain and use energy.

Miller-Urey experiment (also known as the **Urey-Miller experiment**): The first chemical experiment specifically designed to mimic a plausible prebiotic environment. Amino acids and other key biomolecules were synthesized.

nucleotides: The building blocks of DNA, RNA, and other genetic molecules. Each nucleotide consists of a base, a sugar, and a phosphate.

Occam's Razor: The scientific precept that one should accept the hypothesis that requires the fewest assumptions.

oxaloacetate: A 4-carbon molecule that is part of the reverse citric acid cycle. A key step in this cycle is the synthesis of oxaloacetate from pyruvate plus carbon dioxide.

PAH (Polycyclic Aromatic Hydrocarbons): A group of molecules in which carbon atoms form interlocking 5- and 6-member rings.

PNA, or Peptide Nucleic Acid: An alternative genetic molecule in which amino acids, rather than sugar-phosphate, form the backbone.

polymerization: The chemical reaction by which small molecules are linked into a long chain, or polymer.

Principle of Continuity, or Congruity: The principle that the biochemistry we see in today's cells, no matter how intricate it is, must have evolved in a continuous chemical path from primordial geochemistry.

proteins: Polymers of amino acids that fold into complex shapes. Proteins serve as the chemical workhorses of life.

pyrite: A common mineral of iron and sulfur (FeS_2) that serves as a solid foundation for the Iron-Sulfur World hypothesis.

pyrrhotite: A mineral of iron and sulfur (FeS) that is produced at hydrothermal vents. Pyrrhotite provides much of the chemical energy in the Iron-Sulfur World hypothesis.

pyruvate: A 3-carbon molecule in the citric acid cycle. The reaction of pyruvate with carbon dioxide to make oxaloacetate is a key reaction that has not yet been demonstrated for a plausible prebiotic environment.

reverse citric acid cycle (also known as the **reductive citric acid cycle**): A primitive version of the citric acid cycle that runs the opposite way, building larger molecules a step at a time to make 6-carbon citric acid, which then splits to start two cycles.

ribose: A 5-carbon sugar that forms part of the backbone of RNA.

ribosomes: Cellular structures that assemble proteins from the genetic instructions of DNA.

ribozymes: RNA strands that act as catalysts.

RNA, or Ribose Nucleic Acid: A genetic polymer that carries information and helps to assemble proteins. The discovery that RNA can act as a catalyst (a ribozyme) was instrumental in the development of the RNA World hypothesis.

RNA World: The hypothesis that a self-replicating, catalytic strand of RNA was the first life-form.

Second Genesis: A completely independent origin of life from life on Earth. Discovery of a second genesis would imply that life in the universe is common.

sugars: The key molecular building blocks of carbohydrates.

thioester: An energy-rich molecule that contains a carbon-sulfur bond.

Thioester World Model: The origin-of-life model of Christian de Duve, who suggest that thioesters provided the initial metabolic energy source.

TNA, or Threose Nucleic Acid: An alternative genetic polymer that incorporates the 4-carbon sugar threose in its backbone.

Urey-Miller experiment: See **Miller-Urey experiment**.

vitalism: The idea that every organism is imbued with a "life force," intrinsically different from the forces that act on inanimate matter.

Biographical Notes

Louis J. Allamandola (1946–). Chemist at the NASA Ames Research Center in California, whose group has demonstrated the facile production of key biomolecules in experiments that simulate deep space environments. Working with Max Bernstein and others, Allamandola has shown that deep space was a major source of organic molecules on the early Earth.

Gustaf Arrhenius (1922–). Research geochemist and cosmochemist at the Scripps Institution of Oceanography in La Jolla, California. His work has focused on the roles of minerals in the selection, concentration, and catalysis of prebiotic molecules.

Jeffrey L. Bada (1942–). Chemist at the Scripps Institute and head of a NASA-sponsored team that is studying the origin of life. Bada, a student of Stanley Miller, has been an outspoken proponent of the primordial soup hypothesis for life's origin. He and his colleagues, notably Mexican chemist Antonio Lazcano, have questioned the survivability of biomolecules at hydrothermal vents.

John Desmond Bernal (1901–1971). British biophysicist who was one of the earliest proponents of mineral-assisted life. Bernal advocated the special role of clays, which are ubiquitous minerals with regularly layered atomic structures.

A. Graham Cairns-Smith (1931–). Scottish chemist who proposed the clay life hypothesis, by which evolving, self-replicating clay minerals are the first life-forms on Earth. Cairns-Smith's creative ideas are widely respected, though they are speculative and largely untested.

George D. Cody (1960–). A Staff Scientist at the Carnegie Institution's Geophysical Laboratory, Cody is an organic geochemist with a background in coal geology. He has instituted a research program on prebiotic hydrothermal organic synthesis, with an emphasis on the catalytic role of minerals.

Charles Darwin (1809–1882). Darwin wrote the revolutionary book *On the Origin of Species by Means of Natural Selection,* which was published in 1859. His theory of natural selection plays a significant role in modern theories of molecular evolution and the origin of life.

David W. Deamer (1939–). University of California biochemist who has demonstrated the self-organization of lipids from many plausible prebiotic sources into encapsulating vesicles, reminiscent of cell membranes. He views the encapsulation of a self-replicating strand of genetic material, such as RNA, as the most probable scenario for life's origin.

Christian de Duve (1917–). Belgian chemist who won the Nobel Prize for his studies of cellular processes. De Duve proposed the Thioester World model for life's origin. In this model, a primitive *protometabolism* based on energetic carbon-sulfur bonds helps to jumpstart a genetic system based on an RNA-like genetic polymer.

Albert Eschenmoser (1925–). Swiss chemist who has synthesized more than a dozen variants of RNA with different sugar-phosphate backbones. This work reveals alternative chemistries for life's genetic material. Of special importance is his synthesis of TNA, which employs a simpler 4-carbon sugar, threose, instead of the 5-carbon ribose of RNA.

James P. Ferris (1925–). Ferris leads a NASA-sponsored research team at Rensselaer Polytechnic Institute in Troy, New York, where he has demonstrated that clay minerals can adsorb and assemble a variety of organic molecules, including RNA nucleotides.

Sidney W. Fox (1912–1998). Fox, who entered the field of origins-of-life chemistry shortly after the Miller-Urey experiment of 1953, proposed the Protenoid World model—the first comprehensive model of primitive metabolism. His increasingly bizarre claims of lifelike protenoid properties isolated him from the mainstream origins-of-life community. Nevertheless, his vision was influential as the first deterministic, metabolism-first model for life's origin.

John B. S. Haldane (1892–1964). British biochemist and geneticist who wrote a brief, perceptive essay entitled "The Origin of Life." This article introduced the idea of a "hot, dilute soup," what is now commonly called the *primordial soup*.

John H. Holland (1929–). Santa Fe Institute scientist who is considered a founder of the field of emergence. His approach is to model emergent systems, including life, as if they are computer programs with a fixed set of operating instructions.

Gerald F. Joyce (1956–). Molecular biologist at the Scripps Research Institute who is a leading advocate for the RNA World model of life's origin. Joyce crafted the most widely used definition of life as a "self-sustained chemical system capable of undergoing Darwinian evolution."

Stuart A. Kauffman (1939–). Theorist at the Santa Fe Institute who is a leader in the field of emergence and autocatalytic systems. He views the origin of life as the result of a self-replicating, evolving, autocatalytic network of molecules.

Pier Luigi Luisi (1938–). Swiss chemist who developed the Lipid World model for life's origin. Luisi argues that self-organizing lipids can form vesicles that grow and replicate.

Stanley L. Miller (1930–). University of California chemist whose pioneering research began the modern era of experimental research on the origin of life. While a University of Chicago graduate student under Harold Urey in the early 1950s, Miller performed the first of the so-called Miller-Urey experiments, by which amino acids and other key biomolecules were synthesized in a simple glass apparatus that passed electric sparks through a mixture of simple gas molecules.

Harold J. Morowitz (1927–). Professor of biology, formerly at Yale University and now at George Mason University, who has elaborated on the metabolism-first theory of life's origin.

Alexander I. Oparin (1894–1980). Russian biochemist who in his 1924 book, *The Origin of Life*, elaborated on the idea that life arose from a body of water that gradually became enriched in organic molecules. His work was not widely recognized in the West until after an English translation appeared in 1938.

Leslie E. Orgel (1927–). Research professor at the Salk Institute for Biological Studies in San Diego, who is a leading proponent of the RNA World hypothesis. He wrote the classic 1968 paper entitled "Evolution of the genetic apparatus."

Louis Pasteur (1822–1885). Famous French chemist who refuted the theory of the spontaneous generation of life by performing experiments with sealed and unsealed flasks. Ironically, his discovery pushed the actual origin of life to a time and place so remote that it seemed impossible to study in any systematic way.

Ilya Prigogine (1917–2003). Russian-born chemist who recognized that all sorts of complex emergent systems arise when energy flows through a collection of many interacting particles. He won the Nobel Prize for his research.

J. William Schopf (1941–). Professor of paleontology at UCLA and a leading expert on Earth's early life, which is preserved as fossil single cells in rocks more than a billion years old. Schopf's claims of "Earth's oldest fossils" from Australia have come under increased scrutiny in recent years.

Joseph V. Smith (1928–). Emeritus Professor of Mineralogy at the University of Chicago's Department of Geophysical Sciences. Smith has developed models of life's origin in the protective weathered cavities and structural channels of common rock-forming minerals.

Andrew Steele (1966–). A Staff Scientist at the Carnegie Institution of Washington's Geophysical Laboratory, Steele is a microbiologist who is developing a flight-ready life-detection instrument for Mars. This NASA-sponsored effort employs a variety of fluorescent antibodies to detect key biomolecules.

Jack W. Szostak (1952–). Harvard Medical School biologist who is attempting to engineer an evolving life-form in his laboratory. Szostak's group has successfully developed strands of RNA that copy other RNA segments.

Harold C. Urey (1893–1981). Urey won the 1934 Nobel Prize in chemistry for his discovery of the heavy isotope of hydrogen, known as deuterium. He proposed the experiment that came to be known as the Miller-Urey experiment, first performed in 1952 by Stanley Miller.

Günter Wächtershäuser (1938–). German chemist and patent attorney who developed the Iron-Sulfur World model for life's origins. Wächtershäuser is an outspoken advocate for a metabolism-first, autotrophic origin of life. His presentation is the most elaborate and chemically detailed model in the field and, therefore, the most testable by experiments.

Carl R. Woese (1928–). Professor of biology at the University of Illinois who proposed a radically new tree of life based on genetic similarities and differences. He lumped all plants, animals, fungi, and complex single-celled organisms into one "domain" of life called the *Eukarya*. He divided all other primitive single-celled organisms into two domains, the *Bacteria* and the *Archaea*.

Bibliography

Essential Reading:

Dick, Stephen J., and James E. Strick. *The Living Universe: NASA and the Development of Astrobiology*. New Brunswick, NJ: Rutgers University Press, 2004. A scholarly historical account of NASA's role in astrobiology and research on the origin of life.

Fry, Iris. *The Emergence of Life on Earth: A Historical and Scientific Overview*. New Brunswick, NJ: Rutgers University Press, 2000. A thoughtful account of origins research and its philosophical implications.

Hazen, Robert M. *Genesis: The Scientific Quest for Life's Origin*. Washington, DC: Joseph Henry Press, 2005. The companion volume to this lecture series.

Lahav, Noam. *Biogenesis*. Oxford: Oxford University Press, 1999. A comprehensive bibliographic survey of research on the origin of life.

Morowitz, Harold J. *The Emergence of Everything*. New York: Oxford University Press, 2002. A sweeping account of cosmic evolution from the perspective of emergence.

Wills, Christopher, and Jeffrey L. Bada. *The Spark of Life: Darwin and the Primeval Soup*. Cambridge, MA: Perseus, 2000. A popular account of origins research, with special attention paid to Stanley Miller and the prebiotic soup hypothesis.

Supplemental Reading:

Bak, Per. *How Nature Works: The Science of Self-Organized Criticality*. New York: Copernicus Press, 1996. One of several approaches to the study of the emergence of complexity.

Bernal, John D. *The Origin of Life*. London: Weidenfeld and Nicholson, 1967. A classic work on origin-of-life theory, broad and general in content and scope.

Bernstein, Max P., Scott A. Sandford, and Louis J. Allamandola. "Life's Far-Flung Raw Materials." *Scientific American* 281(1)(1999): 42–49. The best popular account of organic molecule synthesis in space.

Boyce, Joseph M. *The Smithsonian Book of Mars*. Washington: Smithsonian Institution Press, 2002. An excellent popular overview of Mars.

Cairns-Smith, A. Graham. *Seven Clues to the Origin of Life*. Cambridge, UK: Cambridge University Press, 1985. Lucid popular account of Cairns-Smith's clay life hypothesis for life's origin.

Chaisson, Eric J. *Cosmic Evolution: The Rise of Complexity in Nature.* Cambridge: Harvard University Press, 2001. Fascinating take on the deterministic emergence of complexity in the universe.

Darwin, Charles. *The Origin of Species.* London: John Murray, 1859. The landmark work that introduced the concept of evolution by natural selection.

Davies, Paul C. W. *The Fifth Miracle: The Search for the Origin and Meaning of Life.* New York: Touchstone, 1999. A competent popular survey of origins research.

Deamer, David W., and Gail R. Fleishaker, eds. *Origin of Life: The Central Concepts.* London: Jones and Bartlett, 1994. A reprint volume with many of the most important papers in the field, plus commentary.

De Duve, Christian. *Vital Dust: Life as a Cosmic Imperative.* New York: BasicBooks, 1995. A popular account of de Duve's Thioester World scenario.

Dyson, Freeman. *Origins of Life.* Cambridge, UK: Cambridge University Press, 1985; 1999, 2nd ed. A short, thoughtful argument in favor of a metabolism-first origin of life.

Eigen, Manfred, and Peter Schuster. *The Hypercycle: A Principle of Natural Self-Organization.* Berlin: Springer-Verlag, 1979. Technical treatment of the mathematics of self-replicating cycles of molecules.

Fox, Sidney W. *The Emergence of Life: Darwinian Evolution from the Inside*. New York: Basic Books, 1988. Fox's popular overview of his controversial Protenoid World model.

Frederickson, John K., and Tullis C. Onstott. "Microbes Deep Inside the Earth." *Scientific American* 275(4)(1996): 68–73. A beautifully illustrated account of drilling for deep microbes.

Gell-Mann, Murray. *The Quark and the Jaguar*. New York: W. H. Freeman, 1994. Yet another approach to the emergence of complexity, written in a popular style.

Gold, Thomas. *The Deep Hot Biosphere*. New York: Copernicus, 1999. Gold's accessible account of his hypothesis that petroleum comes from deep inside Earth.

Goldsmith, Donald. *The Hunt for Life on Mars*. New York: Dutton, 1997. A journalist's fast-paced account of the Martian meteorite story.

Holland, John H. *Hidden Order.* Reading, MA: Helix Books, 1995. See below.

———. *Emergence: From Chaos to Order*. Reading, MA: Helix Books, 1998. These two books by John Holland are classics in the field of emergence. He treats emergent phenomena as simple computer programs.

Kauffman, Stuart A. *The Origins of Order: Self-Organization and Selection in Evolution.* New York: Oxford University Press, 1993. See below.

———. *At Home in the Universe: The Search for Laws of Self-Organization and Complexity.* New York: Oxford University Press, 1995. Kauffman's most detailed accounts of his theory of evolving, autocatalytic cycles. The 1993 book is more technical, while the 1995 book is for a popular audience.

Knoll, Andrew H. *Life on a Young Planet: The First Three Billion Years of Evolution on Earth.* Princeton, NJ: Princeton University Press, 2003. An elegant review of Earth's earliest life by a leading paleontologist.

Lehninger, Albert L., David L. Nelson, and Michael M. Cox. *Principles of Biochemistry*, 2nd ed. New York: Worth Publishers, 1993. The leading biochemistry textbook.

Margulis, Lynn, and Dorian Sagan. *What Is Life?* New York: Simon and Schuster, 1995. A provocative take on the question of life's origin and its definition, aimed at a general audience.

Maynard-Smith, John, and Eörs Szathmáry. *The Origins of Life: From the Birth of Life to the Origin of Language*. New York: Oxford University Press, 1999. Another book that looks at the sweep of evolution in the cosmos. The chapters on the genetic code are concise and accessible.

Monod, Jacques. *Chance and Necessity.* Translated from the French by A. Wainhouse. New York: Knopf, 1971. A classic argument against the deterministic origin of life. Monod claims that life is rare.

Morowitz, Harold J. *The Beginnings of Cellular Life: Metabolism Recapitulates Biogenesis*. New Haven: Yale University Press, 1992. A powerful argument for the inevitability of life, based on the simplest biochemical characteristics of all known life.

Nicolis, Grégoire, and Ilya Prigogine. *Exploring Complexity: An Introduction.* New York: W. H. Freeman, 1989. A popular introduction to Prigogine's theory of complex systems.

Oparin, Alexandre I. *The Origin of Life*. Translated by S. Morgulis. New York: Macmillan, 1938. A classic theoretical work that has influenced the subsequent history of origins research.

Pályi, Gyula, Claudia Zucchi, and Luciano Caglioti, eds. *Fundamentals of Life.* New York: Elsevier, 2002. Proceedings of a conference that explored the question "What is life?"

Popper, Karl R. *Conjectures and Refutations: The Growth of Scientific Knowledge*, 4th ed. London: Routledge and Kegan Paul, 1963. A classic in the philosophy of science. Popper emphasizes that a scientific theory must be falsifiable.

Prigogine, Ilya. *Order out of Chaos: Man's New Dialogue with Nature.* Toronto: Bantam Books, 1984. A popular treatment of Prigogine's ideas about the behavior of complex systems.

Rudwick, Martin J. S. *The Meaning of Fossils: Episodes in the History of Paleontology.* 2nd ed. Chicago: University of Chicago Press, 1976. The preeminent history of paleontology.

Schopf, J. William, ed. *Earth's Earliest Biosphere: Its Origin and Evolution.* Princeton, NJ: Princeton University Press, 1983. Also see below.

———. *Cradle of Life: The Discovery of Earth's Earliest Fossils.* Princeton, NJ: Princeton University Press, 1999. Schopf's works on Precambrian paleontology. The former is a more technical collection of important review articles; the latter is Schopf's idiosyncratic personal account.

Solé, Richard, and Brian Goodwin. *Signs of Life: How Complexity Pervades Biology.* New York: Basic Books, 2000. A wonderful, though at times technical, treatment of complex biological systems.

Strick, James E. *Sparks of Life: Darwinism and the Victorian Debate over Spontaneous Generation.* Cambridge, MA: Harvard University Press, 2000. An excellent history of the spontaneous generation debates of the 19th century.

Ward, Peter D., and Donald Brownlee. *Rare Earth: Why Complex Life Is Uncommon in the Universe.* New York: Copernicus, 2000. A well-written account that addresses the importance of contingency in life's origins.

Wharton, David A. *Life at the Limits: Organisms in Extreme Environments.* Cambridge, UK: Cambridge University Press, 2002. Excellent, accessible overview of research on extremophiles.

Wolfram, Stephen. *A New Kind of Science*. Champaign, IL: Wolfram Media, Inc., 2002. A massive, fascinating tome on the ability of simple mathematical procedures to produce complex patterns.

Internet References

http://www.issol.org. The home page for the International Society for the Study of the Origin of Life (ISSOL). Take a look at the historical archives.

http://nai.arc.nasa.gov/. The home page of NASA's Astrobiology Institute, which is based at the Ames Research Center in California. Links to more than a dozen member teams provide a wide range of information on origins research, as well as efforts to detect life on other worlds.

www.origins.rpi.edu. Web site of the NASA Specialized Center of Research and Training at the New York Center for Studies of the Origins of Life (NSCORT), which represents a partnership between RPI and SUNY-Albany.

http://www.resa.net/nasa/origins_life.htm. An exceptional Web site maintained by the Wayne County (Michigan) Regional Educational Service Agency. Includes an excellent overview of current ideas about the origin of life with numerous links to other sites.

Notes

Notes

Notes

Notes

Notes

Notes

Notes